ELECTRONICS

Circuits,
Amplifiers
and Gates

ELECTRONICS

Circuits, Amplifiers and Gates

BY

D V BUGG

Queen Mary and Westfield College
University of London
and
Rutherford–Appleton Laboratory

Adam Hilger
Bristol, Philadelphia and New York

British Library Cataloguing in Publication Data

Bugg, D.V.
    Electronics: circuits, amplifiers and gates.
    I. Title
    621.3815

    ISBN 0-7503-0109-0
    ISBN 0-7503-0110-4 pbk

Library of Congress Cataloging-in-Publication Data are available

Published under the Adam Hilger imprint by IOP Publishing Ltd
Techno House, Redcliffe Way, Bristol BS1 6NX, England
335 East 45th Street, New York, NY 10017-3483, USA

US Editorial Office: 1411 Walnut Street, Philadelphia, PA 19102

Printed in Great Britain by The Bath Press, Avon

Contents

x   *Contents*

Preface

With a name like mine, it was almost inevitable that I should be assigned the task of teaching electronics. This book is based upon many years of experience coping with the difficulties and questions of both physicists and engineers at many levels of ability and experience. The aim is a rigorous but introductory treatment suitable for first and second year students at Universities and Polytechnics. Though written from a physicist's viewpoint, I would be gratified if electrical engineers also find it helpful. Some topics, such as transformers and three-phase supplies, are largely for their benefit.

It is possible to teach the subject in a wide variety of orderings; students will often be taught two or more topics in parallel. In order to accomodate this, groups of chapters in this book are written to be entirely independent of one another. The figure below shows a flowchart of the interconnections. It is possible, for example, to take Chapters 11–13 independently of 1–10. The pacing of chapters 14 onwards is distinctly higher than the earlier chapters, and is my view of second year material. Sections labelled with an asterisk are more advanced topics which may be omitted safely or delayed.

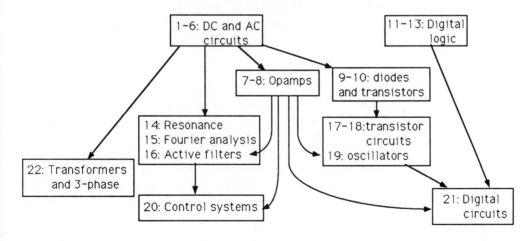

Fig. P.1. Interconnections amongst chapters.

Some of my colleagues have tried to persuade me that students no longer need to understand transistors, because it is easier and cheaper to assemble circuits from integrated circuits. In view of the pace of development of electronics and its uses, I cling to the notion that it is still valuable to understand what goes on in transistors and basic circuits using them. However, I have placed this material fairly late in the book.

Laplace transforms are not used. My own experience is that they are helpful only in elaborate problems beyond the level of this text. However, I have taken the opportunity to use the *s* notation so that students become accustomed to using it and will hopefully make the transition to Laplace transforms more readily. I have adopted the novel approach of considering phase-locked loops (my own hobby) from differential equations rather than Laplace transforms, since I believe it gives more insight.

Practical experience is essential in developing an instinct for electrical circuits and electronics. Most students find it fun after the initial shock of learning how to use instruments. Ideally students should work through every circuit and every exercise of this book, not just on paper but with an oscilloscope. My own students regularly try out their exercise and homework problems experimentally and find this revealing and rewarding. I have included suggestions in the text for the more vital pieces of practical work with realistic component values.

There is an extensive set of exercises, mostly taken from examinations in the University of London. I am indebted to the University for permission to include these questions. All have been triple checked, but I would be grateful to hear of any surviving errors. The solutions are mine and the University disclaims responsibility for them. Questions marked with a single asterisk are lengthy or awkward. Those marked with a double asterisk are particularly revealing and not necessarily difficult. Some material appears *only* in exercises, e.g. attenuators, pulse width modulation and the Cascode amplifier.

Within the text, tricky questions are posed every now and then in italics. Usually no answers are given, deliberately. My students tell me that such unanswered teasers provoke as much thought, discussion and understanding as detailed explanations and I consider them an integral part of the learning process.

I am grateful to Dr E Eisenhandler for suggestions concerning three chapters. I also wish to acknowledge permission from Texas Instruments and Mullards to reproduce characterisitcs of transistors and ICs. I am especially grateful to my wife for her patience and for typing the manuscript in TeX.

David Bugg
November 1990

1

Voltage, Current and Resistance

1.1 Basic Notions

A thunderstorm is a dramatic electrical spectacle. Who could doubt the reality of voltages and currents when confronted by such a display? The storm acts like a giant Wimshurst machine, and separates electrical charges between top and bottom of the cloud (figure 1.1). The energy required for this separation comes from warm moist air which rises and condenses. In a well developed storm, a voltage of $\approx$200 million volts builds up on the thundercloud. Eventually the air breaks down when the electric field at the base of the cloud approaches 10^6 V m^{-1}. In the ensuing lightning stroke the current is 10^4–10^5 amps.

The physics of a thunderstorm is complicated. For present purposes this example serves to focus attention on basic notions of charge, current, voltage and energy. What do the numbers mean? How are voltage and current defined?

First let us make clear the relation between voltage and energy. If a charge Q is moved from point A in the cloud at potential V_A to point B at voltage V_B (figure 1.2), the work done on the charge is $Q \times (V_B - V_A)$. In SI units, it requires one joule (J) to move one coulomb (C) of charge through one volt (V):

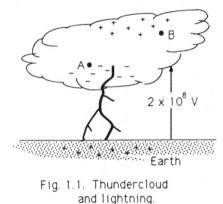

Fig. 1.1. Thundercloud and lightning.

$$\boxed{\text{Energy (J)} = \text{Charge (C)} \times \text{Potential difference (V)}.} \quad (1.1)$$

When the thundercloud discharges, this potential energy is liberated as heat, thunder and electromagnetic radiation. If the cloud carries a charge of 10 C, the energy liberated in a complete discharge is $10 \times 2 \times 10^8$ J $= 2 \times 10^9$ J or 2 GJ (gigajoules).

In an electrical circuit, a battery likewise provides a potential difference and a source of energy. A fully charged 12 V car battery can deliver from chemical reactions a charge of typically 3×10^5 C and an energy of 3.6×10^6 J or 3.6 MJ (megajoules).

A third familiar example will illustrate these concepts. Figure 1.3 shows schematically the layout of an oscilloscope. Electrons are emitted from a heated cathode and are then accelerated

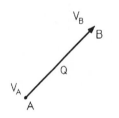

Fig. 1.2. Work done = $Q(V_B - V_A)$.

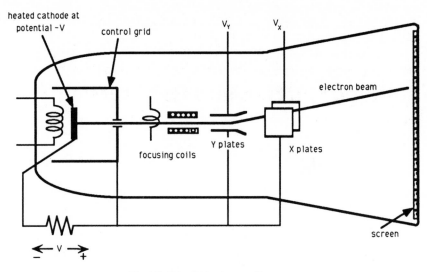

heated cathode at
potential −V

control grid

V_Y V_x

electron beam

Y plates

focusing coils

X plates

electron beam

screen

$\xleftarrow{} V \xrightarrow{}$
−   +

Fig. 1.3. Schematic of an oscilloscope.

through a large voltage V, typically 1–5 kV (kilovolts). They acquire kinetic energy in falling through the potential difference V and reach a velocity v given by

$$eV = \tfrac{1}{2}mv^2 \tag{1.2}$$

where e and m are the charge and mass of the electron. We have to distinguish between the voltage difference V through which the beam is accelerated and the energy eV acquired by each electron. If $V = 5$ kV, this energy is $1.6 \times 10^{-19} \times 5 \times 10^3$ J $= 8 \times 10^{-16}$ J.

Next, what is the relation between current and charge? The moving electrons carry a current, whose magnitude is defined by the charge Q passing through a particular surface per unit time: 1 *Ampere* = 1 *Coulomb per second*. Since the current may vary with time, it needs to be expressed in terms of differential quantities:

$$dQ(C) = I(A) \times dt(s)$$

or

$$\boxed{I = dQ/dt.} \tag{1.3}$$

For example, a beam of 10^{13} electrons/s carries a current of $-1.6 \times 10^{-19} \times 10^{13}$ A $= -1.6 \times 10^{-6}$ A or $- 1.6 \ \mu$A (microamps). The minus sign arises because electrons are negatively charged; conventional current flows in the opposite direction to the electrons. The current may be measured using the force exerted on it by an applied magnetic field. You can demonstrate this force by holding a bar magnet up to an oscilloscope or a TV set and watching the beam deflect. The *Ampere*, the absolute unit of current, is actually defined in terms of the force between two current-carrying coils. Then the *Coulomb* is derived from the *Ampere* using equation (1.3).

A car battery is charged up by a trickle of current of ~4 A over typically 20 hours. It is said to hold a charge of 80 ampere hours, although this is stored as chemical energy rather than as an accumulated charge. In this respect, a battery is different from a capacitor, which stores charge directly (Chapter 3).

Another important electrical quantity is **power** P. It is defined as the rate of change of energy, E:

$$P = dE/dt. \qquad (1.4)$$

It may be related to voltage and current using equation (1.1). A charge dQ falling through potential V gains energy $dE = V dQ$, so a current I flowing through potential difference V generates power

$$\boxed{P = V\frac{dQ}{dt} = VI.} \qquad (1.5)$$

This is the familiar result for power dissipated in a resistor. Power is measured in watts (W): one watt = 1 joule per second. In each stroke of a lightning discharge, power is dissipated for ≈ 100 μs at a rate of about 5×10^{12} W. By comparison, a large power station generates 2000 MW = 2×10^9 W.

1.2   Waveforms

This chapter will be concerned mostly with constant voltages and currents. Such a situation is referred to as DC, meaning **direct current**. However, the principles will carry over to situations where currents and voltages are varying. In the simplest case, figure 1.4(b), they vary with time t as $\sin \omega t$ or $\cos \omega t$. A current of this type is called AC or **alternating current**. Although the phrases AC and DC ought to apply strictly to current, they are often used loosely to refer to voltages too.

Waveforms are readily displayed on the screen of an oscilloscope. Technically, this is done as follows. The beam of electrons of figure 1.3 is collimated through a small hole in the control grid and is then focused by electrostatic or magnetic lenses. Two pairs of plates provide electric fields at right angles to the beam; they deflect it horizontally (X) and vertically (Y). (In a TV set, the plates are replaced by magnetic coils.) The ramp or sawtooth voltage of figure 1.4(c) is applied to the horizontal plates, while the waveform $V(t)$ under study is applied to the vertical plates. The result is a graph of $V(t)$ against t (figure 1.5). When the sawtooth voltage

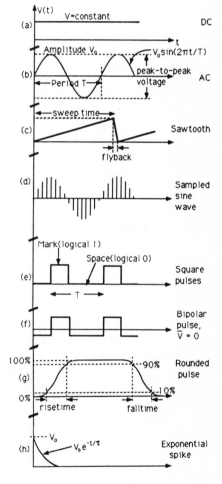

Fig. 1.4. Common waveforms.

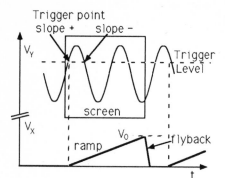

Fig. 1.5. Triggering of an oscilloscope.

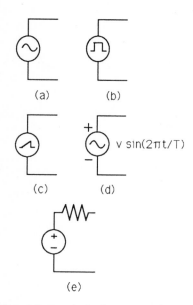

Fig. 1.6. Symbols for generators of sine waves, square waves, ramps. (e) represents a battery explicitly.

reaches a value V_0 such that the display is off the screen, it is restored rapidly to zero; during this 'flyback', the source is blanked off, so as not to confuse the picture.

The virtue of the oscilloscope is its speed. Common oscilloscopes display waveforms on a timescale of ≈ 1 s cm^{-1} to 1 μs cm^{-1}; high-quality oscilloscopes using sampling techniques are capable of displaying features down to 10^{-10} s.

Several other common waveforms are shown in figure 1.4. Square pulses are used in digital circuitry. It is arbitrary whether a low voltage stands for binary 0 and a high voltage for binary 1 or vice versa. Bipolar pulses are square pulses superimposed on a DC level such that the mean voltage is zero averaged over time. Spikes are used to trigger digital circuits.

In circuit diagrams, generators which produce sine waves, square waves, or ramps are drawn as in figure 1.6. The appropriate waveform is shown inside the circle representing the generator. Where it matters, the polarity of the signal is indicated explicitly, as in (d). Strictly speaking, a battery should be represented as in figure 1.6(e) by a DC generator with a resistance in series, to represent the internal resistance of the battery. This formality can frequently be skipped, but you should remember that batteries always have some internal resistance.

1.3   Ohm's Law

Experimentally, the current in many materials, such as metals, depends linearly on applied voltage. This is Ohm's law. The constant of proportionality is defined so that

$$\boxed{V = IR.} \tag{1.6}$$

The **resistance** R is measured in ohms (Ω) if V is in volts and I in amps. For wires of a particular material, it is found that the resistance is proportional to the length L of the wire and inversely proportional to its area A:

$$R = \rho L / A \tag{1.7}$$

where ρ is the resistivity of the material. Can you justify equation (1.7) in terms of resistors in series and parallel? The best conductors are silver and copper with $\rho = 1.5$ and 1.7×10^{-8} Ω m respectively. For aluminium, $\rho = 2.6 \times 10^{-8}$ Ω m. Sometimes it is convenient to use instead **conductivity** $\sigma = 1/\rho$. The **conductance** g of a component is the inverse of its resistance, $g = 1/R$.

Do you know what your resistance is from one hand to the other? Try measuring it with a multimeter or AVOmeter, first touching the leads lightly with one finger of each hand, then making contact by pressing each finger against a coin which rests on a meter lead. Your resistance depends on the area of contact. If you moisten your fingers the resistance decreases, but then the resistance of the rest of the body dominates.

The meter works off a 6 or 9 V battery. Estimate roughly what current flows through your body. Mild shocks are felt from a current of 1 mA, currents of 10 mA are dangerous and 100 mA may be fatal. To what voltages do these currents correspond? When you work with high voltages, make a habit of using only one hand and keep the other in your pocket. That eliminates the risk of a current from hand to hand, affecting the heart, although there is of course still the danger of a current from your hand to your feet if they make good electrical contact with the ground. Make sure you do not grasp anything at high voltage, so that if you do get a shock you jump clear.

1.4   The Resistor Colour Code

In electronics, currents are typically in the range 1 μA to 100 mA and voltages are up to 20 V. Resistances are therefore commonly in the range 10 to 10^6 Ω. Resistors are colour coded according to figure 1.7 and table 1.1. From red to purple, the order follows the colours of the spectrum.

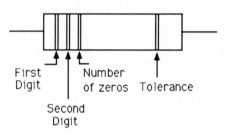

Fig. 1.7. Resistor colour code.

Table 1.1   The resistor colour code.

Colour	Value	Tolerance	Examples
Black	0	silver ±10%	Yellow, purple, red = 4700 Ω
Brown	1	gold ±5%	Brown, black, brown = 100 Ω
Red	2	brown ±1%	Brown, black, green = 1 MΩ
Orange	3		
Yellow	4		
Green	5		
Blue	6		
Purple	7		
Grey	8		
White	9		

Only certain standard values are commonly available: 10, 12, 15, 18, 22, 27, 33, 39, 47, 56, 68 and 82 and these values multiplied by powers of ten. Calculations are usually limited to the sort of accuracy these values permit (±10%), though the tolerances with which resistors are manufactured today are usually ±1%.

An alternative way (the official British Standard) of expressing resistances is

$$4.7\ \Omega = 4R7$$
$$4.7\ k\Omega = 4K7$$
$$2.2\ M\Omega = 2M2.$$

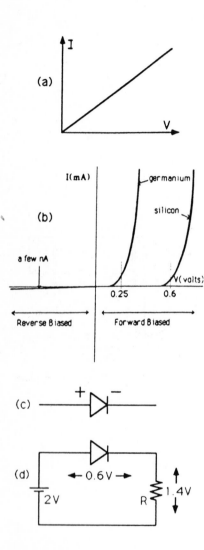

Fig. 1.8. Relation between I and V for (a) a resistor, (b) diodes; (c) the symbol for the diode; (d) a simple circuit including a diode.

1.5   Diodes

For a resistor, V increases linearly with I (figure 1.8(a)). However, not all materials obey Ohm's law. In a lightning discharge, current varies non-linearly with voltage. A second example is provided by the pn diode, where the relation between current and voltage is sketched in figure 1.8(b). This is called the **characteristic curve** of the diode. Current flows much more easily in one direction than the other. When it flows easily the diode is said to be **forward biased** and when the voltage is the other way it is said to be **reverse biased**. The symbol for the diode is shown in figure 1.8(c); the arrow denotes the direction in which the diode conducts.

We shall find later that the current in a diode rises exponentially with voltage for forward bias. Typical applications use currents of 1–100 mA. For silicon diodes, the current rises steeply through this range as V increases from about 0.55 to 0.65 V. When the diode conducts to a significant extent, the voltage across it is usually close to 0.6 V; this voltage may be regarded as the value required to switch the diode on. For lower or negative voltages, little current flows. In the circuit of figure 1.8(d), the diode is forward biased and the current is governed by the voltage $(2.0 - 0.6)$ V $= 1.4$ V across the resistor R. To determine experimentally which way a diode conducts, you can measure its resistance with a multimeter. However, you should connect a resistor of roughly 1 kΩ in series with the diode to prevent damaging it with the 6 or 9 V supplied by the battery of the multimeter.

For reverse bias, the diode current is very small, a few nA for silicon, and it is usually adequate to think of the diode as non-conducting. For a germanium diode, currents are higher and under reverse bias the current is typically 1 μA. Diodes are rated according to the reverse voltage they stand before breaking down. It is possible to control the chemical doping of the material so that this breakdown occurs at a precisely controlled voltage which can provide a reference value in electrical circuits. This kind of diode is called a **Zener diode**. Its characteristic curve is shown in figure 1.9(a) and its circuit symbol in figure 1.9(b); again the arrow indicates the direction of current flow when it is forward biased, although it is mostly used reverse biased, to make use of the threshold. Typical values of this threshold are 2.7 to 30 V.

How can you distinguish between a deviation from Ohm's law and a non-linearity in the meters you use for measuring voltage and current? As a clue, consider how you can double a voltage and how you can double a current.

1.6   Kirchhoff's Laws

The remainder of the chapter develops the methods needed to calculate currents and voltages round a circuit, such as that in figure 1.10(a). Kirchhoff's laws are the foundation. They work for any components: resistors, capacitors, inductors, diodes and transistors, whether or not Ohm's law is obeyed. They reduce any circuit, however complicated, to a set of simultaneous equations. Once this is achieved, the rest is algebra. Sometimes it is convenient to express the equations in terms of currents, sometimes in terms of voltages. We shall explore both.

Kirchhoff's laws express two of the fundamental laws of physics: conservation of charge and conservation of energy. The first is **Kirchhoff's current law**. At any junction or **node**, charge is conserved and so is current (figure 1.10(b)); remember $I = \mathrm{d}Q/\mathrm{d}t$. This conservation law allows us to reduce the number of variables by using the currents I_4 and I_5 which flow round closed loops of figure 1.10(a). These are called **mesh currents**. They are related to I_{1-3} by

$$I_1 = I_4; \qquad I_2 = I_5 \qquad \text{and} \qquad I_3 = I_4 - I_5.$$

Using mesh currents automatically disposes of Kirchhoff's current law.

Next we apply **Kirchhoff's voltage law**. Round any closed loop of the circuit, energy conservation requires that the net change of potential is zero. The changes in potential summed over individual components of a loop give zero resultant. Care is required over signs in working out which voltage changes are positive and which negative. If in figure 1.10(a) we start at the point Y and move clockwise round each loop, the voltages across each component give

$$9 - 3I_4 - 6(I_4 - I_5) = 0$$

$$6(I_4 - I_5) - 2I_5 + 2 = 0$$

where I is expressed in mA (milliamps). There are two equations for the two unknowns I_4 and I_5 with the solution $I_4 = \frac{7}{3}$ mA and $I_5 = 2$ mA. After solving the equations, it is a good idea to evaluate the potential drops across individual resistors as a check, particularly on signs.

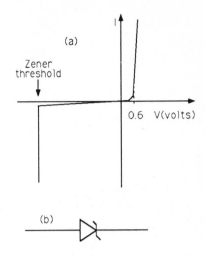

Fig. 1.9(a). Zener diode characteristics, (b) the symbol for the Zener diode.

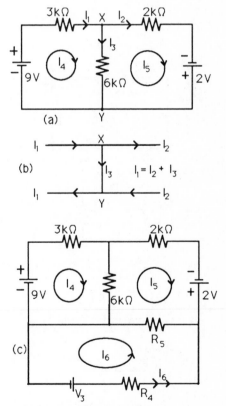

Fig. 1.10(a) Illustrating Kirchhoff's laws, (b) current conservation, (c) 3 loops.

If more loops containing resistors and batteries are added, as in figure 1.10(c), each new loop adds one further current but one further equation. For figure 1.10(c), the equations are

$$9 = 3I_4 + 6(I_4 - I_5) \qquad \text{as before}$$
$$2 = R_5(I_5 + I_6) + 6(I_5 - I_4) + 2I_5$$
$$V_3 = R_4 I_6 + R_5(I_5 + I_6).$$

Kirchhoff's laws may be used to find equations for networks of any degree of complexity, though solving the equations may be tedious.

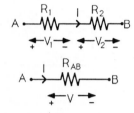

FIG. 1.11. Resistors in series.

Series resistors

As elementary examples of Kirchhoff's laws, the familiar formulae for resistances in series and parallel will be derived. In figure 1.11, the voltages and resistances are **in series**. Kirchhoff's voltage law gives

$$V = V_1 + V_2 = IR_1 + IR_2 = I(R_1 + R_2).$$

The relation between V and I is just the same as for an equivalent resistor R_{AB} where

$$\boxed{R_{AB} = R_1 + R_2 \qquad \text{(series).}} \tag{1.8}$$

This equivalence also holds for power dissipation: in R_1 and R_2 the power dissipated is

$$VI = I^2 R_1 + I^2 R_2 = V^2/(R_1 + R_2)$$

and this is equal to the power dissipated in R_{AB}, namely V^2/R_{AB}.

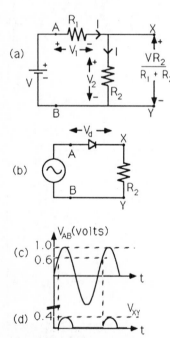

The potential divider

A common arrangement of resistors is the potential divider shown in figure 1.12(a). It fixes V_X at a value intermediate between V and V_Y:

$$V = V_1 + V_2 = I(R_1 + R_2)$$

$$\boxed{V_{XY} = IR_2 = VR_2/(R_1 + R_2).} \tag{1.9}$$

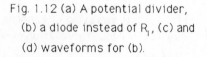

Fig. 1.12 (a) A potential divider, (b) a diode instead of R_1, (c) and (d) waveforms for (b).

You can use this result to verify Ohm's law. Suppose the battery is replaced by a sine-wave generator. The equation works for AC voltages as well as DC: the applied voltage $V(t)$ varies with time, but at any instant Ohm's law still applies. If you display V_{XY} and V_{AB} on two traces of an oscilloscope, you can verify that V_{XY} is a scaled-down replica of V_{AB}. Most oscilloscopes have a knob allowing you to vary continuously the vertical scale of each trace, so with a little dexterity you can superpose one waveform on top of the other. This demonstrates that $V = IR$.

If, however, you replace R_1 with a diode, as in figure 1.12(b), and set the amplitude of the sine-wave generator to ~ 1 V, you get a result more like that in figure 1.12(c); V_{XY} is quite different in shape from V_{AB}. When the diode conducts, the voltage V_d across it is roughly 0.6 V and $V_{XY} = V_{AB} - V_d$; when $V_{AB} < 0.6$ V, the diode does not conduct, so $V_{XY} \simeq 0$. Ohm's law fails here because the diode behaves non-linearly.

It has probably not escaped your attention that the oscilloscope has a resistance of typically $10^6 \ \Omega = 1$ MΩ. This is called its **input resistance**, R_{in}, or **input impedance**. If this is comparable with R_2, you must allow for it. It is a good idea to measure the resistances of the oscilloscope and multimeter you use in the lab, so that you can allow for them when necessary. This is easily done with the circuit of figure 1.13 by measuring the current $I = V/R_{in}$.

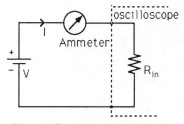

Fig. 1.13. Measuring R_{in}.

When you use a multimeter to measure a resistance, the circuit is identical to that of figure 1.13, with an internal battery driving current through the meter and resistance. If the resistor is part of a circuit, you will need to disconnect one of its ends to make sure that (a) other components in the circuit do not appear in parallel, (b) voltage sources in the circuit do not drive current through the meter and (c) the battery in the meter does not affect the way the circuit operates.

Parallel resistors

As another exercise, the formulae for resistors in parallel (figure 1.14) will be derived. The same potential V appears across R_1 and R_2, so

$$V = I_1 R_1 = I_2 R_2. \qquad (1.10)$$

By Kirchhoff's current law, the total current is

$$I = I_1 + I_2 = (V/R_1) + (V/R_2) \equiv V/R_{EQ}. \qquad (1.11)$$

Hence

$$\boxed{\frac{1}{R_{EQ}} = \frac{1}{R_1} + \frac{1}{R_2} = \frac{R_2 + R_1}{R_1 R_2}} \qquad (1.12)$$

or

$$\boxed{R_{EQ} = \frac{R_1 R_2}{R_1 + R_2} \quad \text{(parallel)}.} \qquad (1.12a)$$

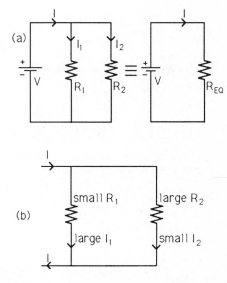

Fig. 1.14. Resistors in parallel.

Both forms (1.12) are worth memorising.

It is useful to make approximations if the resistors have widely different values. Suppose R_2 is much the larger. It carries little current and $R_{EQ} \simeq R_1$. More exactly,

$$R_{EQ} = \frac{R_1}{1 + R_1/R_2} \simeq R_1 \left(1 - \frac{R_1}{R_2}\right) \qquad (1.12b)$$

where the binomial theorem has been used to expand $(1 + R_1/R_2)^{-1}$. For example, if $R_1 = 1$ kΩ and $R_2 = 33$ kΩ, $R_{EQ} \simeq 1 \times (1 - 0.03)$ kΩ $= 0.97$ kΩ. Get used to doing this rough arithmetic in your head.

Parallel resistors act as a **current divider** and it is worth remembering expressions for I_1 and I_2. From (1.10) and (1.11):

$$I = I_1(1 + R_1/R_2)$$

so

$$I_1 = \frac{I R_2}{R_1 + R_2} \qquad \text{and} \qquad I_2 = \frac{I R_1}{R_1 + R_2}. \qquad (1.13)$$

The larger current flows in the smaller resistor (figure 1.14(b)). Again the total power dissipated in R_1 and R_2 is the same as in R_{EQ}:

$$P = VI_1 + VI_2 = \frac{V^2}{R_1} + \frac{V^2}{R_2} = \frac{V^2}{R_{EQ}}.$$

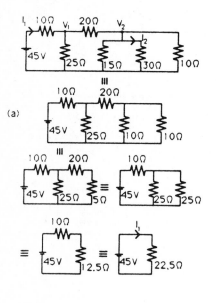

(a)

Worked example

It is often straightforward to reduce a network by combining resistors in series and parallel. In doing this, make sure you solve for the quantity you want to know. Suppose, for example, you require the total power dissipated in the circuit shown in figure 1.15(a). You need to find the current I_1 delivered by the battery. This is easily done by the steps shown in the figure. From the final diagram, $I_1 = 2$ A and the power supplied is $P = 45 \times 2$ W $= 90$ W.

Suppose, however, you want to know instead the current I_2 in the 30 Ω resistor and the power dissipated there. Instead of solving for I_1 and then back-tracking to I_2, it is convenient to evaluate all voltages and currents in terms of I_2 from the outset. This is illustrated in figure 1.15(b). Firstly $V_2 = 30I_2$. The currents in the 15 Ω and 10 Ω resistors are $2I_2$ and $3I_2$, so the total current through the 20 Ω resistor is $6I_2$ and

$$V_1 = V_2 + 120I_2 = 150I_2.$$

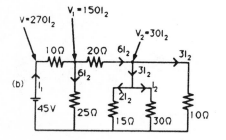

(b)

Fig 1.15 (a) Reduction of a network by series and parallel equivalents, (b) solution in terms of I_2.

This procedure continues in the same way back to the battery, which is known to be 45 V. The current through the 25 Ω resistor is $6I_2$ and finally $I_1 = 12I_2$; so the voltage at the battery is given by $V_1 + 120I_2 = 270I_2$. Hence $I_2 = \frac{1}{6}$ A. Then the power in the 30 Ω resistor is $R_2 I_2^2 = 30I_2^2 = \frac{5}{6}$ W.

You should become fluent at manipulations such as these, since they crop up endlessly when analysing electrical and electronic circuits. Examples at the end of the chapter develop the tricks you frequently need. Note, however, that it is not always possible to reduce networks to series and parallel combinations. A counter-example is shown in exercise 9. There, you must use Kirchhoff's laws directly.

It is always possible to find the voltages and currents in a circuit from Kirchhoff's laws. It is a safe method. However, it is often unnecessarily long-winded. In the worked example we have just done, blind application of Kirchhoff's laws leads to four mesh currents and a set of four simultaneous equations. The remainder of this chapter and the next will explore a variety of shortcuts which often give the required answer more directly.

1.7   Node Voltages

In the worked example, it came in useful to choose the voltages V_1 and V_2 at nodes as unknowns. This often reduces the number of equations. Figure 1.10(a) serves as a second example. There are two nodes X and Y. Suppose voltages are measured relative to the point Y by taking $V_Y = 0$. This leaves only one unknown, the voltage V_X at X. Then the current through the 3 kΩ resistor is $(9 - V_X)/3$ mA, that through the 6 kΩ resistor is $(V_X/6)$ mA and the current through the 2 kΩ resistor is $(V_X+2)/2$ mA. Kirchhoff's current law applied to the node X gives

$$\frac{9 - V_X}{3} = \frac{V_X}{6} + \frac{V_X + 2}{2}$$

or $V_X = 2$ V. From this, I_5 and I_4 follow immediately; the results of course agree with the previous solution.

In this method, only a single equation has to be solved instead of the two simultaneous equations of the previous section. This is a useful simplification.

1.8   Earths

You have probably noticed that only voltage differences appear in Kirchhoff's laws. The absolute voltages of points in the circuit are irrelevant. In fact, one of the subtleties of electromagnetism is that absolute voltages can *never* be measured.

It is very convenient in analysing a circuit to define $V = 0$ at some point such as Y of the previous section. In practical work, it is likewise convenient to measure voltages from that of the Earth. Figure 1.16(a) shows a widespread convention in electrical equipment. The earth terminal of the three-pin plug is connected internally to the metal instrument case and any earth terminals on it; they are all at the same potential, called **earth** or **ground**. This way, anything you touch should be at earth potential and should be incapable of giving you a shock. The earthing point is denoted

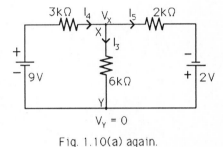

Fig. 1.10(a) again.

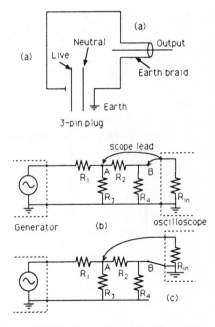

Fig. 1.16 (a) Earthing convention, (b) correct and (c) incorrect ways to connect earth leads when measuring a circuit.

by the symbol shown in figure 1.16(a). It will be explained later
that the **neutral** differs from Earth by at most 1 or 2 V under all
ordinary circumstances. The **live** wire carries high voltage from
the mains.

The output (or input) signal is developed between the central
wire and earth. When you connect instruments to a circuit, e.g.
a sine-wave generator and an oscilloscope, you need to remember
this convention and connect their earths to a common point, as in
figure 1.16(b). If instead you mistakenly connect one of the earth
terminals to a point which is not meant to be earthed, as in figure 1.16(c), you will short out part of the circuit via the mains leads
and upset its operation. If you want to find the signal across R_2 using an oscilloscope, you will need to make separate measurements
at points A and B and take the difference. Most oscilloscopes have
two traces to make this easy. The more expensive oscilloscopes will
actually take the difference for you.

For this particular measurement, a multimeter may be more
convenient. Its case is not earthed, but is left **floating**. It is
able to measure the voltage *difference* between any two points in
a circuit.

1.9   Superposition

Let us return to the circuit of figure 1.10(a), reproduced in figure 1.17, and work out what happens if each battery in turn is
reduced to zero, but otherwise the circuit is left connected. The
currents for these two cases are shown in (b) and (c).

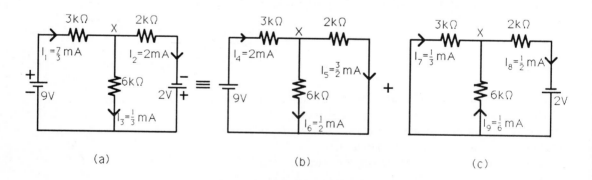

Fig. 1.17. The superposition principle.

If we add the results for (b) and (c), the total currents are the
same as in (a), i.e. the currents produced by the individual batteries simply add. This is called the **superposition principle**. It is
a consequence of the linear relation between voltage and current,

and holds in general when network elements obey this linear relation. What it implies is that the currents produced by one battery
split at nodes in a way which depends only on the resistances in
the network, and is independent of other batteries. This can be
handy (a) in thinking about the way circuits behave, and (b) in
doing arithmetic on individual circuits. For example, in dealing
with figure 1.10(c), we wrote down three simultaneous equations,
but were then faced with the algebraic problem of solving them.
It may be more convenient to find the currents generated by each
of the three batteries in turn. Further examples are given in the
worked example which follows and in the exercises at the end of
the chapter.

A nice demonstration is to replace one battery of figure 1.17 by
a sine-wave generator. Although the applied voltage varies with
time, the superposition principle applies at any instant. The voltage across any resistor may be displayed on the oscilloscope, and
is a superposition of a DC voltage from the battery and an AC component from the generator (figure 1.18). Running the generator
down to zero leaves the DC component unchanged; taking out the
battery leaves the AC component unchanged.

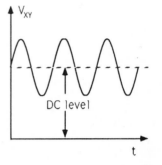

Fig. 1.18. Superposition of
DC and AC voltages.

Another worked example

Figure 1.19(a) shows a rather complicated circuit. As an illustration of all the methods which have been developed in this chapter,
we shall find all the currents and the voltages V_A and V_B at nodes
by three different methods.

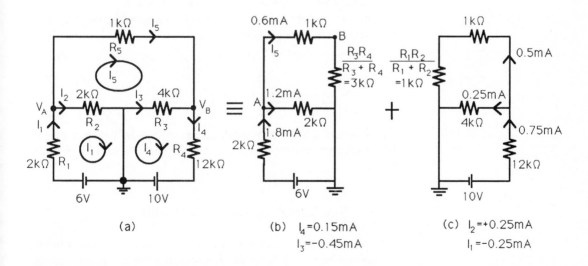

Fig. 1.19 Worked example.

Using superposition, the currents may be obtained from those due to the individual batteries. In (b), the 10 V battery is shorted out and R_3 and R_4 appear in parallel. The top three resistors of (b) provide 2 kΩ in parallel with 4 kΩ between A and earth, i.e. $\frac{4}{3}$ kΩ, so the contribution to $I_1 = 1.8$ mA. It is easy to see how this divides at A; then I_5 splits between 0.15 mA through R_4 and a contribution of -0.45 mA to I_3. In (c), the 6 V battery is shorted, so R_1 and R_2 appear in parallel. The arithmetic of the resulting currents is shown in the figure. The signs of the contributions to I_1 and I_2 are easy to follow from (a) and the sense in which the 10 V battery drives currents. Adding currents from (b) and (c), $I_1 = 1.55$ mA, $I_2 = 1.45$ mA, $I_3 = -0.7$ mA, $I_4 = -0.6$ mA and $I_5 = 0.1$ mA. From these currents, it is simple to find $V_A = 6 - 3.1 = 2.9$ V and $V_B = 10 + I_4 R_4 = 2.8$ V.

Suppose instead the problem is to be solved using mesh currents. The ones to choose would be I_1, I_4 and I_5, as shown in (a). Applying Kirchhoff's current law, $I_1 = I_2 + I_5$ and $I_4 = I_3 + I_5$. The values given in the previous paragraph satisfy these relations. Then applying Kirchhoff's voltage law to each loop in turn:

$$6 = 2I_1 + 2(I_1 - I_5) = 4I_1 - 2I_5$$
$$10 = -12I_4 + 4(I_5 - I_4) = 4I_5 - 16I_4$$
$$0 = 1I_5 + 4(I_5 - I_4) + 2(I_5 - I_1) = 7I_5 - 4I_4 - 2I_1$$

with currents in mA. Solving these three simultaneous equations is tedious. It is however straightforward to substitute the values derived above and demonstrate that the equations are correctly satisfied. Using superposition is really a graphical way of eliminating variables from the simultaneous equations.

The third alternative is to use node voltages V_A and V_B. Then current conservation at these nodes gives

$$\frac{6 - V_A}{2} = \frac{V_A - V_B}{1} + \frac{V_A}{2}$$
$$\frac{10 - V_B}{12} = \frac{V_B}{4} + \frac{V_B - V_A}{1}.$$

The solution of these two simultaneous equations is easy; a check is that the equations are satisfied by the values of V_A and V_B obtained above.

1.10*   Non-linear Elements in a Circuit

Superposition is a valuable shortcut, but (as demonstrated below) it only works exactly for circuits containing linear components like resistors, where $V \propto I$. The next chapter develops other powerful shortcuts which again depend on linearity. However, many electronic devices such as diodes and transistors do not obey Ohm's law, but follow a non-linear relation between current and voltage like figure 1.20. The resulting circuit equations can only be solved by graphical or approximate methods. The details of this awkward problem will be deferred until Chapter 17, when transistor circuits will be discussed quantitatively. Nonetheless, it is valuable to have a general idea of what to do about non-linearity from an early stage. It can then be pushed into the background while other more fundamental concepts are being developed.

For any network containing non-linear elements, Kirchhoff's laws still apply, because they depend only on the fundamental laws of energy conservation and charge conservation; they are quite general and do not depend on Ohm's law. Problems may be solved using mesh currents or node voltages, whichever happen to be convenient. We shall find that exact superposition breaks down, since its validity depends on Ohm's law. However, it can be recovered in a useful approximate form. Consequently, all the methods developed so far can be carried over to electronic circuits, at least in an approximate way.

In order to develop these approximations, consider the simple circuit of figure 1.21(a). The voltage V_0 is large enough to switch on the diode, so that a significant current flows. From Kirchhoff's voltage law,

$$V_0 = V_d + V_R = V_d(I) + RI. \tag{1.14}$$

Earlier, the simplification $V_d \simeq 0.6$ V was adopted (or 0.25 V for germanium-based diodes). In this case, the solution of equation (1.14) is simply

$$I = (V_0 - 0.6)/R. \tag{1.15}$$

This approximation is very frequently good enough.

Next suppose that V_0 is made up of two components V_1 and V_2 (figure 1.21(b)). Blind application of the superposition principle would give

$$I = I_1 + I_2$$

where

$$I_1 = (V_1 - 0.6)/R$$

and

$$I_2 = (V_2 - 0.6)/R.$$

This disagrees with (1.15). Clearly the superposition principle no longer applies. The difficulty lies with the diode, which has been replaced with a fixed voltage drop of 0.6 V. We have to tread more carefully.

Suppose equation (1.14) has to be solved exactly. Figure 1.22(a) shows a graphical method. The equation is rewritten as

$$V_d(I) = V_0 - RI.$$

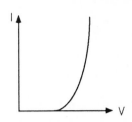

Fig. 1.20. The diode characteristic again.

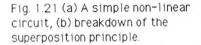

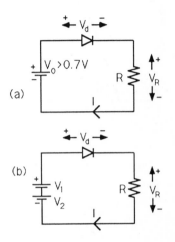

Fig. 1.21 (a) A simple non-linear circuit, (b) breakdown of the superposition principle.

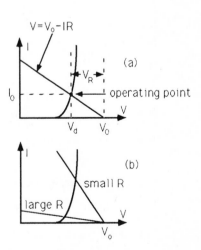

Fig. 1.22 (a) Finding I_0 and $V_d(I_0)$ graphically, (b) variations with R.

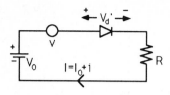

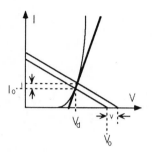

Fig. 1.23. Superposition of a small voltage v on V_0.

The right-hand side gives a straight line against I and its intersection with the diode curve determines the solution I_0. This intersection is called the **operating point**. The straight line is called the **load line**. If you want to measure the characteristic curve experimentally, you can vary V_0 or R and plot out the variation of V_d with I, as in figure 1.22(b).

Suppose equation (1.14) has been solved by the graphical method. What happens if V_0 changes a little? Suppose, in fact, a *small* voltage v is added to the circuit, as in figure 1.23. Correspondingly, the current changes by a small amount i. Rather than going through the whole graphical solution again, it is a reasonable approximation to replace the diode characteristic by a straight line, as in figure 1.24. Providing v does not move the operating point too far, this gives a first approximation to the diode characteristic near the operating point. The voltage V_d' across the diode is now

$$V_d' = V_d + \frac{dV_d}{dI}i \tag{1.16}$$

and with the inclusion of the voltage v in figure 1.24, Kirchhoff's voltage law gives

$$V_0 + v = V_d + \frac{dV_d}{dI}i + R(I_0 + i). \tag{1.17}$$

When $v = 0$,

$$V_0 = V_d + RI_0. \tag{1.18}$$

Subtracting this equation from (1.17), the equation for i reads

$$v = \frac{dV_d}{dI}i + Ri = \rho i + Ri \tag{1.19}$$

where $\rho = (dV_d/dI)$ is called the **slope resistance** of the diode. It is quite different from the DC resistance $V_d(I_0)/I_0$, since the straight line approximating the diode characteristic does not go through the origin.

Figure 1.25 interprets equations (1.17)–(1.19). The full circuit (a) may be expressed as the superposition of (b) for the original equation (1.18) and the perturbation (c) relating the change v in applied voltage to the change i in current, equation (1.19). The superposition principle has been recovered by replacing the diode characteristic by a straight line. Of course, this is only an approximation and is only as good as the straight line approximation to the diode characteristic. It is nonetheless a very valuable approximation in dealing with circuits involving diodes and transistors. The circuit describing the small voltage v and current i (figure 1.25(c)) is known as the **small-signal equivalent circuit**. There, the behaviour of the diode is approximated by an equivalent resistor ρ,

Fig. 1.24. Movement of the operating point by a small voltage v.

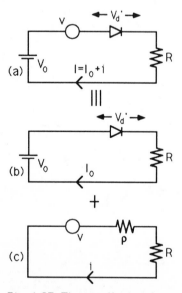

Fig. 1.25. The small-signal equivalent circuit for a diode circuit.

whose value represents the fact that the small signal v moves the operating point up and down the slope of the characteristic curve.

Consider a numerical example. Suppose the operating point is such that $V_d(I_0) = 0.6$ V when $I_0 = 20$ mA; and suppose that changing I to 25 mA changes $V_d(I)$ to 0.606 V. The DC resistance is $0.6/0.020 = 30$ Ω. The slope resistance ρ is $(0.606 - 0.60)/(0.025 - 0.020)$ $\Omega = 0.006/0.005$ $\Omega = 1.2$ Ω.

This algebra is successful only if the perturbation v is small enough that the diode characteristic may be approximated by a straight line over the range of voltages involved. If instead V_0 and v are both batteries providing quite large voltages, the superposition principle breaks down. In particular, if v drives the diode down to the bottom of the characteristic curve and turns it off, the approximation is hopelessly bad.

You can experiment with these phenomena with the circuit of figure 1.26(a). For *small* AC signals, the behaviour of the circuit may be analysed as the superposition of (b) and (c). If the AC voltage is turned down to zero, $V_{XY} \simeq 2.4$ V and the current I_0 through the diode is $\simeq (2.4/1.5)$ mA $= 1.6$ mA. An ammeter placed in series with the diode will confirm this. If the AC voltage is small, the voltage V_{XY} has a DC component $\simeq 2.4$ V and a sine-wave component superimposed, like figure 1.18. The AC component is very small because the slope resistance of the diode is small for AC signals, shorting out the 6 kΩ resistor. Exercise 16 examines its magnitude quantitatively. As you increase the AC voltage, the sine-wave component of V_{XY} becomes distorted when the diode begins to switch off. Exercise 16 works out how large v must be for this to happen.

In conclusion, the superposition principle works successfully when a *small* signal v is added to a DC signal, as in figure 1.26, providing the operating point does not move off the linear part of the characteristic. If you fancy your hand at algebra, show how the superposition principle depends on linearity by taking $V_d = aI + bI^2$ and following the algebra through the circuit of figure 1.26(a) using Kirchhoff's laws; solve for I_1 firstly with $b = 0$, then see what happens when $b \neq 0$.

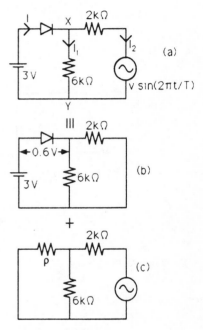

Fig. 1.26 (a) AC and DC signals applied to a circuit containing a diode, (b) and (c) DC and AC equivalent circuits for small AC signals.

1.11   Summary

Kirchhoff's voltage and current laws apply to any circuit, whether components are linear or not. Series resistors make a potential divider, parallel resistors a current divider. Networks may be solved using mesh currents or node voltages; either may be convenient under different circumstances, and it is a matter of practice to spot which gives the easier solution. When the circuit components are linear, superposition offers a shortcut; it fails however when the

voltage–current relation for any component moves over a significantly non-linear region.

1.12   Exercises

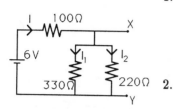

Fig. 1.27.

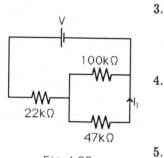

Fig. 1.28

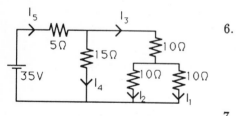

Fig. 1.29

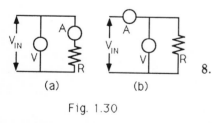

(a)   (b)

Fig. 1.30

1. *Series resistance and power.* The lights on a Christmas tree
    are 20 × 1 W bulbs in series connected to the 240 V mains.
    One bulb fuses, but still conducts. What current flows and
    what power is now dissipated in each bulb? (Ans: 88 mA,
    1.11 W.)

2. *Potential and current dividers.* Find V_{XY} in figure 1.27 and
    also the currents I_1 and I_2. (Ans: $V_{XY} = 3.41$ V, $I_1 =$
    10.34 mA, $I_2 = 15.5$ mA.)

3. *Parallel resistances.* Two resistors R_1 and R_2 in parallel have
    an equivalent resistance of 30 Ω. Current divides between
    them in the ratio 3:1. Determine R_1 and R_2. (Ans: 40 Ω and
    120 Ω.)

4. *Approximations.* A 33 kΩ resistor is in parallel with 330 kΩ.
    Find their equivalent resistance (a) exactly, and (b) using the
    binomial theorem and equation (1.12b). By what percentage
    do these results differ? (Ans: (a) 30 kΩ, (b) 29.7 kΩ, (c) 1%.)

5. *Parallel resistances and currents.* In the circuit in figure 1.28,
    $I_1 = 50$ μA. What is V? What current flows through the
    100 kΩ resistor? (Ans: 3.97 V; 23.5 μA.)

6. *Parallel resistances and power.* Determine the currents I_1
    and I_2 in figure 1.29. Then find I_3, I_4 and I_5. Find the
    total power loss in the resistors and verify that it equals the
    power delivered by the battery. (Ans: $I_1 = I_2 = 0.7$ A;
    $I_3 = I_4 = 1.4$ A; $I_5 = 2.8$ A; $P = 98$ W.)

7. *Instrument impedance.* Figures 1.30(a) and (b) show two
    ways of measuring R. The voltmeter V has a resistance of
    20 kΩ and the ammeter A a resistance of 100 Ω. If the applied
    voltage V_{IN} is 12 V and $R = 4.7$ kΩ, what current and voltage
    are measured in each case? (Ans: (c) 12 V, 2.5 mA, (d) 11.69
    V, 3.07 mA.)

8. *Mesh currents.* Use mesh currents in the circuit of figure 1.28
    to find I_1 and the current supplied by the battery if $V = 10$ V.
    Check by combining resistors in series and parallel. (Ans:
    $I_1 = 126$ mA, 185 mA.)

9. *Mesh currents and node voltages.* Find the node voltages
    V_1 and V_2 in figure 1.31. Hence find the mesh currents I_1,
    I_2 and I_3 and check that your solution satisfies Kirchhoff's
    voltage law round each loop. (Ans: $V_1 = 35/6$ V, $V_2 = 5$ V;
    $I_1 = 15/144$ A; $I_2 = 45/144$ A; $I_3 = 115/144$ A.)

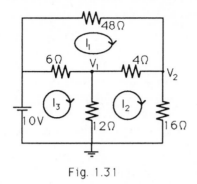

Fig. 1.31

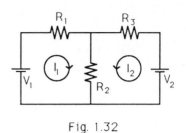

Fig. 1.32

10. *Superposition.* Find I_1 and I_2 in figure 1.32 due to battery V_1 acting alone and also due to V_2 acting alone. Show that the sum of these currents satisfies Kirchhoff's voltage law for the loops. (Ans: V_1 alone gives $I_1 = V_1(R_2 + R_3)/X$, $I_2 = -V_1 R_2/X$, where $X = R_1 R_2 + R_1 R_3 + R_2 R_3$; V_2 alone gives $I_1 = -V_2 R_2/X$, $I_2 = V_2(R_1 + R_2)/X$.)

11*. *Node voltage and superposition.* (QMC 1985) Find the currents I_{1-4} in figure 1.33 and the voltage V_X at node X. You are advised to consider separately the currents due to the 1 V and 2 V batteries, solving in each case for I_1 and V_X. Check at the end that your results satisfy Kirchhoff's voltage law round each loop. (Ans: $V_X = 1.75$ V; $I_1 = 30$ mA, $I_2 = 17.5$ mA, $I_3 = -7.5$ mA, $I_4 = 25$ mA.)

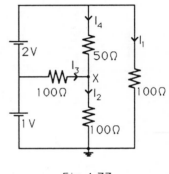

Fig. 1.33

12. *Attenuator, characteristic impedance.* The network shown in figure 1.34 attenuates an input voltage V_{in}, producing an output voltage $V_{out} = V_{in}/A$ across a load Z. The value of n can be chosen so that the impedance measured across the input terminals AB is also Z when the load is connected. In this case, the network is said to have **characteristic impedance** Z, and m elements like figure 1.34 can be joined in sequence to make a ladder network with (a) input impedance Z which is independent of m and (b) output signal V_{in}/A^m across the terminating load Z. Each step of the ladder from AB to CD acts as an **attenuator**, reducing the signal by a constant factor A. Show that the required value of n is $0.5(x^2 - 1)$ where $x = Z/R$ and that A is then $(x + 1)/(x - 1)$. Find n and R if $Z = 100$ Ω and A is (a) 2, (b) 10. Comment on the stability of the components R in the two cases. (Ans: (a) $n = 4$, $x = 3$, (b) $n = 20/81$, $x = 11/9$; A sensitive to n and R in the latter case.)

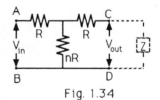

Fig. 1.34

13. *Diode slope resistance.* The diode in the circuit of figure 1.35 has characteristic curve $V = AI + BI^2$ where $A = 50$ Ω and $B = 100$ μΩ/A. Find the DC current I, the slope resistance

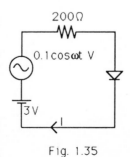

Fig. 1.35

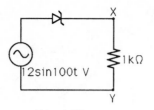

12sin100t V

Fig. 1.36

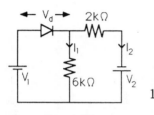

Fig. 1.37.

ρ of the diode, and the AC voltage across the diode. (Ans: 12 mA, 52.4 Ω, 0.0208 $\cos \omega t$ V.)

14. *Diode circuit.* Assume that the Zener diode in figure 1.36 has a characteristic curve given by figure 1.9(a) with a Zener threshold at -6 V. Find the current through the resistor and V_{XY} as a function of t (which is in s). Sketch $V_{XY}(t)$ and $V_d(t)$. (Ans: For $\sin 100t > 0.05$, $I = 11.4 \sin \omega t$ mA and $V_{XY} = 12 \sin 100t - 0.6$ V. For $\sin 100t < 0.5$, $I = -6 \sin 100t$ mA and $V_{XY} = 12 \sin 100t + 6$ V.)

15. *Diode and superposition.* Suppose that in figure 1.37, $V_1 = 9$ V and $V_2 = 2$ V. Find I_1 and I_2, assuming that the voltage V_d across the diode is 0.6 V. Now find I_1 and I_2 if (a) $V_1 = 9$ V and $V_2 = 0$, (b) $V_1 = 0$ and $V_2 = 2$ V. Is the superposition principle satisfied?(Ans: $I_1 = 1.4$ mA, $I_2 = 5.2$ mA; (a) $I_1 = 1.4$ mA, $I_2 = 4.2$ mA; (b) $I_1 = 0.1$ mA, $I_2 = 0.7$ mA. No.)

16*. *Diode and superposition.* In figure 1.26, find the slope resistance of the diode given $\rho = 25 \ \Omega/I$ (mA), assuming $v = 0$. Hence find the DC and AC components of V_{XY} for small v. How large does v have to be to begin to switch off the diode? (Ans: $\rho = 15.6 \ \Omega$; $V_{XY} = 2.4 + (v/129) \sin(2\pi t/T)$; 3.2 V.)

2

Thevenin and Norton

2.1 Thevenin's Theorem

In this chapter we discuss a very important method which will be extended later in the book to circuits containing capacitors, inductors and transistors. It is important in two quite different respects: firstly, it is often a shortcut, and secondly, it leads to conceptual simplifications in thinking about complicated circuitry.

Very frequently, what is of interest is the current or voltage waveform for only one component of a circuit. It is a waste of time solving for all the currents or voltages in the circuit, and mistakes can easily arise in solving many equations. Thevenin's theorem bypasses this drudgery. It will be introduced by means of a simple example, shown in figure 2.1(a).

Suppose we want to know the current I_2 between the terminals AB through a load resistor R_L. Thevenin's theorem states that I_2 is given by the equivalent circuit of figure 2.1(b). The circuit enclosed by the broken lines behaves simply as a source of voltage V_{EQ} in series with an internal resistance R_{EQ}. A straightforward way of deriving V_{EQ} and R_{EQ} will emerge below. Then the voltage V_{AB} across the load resistor is

$$\boxed{V_{AB} = V_{EQ} - R_{EQ}I_2 = I_2R_L.} \qquad (2.1)$$

This theorem works no matter how complicated the network of batteries and resistors inside the broken lines, with the sole proviso that all the elements obey a linear relation between voltage and current. If all the components in the circuit are linear, it is very reasonable that V_{AB} should be linearly related to I_2. What is less obvious is that the values of V_{EQ} and R_{EQ} depend only on the values of components within the broken lines *and are independent of R_L and hence I_2*. For practical purposes, the broken lines could represent a black box, with two output terminals; all that matters are V_{EQ} and R_{EQ}. For a complicated circuit, this is a great conceptual and practical simplification. Let us first check it for the simple circuit of figure 2.1(a).

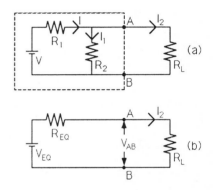

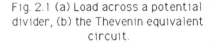

Fig. 2.1 (a) Load across a potential divider, (b) the Thevenin equivalent circuit.

To do this, the equations describing figure 2.1(a) need to be arranged so as to display current I_2 and the voltage V_{AB}:

$$I_2 = I - I_1 = \frac{V - V_{AB}}{R_1} - \frac{V_{AB}}{R_2} = \frac{V}{R_1} - V_{AB}\left(\frac{1}{R_1} + \frac{1}{R_2}\right)$$

Rearranging,

$$V_{AB} = \frac{V R_2}{R_1 + R_2} - I_2\frac{R_1 R_2}{R_1 + R_2}.$$

This indeed looks like equation (2.1) if

$$V_{EQ} = \frac{V R_2}{R_1 + R_2} \tag{2.2}$$

$$R_{EQ} = \frac{R_1 R_2}{R_1 + R_2}. \tag{2.3}$$

Notice that V_{EQ} and R_{EQ} do not depend on R_L, but only on quantities within the broken lines.

This example verifies Thevenin's theorem in one particular case. The general case, where there is a large number of loops in the network, goes the same way and is given in appendix A.

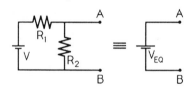

Fig. 2.2. Finding V_{EQ}.

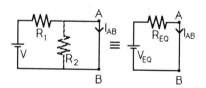

Fig. 2.3. Shorting AB.

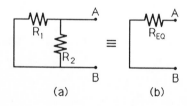

(a)   (b)

Fig. 2.4. Finding R_{EQ}.

Finding V_{EQ} and R_{EQ}

Given Thevenin's theorem, there is a quicker way of arriving at V_{EQ} and R_{EQ}. Remember that these two quantities are independent of R_L. It is convenient to consider three special values of R_L. The first two give V_{EQ} and R_{EQ} and the third acts as a cross-check against mistakes.

(i) If $R_L \rightarrow \infty$ (i.e. an open circuit across AB), no current flows through AB, so there is no voltage drop in R_{EQ} and the value of V_{EQ} is equal to V_{AB} when AB is open circuit. Then, from figure 2.2,

$$V_{EQ} = V_{AB} = V R_2/(R_1 + R_2)$$

in agreement with (2.2).

(ii) If terminals AB are short-circuited (figure 2.3), no current flows through R_2, so $I_{AB} = V/R_1$. But in this case the short-circuit current I_{AB} is equal to V_{EQ}/R_{EQ}, hence

$$R_{EQ} = \frac{V_{EQ}}{I_{AB}} = \frac{V R_2}{R_1 + R_2} \times \frac{R_1}{V} = \frac{R_1 R_2}{R_1 + R_2}$$

in agreement with (2.3).

(iii) Imagine that the voltage V is scaled down by a large factor f. All the currents in the circuit scale down by this factor; resistances remain unchanged in both figure 2.1(a) and its equivalent circuit. Ultimately, as $V \rightarrow 0$, the equivalent circuit contains just R_{EQ} to the left of the terminals AB (figure 2.4(b)). Thus

R_{EQ} is found by reducing all voltage sources to zero, i.e. by short-circuiting them. Figure 2.4(a) contains R_1 and R_2 in parallel to the left of AB:

$$R_{EQ} = \frac{R_1 R_2}{R_1 + R_2}$$

in agreement with (ii) and (2.3).

Method (iii) is applicable only if voltage sources in the circuit are independent of current, as in the example discussed above. Later we shall encounter voltage sources (and current sources) whose magnitude is proportional to another voltage or current in the circuit. In this case, it is no longer permissible to scale voltages and currents down to zero and method (iii) is no longer applicable, though (i) and (ii) are. We shall return to this refinement in later chapters.

A second warning is that the equivalent circuit constructed here refers specifically to the terminals AB across resistor R_L. The equivalent circuit across R_1 or R_2 will be different. Suppose, for example, figure 2.1 is redrawn as in figure 2.5(a) to display the equivalent circuit across R_2; it is an exercise for the student to show that the new equivalent circuit has $R_{EQ} = R_1 R_L/(R_1 + R_L)$ and $V_{EQ} = V R_L/(R_1 + R_L)$. These have different values from those for the previous equivalent circuit across R_L.

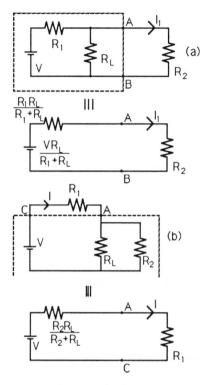

Fig. 2.5. Equivalent circuits across (a) R_2 (b) R_1.

Examples

A familiar example of a Thevenin equivalent circuit is that for a battery: a voltage source with an internal resistance. The concept is however quite general. Consider a pulse generator. It is a very complicated box of electronics. But it behaves just like a voltage source $V_{EQ}(t)$ with an output resistance R_{EQ}. The voltage source may have a square-wave dependence on time t or some other t dependence (e.g. a sawtooth). However, providing the circuit is linear, R_{EQ} does not vary with V_{EQ} (or equivalently with current). If it does, the circuit is non-linear. Even if R_{EQ} varies slightly with current, the approximate notion of a Thevenin equivalent circuit may prove a useful guide.

Suppose the pulse generator is connected to an oscilloscope (figure 2.6). The oscilloscope may again be thought of in terms of its Thevenin equivalent circuit. But for the oscilloscope V_{EQ} is zero, i.e. there is no 'back-EMF'. The oscilloscope simply has an input resistance.

In this way a very complicated set of circuits has been reduced to two simple black boxes. A manufacturer will normally quote the output or input resistance of his product in its specification. For most applications, he goes to considerable trouble to make the performance of the circuit linear, so that Thevenin's theorem applies.

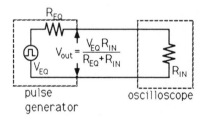

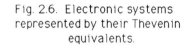

Fig. 2.6. Electronic systems represented by their Thevenin equivalents.

The pulse generator and oscilloscope are examples of **one-port systems**: they each have a single pair of terminals, an output or input. Later we shall meet more complicated devices with more than one set of terminals. For example, an amplifier has both an input and an output. This is a **two-port system**.

2.2   How to Measure V_{EQ} and R_{EQ}

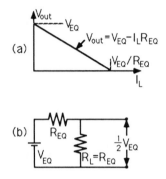

(a)

(b)

Fig. 2.7 (a) straight-line relation between V_{out} and I_L, (b) $V_{out} = V_{EQ}/2$ when $R_L = R_{EQ}$

Suppose we are presented with a gadget which measures the electrical voltage produced by a nerve cell. We know nothing about the internal workings of the gadget, only that it can be represented by a Thevenin equivalent circuit. How do we find V_{EQ} and R_{EQ}? If the gadget is connected to an oscilloscope, as in figure 2.6, the voltage measured by the oscilloscope is $V_{EQ} R_{IN}/(R_{IN} + R_{EQ})$. If $R_{IN} \gg R_{EQ}$, the oscilloscope measures V_{EQ} directly. Oscilloscopes usually have a large input resistance, typically 1 MΩ. If this resistance is inadequate, special probes are available to raise the input resistance to 10 or 20 MΩ.

The value of R_{EQ} may be determined by substituting various loads R_L for the oscilloscope and measuring the output voltage V_{out}. The current I_L through the load is V_{out}/R_L, and should follow the straight-line relation of figure 2.7(a). If not, the components inside the black box are behaving non-linearly, leading to a breakdown of Thevenin's theorem. The straight line on the figure is called the **load line** of the device, since it describes how it responds to a load. Its slope determines R_{EQ}.

Often R_{EQ} may be found by adjusting R_L until $V_{out} = \frac{1}{2}V_{EQ}$, as in figure 2.7(b); in this case, $R_L = R_{EQ}$. However, care should be taken not to draw too large a current and damage the source; for example, a car battery has a resistance which is a fraction of an ohm and the battery would be damaged by applying such a small resistance across the terminals.

A torchlight battery produces 1.5 V. Why will eight such batteries in series not start a car engine?

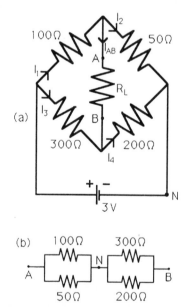

(a)

(b)

Fig. 2.8 (a) Wheatstone bridge, (b) R_{EQ}

Second worked example on Thevenin's theorem
Consider the Wheatstone bridge of figure 2.8. Suppose we want to find the current through the galvanometer, of resistance R_L, between A and B. It is necessary to construct the Thevenin equivalent circuit across terminals AB.

(i) To find V_{EQ}, let $R_L \to \infty$. Taking the negative terminal N of the battery as the zero of voltage,

$$V_A = \frac{3 \times 50}{100 + 50} = 1 \text{ V}$$

$$V_B = \frac{3 \times 200}{200 + 300} = 1.2 \text{ V}.$$

So

$$V_{\text{EQ}} = V_{\text{A}} - V_{\text{B}} = -0.2 \text{ V}.$$

(ii) To find R_{EQ}, suppose the battery is replaced by a short circuit. Then the resistors between A and B may be redrawn as in figure 2.8(b), so

$$R_{\text{EQ}} = \frac{100 \times 50}{100 + 50} + \frac{300 \times 200}{300 + 200} = 33\tfrac{1}{3} + 120 = 153\tfrac{1}{3} \text{ } \Omega.$$

(iii) In this particular example, finding I_{AB} when $R_{\text{L}} = 0$ is a little lengthy, but still a worthwhile check. It is equal to $(I_1 - I_2)$ and also $(I_4 - I_3)$, so it is necessary to solve for these currents:

$$V_{\text{A}} = 50I_2$$
$$I_1 = (3 - V_{\text{A}})/100 = 0.03 - 0.5I_2 = I_2 + I_{\text{AB}}$$

so

$$I_{\text{AB}} = 0.03 - 1.5I_2.$$

Also

$$I_4 = V_{\text{A}}/200 = I_2/4$$
$$I_3 = (3 - V_{\text{A}})/300 = 0.01 - I_2/6.$$

Also

$$I_3 = I_4 - I_{\text{AB}} = 1.75I_2 - 0.03.$$

Equating these two expressions for I_3,

$$0.04 = (I_2/12)(2 + 21) = 23I_2/12.$$

Finally

$$I_{\text{AB}} = (0.69 - 0.72)/23 = -0.0\dot{3}/23 \text{ A}.$$

This is consistent with $V_{\text{EQ}}/R_{\text{EQ}}$.

2.3 Current Sources

So far we have considered voltage sources such as batteries and waveform generators. The ideal battery produces a constant output voltage, regardless of current, as shown in figure 2.9(a). There do exist sources, however, which behave much more nearly like figure 2.9(b); the output current is constant, regardless of voltage. Such a source is called a **constant current source**. It is represented in circuit diagrams by the symbol of figure 2.9(c), for both AC and DC situations. An example is the photodiode, described in more detail in Chapter 9. When light falls on this diode, electrons are liberated and the current through the diode is proportional to the light intensity.

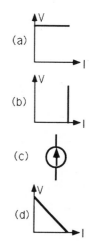

Fig. 2.9 (a) Ideal voltage source, (b) ideal current source, (c) the symbol for a current source, (d) non-ideal source.

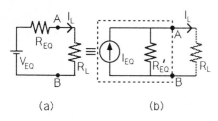

(a)   (b)

Fig.2 10. Norton's equivalent circuit.

Non-ideal voltage sources produce an output voltage which drops with current, as in figure 2.7(a) and figure 2.9(d); for small output resistance R_{EQ}, the slope is small. A non-ideal current source also produces a diagram like figure 2.9(d), but with large slope (large output resistance). There is no clear distinction between non-ideal voltage and non-ideal current sources, and we shall now demonstrate that the circuits of figure 2.10 are interchangeable, in the sense that they produce the same straight-line relation between V_{AB} and I_L. The circuit within the broken lines of figure 2.10(b) is known as Norton's equivalent circuit and is an alternative to Thevenin's equivalent circuit, figure 2.10(a).

2.4   Norton's Theorem

Norton's theorem states that any network with two output terminals is equivalent to a current source I_{EQ} with R'_{EQ} in parallel (figure 2.10(b)). As with Thevenin's theorem, it depends on components in the network obeying a linear relation between current and voltage. It follows very simply from Thevenin's theorem. To see the equivalence between the two circuits of figure 2.10, suppose current I_L flows through a load resistor R_L attached to the terminals AB. Then, in the Norton circuit, current $(I_{EQ} - I_L)$ flows through R'_{EQ} and

$$V_{AB} = R'_{EQ}(I_{EQ} - I_L) = R'_{EQ}I_{EQ} - R'_{EQ}I_L.$$

For Thevenin's circuit, $V_{AB} = V_{EQ} - R_{EQ}I_L$. These two results agree if

$$\boxed{R'_{EQ} = R_{EQ}} \tag{2.4}$$

and

$$\boxed{V_{EQ} = R_{EQ}I_{EQ}.} \tag{2.5}$$

These two relations may be checked very simply from the fact that when AB is open circuit $V_{EQ} \equiv R'_{EQ}I_{EQ}$ and on short circuit $I_L = I_{EQ} \equiv V_{EQ}/R_{EQ}$.

The value of R_{EQ} in the Thevenin equivalent circuit was obtained in figure 2.4 from the internal resistance with the batteries short circuited. This is clear from the left-hand side of figure 2.10. If there are current sources present, these are replaced by open circuits when finding R_{EQ}; this follows from an inspection of the right-hand side of figure 2.10.

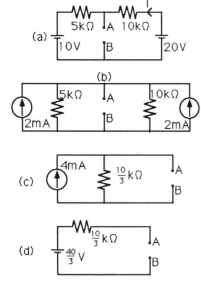

Fig. 2.11 (a) Example for Norton's equivalent circuit, (b) - (d) ways of simplifying it.

Example using Norton's theorem

(i) In figure 2.11(a), if terminals AB are shorted, the current through them is 10/5 mA from the 10 V battery and 20/10 mA from the 20 V battery. So $I_{EQ} = 4$ mA.

(ii) With the batteries replaced by short circuits, R_{EQ} is the resistance across AB, namely 5 kΩ in parallel with 10 kΩ. So $R_{EQ} = 10/3$ kΩ.

(iii) As a check, consider the situation with AB open circuit. In this case, there is a net voltage of 10 V in (a), driving current I in the direction of the arrow through the 5 and 10 kΩ resistors; $I = 10/15$ mA and $V_{AB} = 10 + 5I = 10 + 10/3 = 40/3$ V. This agrees with $I_{EQ} R_{EQ}$ from (i) and (ii).

A circuit can often be simplified quickly and neatly by swopping backwards and forwards between Thevenin and Norton equivalent forms. This is illustrated in figures 2.11(b)–(d). It is a trick worth practising, since it often saves a great deal of algebra. The batteries and resistors of (a) are replaced by equivalent Norton circuits in (b); these are combined in parallel in (c) and then (d) converts back to the Thevenin equivalent form. An important warning is that you must not include the load resistor between terminals A and B in these manipulations: Thevenin's and Norton's theorems apply to the circuits *feeding* terminals AB.

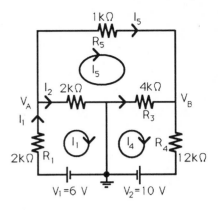

Fig. 2.12. A worked example.

Another worked example

Figure 2.12 reproduces a fairly complicated example from Chapter 1, figure 1.19(a). If all currents and voltages in the circuit are required, it is best to use one of the methods from Chapter 1. Suppose, however, only current I_2 is to be found. It can be obtained straightforwardly by application of Thevenin's and Norton's theorems. The steps are shown in figure 2.13. In (b), V_2 and R_4 are replaced by their Norton equivalent. Then R_3 and R_4 are combined in parallel and (c) returns to the Thevenin equivalent form.

With AB open circuit,

$$V_{EQ} = V_1 - \left(V_1 - \frac{V_2 R_3}{R_3 + R_4}\right) R_1 \left(R_1 + R_5 + \frac{R_3 R_4}{R_3 + R_4}\right)^{-1}$$

$$= 6 - (6 - 2.5)2/(2 + 1 + 3)$$

$$= \frac{29}{6} \text{ V}.$$

With the batteries shorted out, R_{EQ} is given by the parallel combination of R_1 with

$$R_5 + R_3 R_4/(R_3 + R_4)$$

i.e. 2 kΩ in parallel with 4 kΩ, so $R_{EQ} = (4/3)$ kΩ.

As a check, the current through AB when shorted is I_{EQ}:

$$I_{EQ} = \frac{V_1}{R_1} + \frac{V_2 R_3}{R_3 + R_4} \left(R_5 + \frac{R_3 R_4}{R_3 + R_4}\right)^{-1}$$

$$= 3 + 2.5/4 = 29/8 \text{ mA}.$$

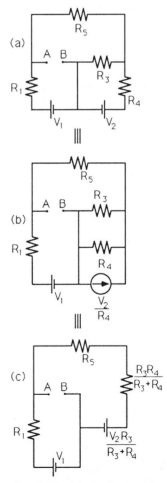

Fig. 2.13. Worked example.

This agrees with V_{EQ}/R_{EQ} as it should. The arithmetic and algebra are sufficiently tortuous that this is a valuable crosscheck.

Finally, the current I_2 of figure 2.12 is

$$I_2 = V_{EQ}/(R_{EQ} + R_2) = \tfrac{29}{6}\left(\tfrac{4}{3} + 2\right)^{-1} = 1.45 \text{ mA}$$

in agreement with the value obtained in the previous chapter.

Further examples are given in the exercises at the end of the chapter. If you can do question 6, you have mastered the vital points of Chapters 1 and 2 up to here.

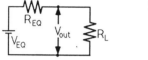

Fig. 2.14. For constant V_{out}, $R_{EQ} \ll R_L$.

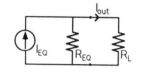

Fig. 2.15. For constant I_{out}, $R_{EQ} \gg R_L$.

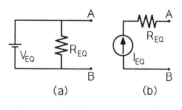

(a)   (b)

Fig. 2.16. Wrong choices for equivalent circuits.

2.5   General Remarks on Thevenin's and Norton's Theorems

(1) Suppose a constant voltage is required across a load R_L, with as little variation as possible when R_L is changed. From figure 2.14, R_{EQ} needs to be small compared with R_L, so that most of V_{EQ} appears across R_L. Thus a constant voltage source should have a low output resistance or **output impedance**, as it is often called.

(2) Conversely, suppose a constant output current is required, independent of load; this is the case, for example, in supplying a magnet or a motor. From figure 2.15, this demands $R_{EQ} \gg R_L$, or high output resistance.

(3) When a circuit is measured with an oscilloscope or voltmeter, it is desirable to disturb the circuit as little as possible, i.e. draw very little current. This requires the detector to have a high input resistance or **input impedance**. Oscilloscopes and multimeters typically have input resistances of 10^6–10^7 Ω. On the other hand, if an ammeter is inserted into a circuit in order to measure current, we want to disturb the current as little as possible. Therefore an ammeter should have a low resistance.

(4) Although Thevenin's and Norton's circuits are equivalent to any network in the sense of giving the same output voltage and current, they are *not* equivalent as regards power consumption *within* the equivalent circuit. You may easily verify that the power dissipated in the Norton equivalent circuit of figure 2.10(b) is different from that dissipated in the Thevenin equivalent circuit (a). This is because power is non-linear in V or I.

(5) Common student howlers are to draw equivalent circuits in the forms shown in figure 2.16. It is worth a moment's thought to see as to why these must be wrong. In the former case, $V_{AB} = V_{EQ}$ independent of load, which gives an absurd result if the terminals are shorted. In the second circuit, $I_{AB} = I_{EQ}$ independent of load, and this is absurd if the terminals are open.

(6) If you encounter a circuit like that in figure 2.16(a) where a resistor is applied directly across a battery, you can ignore the

resistor in forming an equivalent circuit, since V_{AB} is equal to the voltage of the battery, regardless of the resistor value. The resistor dissipates power but has no effect on the equivalent circuit. Likewise, if you meet a circuit like that in figure 2.16(b), you can ignore the resistor in forming an equivalent circuit, since $I_{AB} = I_{EQ}$ regardless of the resistor value.

2.6 Matching

Suppose a given voltage source has fixed characteristics V_{EQ} and R_{EQ}. The question may arise of how to deliver maximum power to a load R_L (figure 2.17(a)). Suppose there is the freedom to vary R_L. The current flowing into the load is

$$I = V_{EQ}/(R_{EQ} + R_L)$$

and the power P dissipated in the load is

$$P = I^2 R_L = V_{EQ}^2 R_L/(R_{EQ} + R_L)^2.$$

This is zero when $R_L = 0$ or when $R_L \to \infty$. In between, there must be a maximum (figure 2.17(b)). When driving a loudspeaker of a discotheque, for example, the aim may be to make maximum noise. Expressed algebraically, this requires $dP/dR_L = 0$. Now

$$\frac{dP}{dR_L} = \frac{V_{EQ}^2}{(R_{EQ} + R_L)^2} - \frac{2V_{EQ}^2 R_L}{(R_{EQ} + R_L)^3} = \frac{V_{EQ}^2(R_{EQ} - R_L)}{(R_{EQ} + R_L)^3}$$

so the power is a maximum when $R_L = R_{EQ}$. In this case, R_{EQ} is said to be **matched** to the load R_L.

There is however a snag with this arrangement. Half of the power is dissipated in the source itself. This is wasteful. If R_{EQ} can be varied, it is still desirable to minimise it, so that most of the power is dissipated in the load.

2.7* Non-linearity

In order to get some insight into Thevenin's theorem, let us find I_2 for the circuit of figure 2.18(a), first algebraically and then by manipulation of Thevenin and Norton equivalents in the network to the left of AB.

Kirchhoff's laws give

$$V_1 = R_1 I_1 + R_3(I_1 - I_2) = (R_1 + R_3)I_1 - R_3 I_2 \qquad (2.6)$$
$$V_2 = R_3(I_2 - I_1) + R_2 I_2 = (R_3 + R_2)I_2 - R_3 I_1. \qquad (2.7)$$

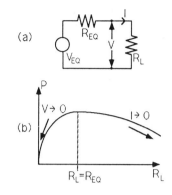

(a)

(b)

Fig. 2.17 (a) A load R_L connected to a voltage source, (b) P as a function of R_L.

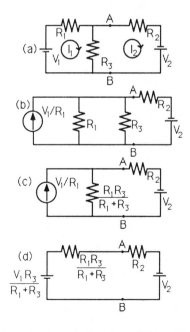

(a)

(b)

(c)

(d)

Fig. 2.18. A new form of Fig. 1.10.

To eliminate I_1 multiply (2.6) through by $R_3/(R_1 + R_3)$:

$$\frac{V_1 R_3}{R_1 + R_3} = R_3 I_1 - \frac{R_3^2 I_2}{R_1 + R_3}. \tag{2.8}$$

Adding to (2.7)

$$V_2 + \frac{V_1 R_3}{R_1 + R_3} = \left((R_3 + R_2) - \frac{R_3^2}{R_1 + R_3}\right) I_2$$

$$= \left(\frac{R_1 R_3}{R_1 + R_3} + R_2\right) I_2. \tag{2.9}$$

In this algebra, V_2 and R_2 were left untouched in the process of eliminating I_1, and this is why Thevenin's theorem works for the network to the left of AB.

The same result may be achieved by manipulation of Thevenin and Norton equivalent circuits as in figure 2.18(b)–(d). The steps in this procedure have a one to one correspondence with the algebra and lead to a circuit (d) directly equivalent to equation (2.9). The diagrammatic method amounts to a visual way of doing the multiplication which derives (2.8) and the addition giving (2.9).

Now let us examine the effect of non-linearity. Suppose a term αI_1^2 were added to the right-hand side of equation (2.6). This could be eliminated by substituting from equation (2.7),

$$I_1 = [(R_2 + R_3)I_2 - V_2]/R_3.$$

It would add to (2.9) a term involving I_2^2, giving a quadratic equation of the form

$$-c = bI_2 + aI_2^2.$$

The solution of this equation involves a term $(b^2 - 4ac)^{1/2}$, where the left-hand side of equation (2.9), $V_2 + V_1 R_3/(R_1 + R_3)$, appears within a square root. (Try it.) Thus the non-linear term αI_2^2 destroys the superposition principle (linear dependence on V_1 and V_2) and also Thevenin's theorem and Norton's theorem.

2.8* Reciprocity

Equations (2.6) and (2.7) have an interesting symmetry. If they are written

$$V_1 = Z_{11} I_1 + Z_{12} I_2 \tag{2.10}$$
$$V_2 = Z_{21} I_1 + Z_{22} I_2 \tag{2.11}$$

then $Z_{12} = Z_{21} = -R_3$. If further loops involving other voltages and currents are added to the circuit, as in figure 2.19, we arrive at more simultaneous equations, but always

$$\boxed{Z_{ij} = Z_{ji}.} \tag{2.12}$$

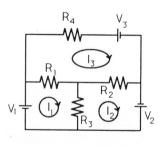

Fig. 2.19. A third loop
added to Fig. 2.18.

This relation is known as **reciprocity**. The origin of the symmetry given by equation (2.12) is easily traced. Whenever a resistor such as R_3 is common to two loops, both currents create a potential across it and the same magnitude of voltage appears in each loop. Specifically, $R_3(I_1 - I_2)$ appears in equation (2.6) and $R_3(I_2 - I_1)$ in (2.7). With the addition of further loops, the same symmetry persists for all pairs of mesh currents.

From (2.9),

$$I_2 = \frac{(R_1 + R_3)V_2 + V_1 R_3}{R_1 R_3 + R_1 R_2 + R_2 R_3} = Y_{21}V_1 + Y_{22}V_2. \qquad (2.13)$$

Then I_1 is found by interchanging indices 1 and 2:

$$I_1 = \frac{(R_2 + R_3)V_1 + V_2 R_3}{R_2 R_3 + R_2 R_1 + R_1 R_3} = Y_{11}V_1 + Y_{12}V_2. \qquad (2.14)$$

So

$$Y_{21} = Y_{12} = R_3/(R_1 R_3 + R_1 R_2 + R_2 R_3) \qquad (2.15)$$

and the symmetry

$$\boxed{Y_{ij} = Y_{ji}} \qquad (2.16)$$

applies to the solution too.

These symmetries are valuable as checks against mistakes and this is the main motivation for pointing them out. They can also be used to simplify the computer programs used to solve the equations: nearly half the Z_{ij} are redundant. The implication of the result $Y_{12} = Y_{21}$ is that the current I_1 originating from V_2 is the same as the current I_2 originating from the *same* voltage V_1. Very occasionally, this can be used to solve problems directly. Suppose, for example, a battery V is applied across the Wheatstone bridge of figure 2.20(a), producing a certain current I_G through the galvanometer. If the battery and galvanometer could be interchanged without changing *any* resistances in the circuit (including R_0 and R_G), the current through the galvanometer would be the same.

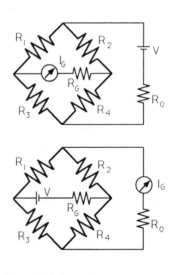

Fig. 2.20 (a) Wheatstone bridge, (b) battery and galvanometer interchanged.

2.9   Amplifiers

This section takes a brief look at transistor amplifiers where Thevenin and Norton equivalent circuits play a vital role. An amplifier is an essential element of any electronic system. It amplifies voltage or current or both. Much later, in Chapter 17, it will be shown that a voltage amplifier may be represented by the circuit shown in figure 2.21.

Internally, the amplifier may be very complicated. Whatever the complexity, the input and output may be represented by Thevenin

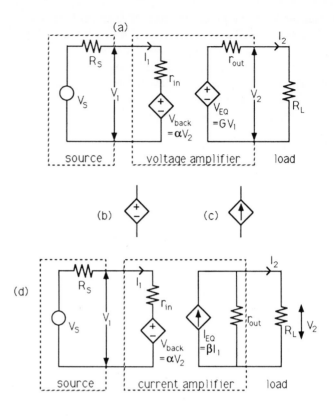

Fig. 2.21 (a) A voltage amplifier, (b) a dependent
voltage source, (c) a dependent current source,
(d) a current amplifier.

equivalent circuits, providing voltage and current are linearly related. In reality, the back-EMF V_{back} in the input side of the amplifier is often negligibly small and the input then behaves simply as an input resistance r_{in} (although this is often the slope resistance of a transistor rather than an actual resistor).

The output circuit consists of a voltage generator $V_{\text{EQ}} = GV_{\text{in}} = GV_1$ amplifying the input voltage, together with an impedance r_{out}. The voltage generator is called a **dependent voltage source** because it depends on another voltage. The symbols for a dependent voltage source and a dependent current source are shown in figures 2.21(b) and (c). The circuit of figure 2.21(a) obeys the equations

$$V_2 = GV_1 - r_{\text{out}} I_2 \qquad (2.17a)$$

$$V_1 = r_{\text{in}} I_1 + \alpha V_2. \qquad (2.17b)$$

For efficient coupling of source to load, r_{in} should be large compared with R_{S}; and r_{out} should be small compared with R_{L}. Then

$V_1 \simeq V_S$ and $V_2 \simeq V_{EQ}$. Figure 2.21(a) is a good representation of a field effect transistor (FET) at low frequencies and in this case $V_{\text{back}} = 0$.

A current amplifier instead amplifies current (figure 2.21(d)). This is obtained from figure 2.21(a) simply by replacing the output Thevenin equivalent circuit by the Norton form. Technically there is no difference between these two forms. Which is more convenient depends on whether G is approximately constant as operating conditions vary (as is the case for the FET) or β (as is the case for the bipolar transistor). For figure 2.21(d),

$$I_2 = \beta I_1 - (1/r_{\text{out}})V_2 \qquad (2.18a)$$

$$I_1 = (V_S - \alpha V_2)/(r_{\text{in}} + R_S). \qquad (2.18b)$$

In this case, efficient coupling from source to load demands that r_{in} is small (I_1 large) and r_{out} large compared to R_L (I_2 large).

For both types of amplifier, the equations can be rearranged by simple substitutions to read in the form of equations (2.10)–(2.14):

$$V_2 = Z_{21}I_1 + Z_{22}I_2 \qquad (2.11)$$

$$V_1 = Z_{11}I_1 + Z_{12}I_2 \qquad (2.10)$$

or

$$I_2 = Y_{21}V_1 + Y_{22}V_2 \qquad (2.13)$$

$$I_1 = Y_{11}V_1 + Y_{12}V_2. \qquad (2.14)$$

However, reciprocity is no longer obeyed: $Z_{12} \neq Z_{21}$ and $Y_{21} \neq Y_{12}$. This can be traced to the presence of transistors in the amplifier and hence to non-linearity.

The general appearance of these equations takes the form of two simultaneous equations relating output voltage and current to input voltage and current. This is the general form of a **two-port network**, having input voltage and current and output voltage and current. By Thevenin's or Norton's theorems, any linear amplifier will behave in this way, regardless of what components are used to build it: valves, bipolar transistors or FETs.

In section 2.1, one simple method of finding R_{EQ} for Thevenin's equivalent circuit was to run down voltage sources to zero (i.e. short circuit them) and replace current sources by open circuits. This method fails if the circuit contains dependent sources. To illustrate this point, consider as an example the circuit of figure 2.22. On open circuit, $V_2 = V_1 - \alpha V_2$, so across terminals A and B $V_{EQ} = V_1/(1 + \alpha)$. On short circuit $I_{EQ} = V_1/R_1$. Hence $R_{EQ} = V_{EQ}/I_{EQ} = R_1/(1 + \alpha)$. This is the correct result. If, however, one imagines the batteries shorted, method (iii) of section 2.1 would lead to the incorrect result that $R_{EQ} = R_1$. The discrepancy arises because the circuit contains a dependent source and it is no longer

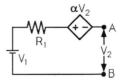

Fig. 2.22. Finding R_{EQ} with a dependent source.

permissible to scale voltages and currents down to zero. For further examples demonstrating this point, see exercises 15–17 at the end of the chapter.

2.10   Systems

Amplifiers are examples of building blocks which go into the construction of an electronic **system**. It is convenient to be able to break down the performance of a large system, such as that of figure 2.23, into subsystems, each of which performs a specific function. Each unit produces an output which is some function of its inputs, sometimes simply proportional to the input but sometimes involving differentiation or integration or perhaps a time delay. This function is called the **transfer function** T or T' of the unit. Each unit has inputs and outputs expressed in Thevenin or Norton equivalent form.

In Chapter 7 we will demonstrate that the properties of the system can be radically altered by **feedback** of a signal to an early stage from a later one. For example, the output V_3 in figure 2.23 might be a signal controlling the speed of a motor; if V_{in} is a signal describing the required speed, it may be convenient to drive the electronics with a signal $(V_{in} - V_3)$ describing the discrepancy between required speed and actual speed. In this case, $\beta = -1$. This is an example of a servosystem. There are also situations where it is convenient to feed a signal forward, skipping intermediate stages; this is illustrated by αV_{in} in the figure.

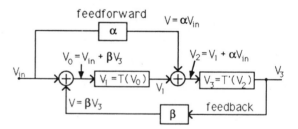

Fig. 2.23. An electronic system including feedback and feedforward.

The behaviour of figure 2.23 is described algebraically by a set of network equations involving the various voltages and transfer functions. Systems design concerns the analysis of these network equations, for example their time dependence and stability. Building a system is nowadays largely a matter of assembling units available commercially: power supplies, amplifiers, digital gates, analogue to digital converters and so forth. The following chapters will be concerned with understanding the performance possible from each type of unit.

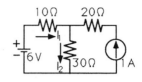

Fig. 2.24.

2.11   Summary

When circuits are linear, Thevenin's and Norton's theorems offer shortcuts. The Thevenin equivalent voltage across two terminals is the open-circuit voltage, figure 2.2, and the Norton equivalent current source is equal to the short-circuit current, figure 2.3. For either, $R_{\rm EQ}$ is obtained by shorting batteries and replacing current sources by open circuits, figure 2.4, unless the circuit contains dependent voltage or current sources. Reciprocity is a useful symmetry check in resistive networks against mistakes: $Z_{ij} = Z_{ji}$ and $Y_{ij} = Y_{ji}$. Amplifiers may be represented by Thevenin or Norton input and output circuits, with the output generator proportional to input current or voltage.

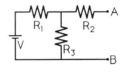

Fig. 2.25.

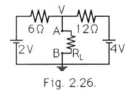

Fig. 2.26.

2.12   Exercises

1.   *Thevenin's theorem.* A circuit produces voltage V on open circuit. When it is measured with a voltmeter having an impedance of 20 kΩ, the voltmeter reads 15 V; when the voltmeter is switched to a range having an impedance of 50 kΩ, it reads 18 V. What is V? (Ans: 20.8 V.)

2.   *Thevenin's theorem.* (QMC 1987) Calculate the currents I_1 and I_2 in the circuit of figure 2.24. Calculate also the potential difference across the current source and indicate its sign. (Ans: $I_1 = -0.6$ A, $I_2 = 0.4$ A, 32 V.)

3.   *Thevenin and Norton.* Find the Thevenin and Norton circuits equivalent to figure 2.25. (Ans: $V_{\rm EQ} = V R_3/(R_1 + R_3)$; $R_{\rm EQ} = (R_1 R_2 + R_2 R_3 + R_3 R_1)/(R_1 + R_3)$; $I_{\rm EQ} = V R_3/(R_1 R_2 + R_2 R_3 + R_3 R_1)$.)

4.   *Thevenin, Norton and superposition.* Find from first principles the voltage V across the load resistor $R_{\rm L}$ in figure 2.26. Find the Thevenin equivalent of the circuit supplying terminals AB and hence check V. (Ans: $V = 8 R_{\rm L}/[3(4 + R_{\rm L})]$; $V_{\rm EQ} = 8/3$ V, $R_{\rm EQ} = 4$ Ω; $I_{\rm EQ} = 2/3$ A.)

5.   *Norton's theorem.* (RHBNC 1988) Find the Norton equivalent circuit across terminals AB of figure 2.27. (Ans: $I_{\rm EQ} = 0.75$ A, $R_{\rm EQ} = 4$ Ω.)

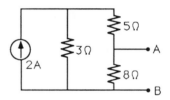

Fig. 2.27.

6**.   *Series and parallel resistors, mesh currents, node voltages, Thevenin and Norton.* (a) Write down the equations for the mesh currents I_{1-4} in figure 2.28, but do not solve them. (b) Find the current $I_3 + I_4$ supplied by the battery by simplifying

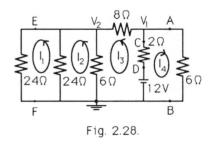

Fig. 2.28.

series and parallel combinations of resistors. (c) Find V_1 and V_2. (d) Hence find I_{1-4} and check them against (a). (e) Find the Thevenin equivalent of the circuit to the left of AB; use this to check I_4. (f) Find the Norton equivalent of the circuit across the terminals CD. Use this to check $I_3 + I_4$. (g) Find the Norton equivalent of the circuit to the right of EF; use it to check I_1. (Ans: (a) $I_3 + I_4 = 2$ A; (c) $V_1 = 8$ V; $V_2 = 8/3$ V; (d) $I_1 = 1/9$ A; $I_2 = 2/9$ A; $I_3 = 2/3$ A; $I_4 = 4/3$ A; (e) $V_{EQ} = 144/14$ V, $R_{EQ} = 24/14$ Ω; (f) $I_{EQ} = 3$ A, $R_{EQ} = 4$ Ω; (g) $I_{EQ} = 18/19$ A, $R_{EQ} = 24 \times 19/143$ Ω.)

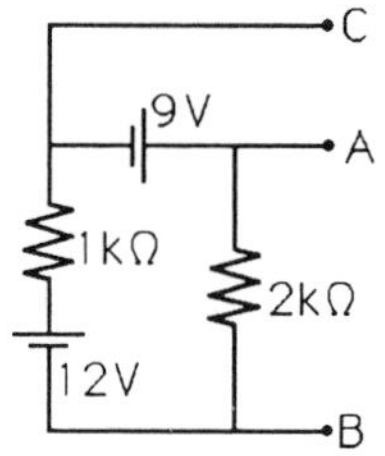

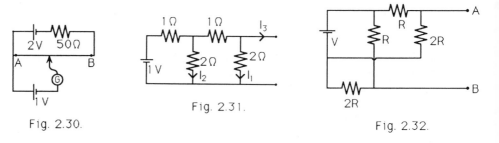

Fig. 2.29.

7. *Thevenin and Norton.* Obtain the Thevenin and Norton circuits equivalent to figure 2.29, considering as output terminals first AB then AC. (Ans: For AB, $V_{EQ} = 14$ V, $R_{EQ} = 2/3$ kΩ, $I_{EQ} = 21$ mA; for AC, $V_{EQ} = 9$ V, $R_{EQ} = 0$, $I_{EQ} = \infty$.)

Fig. 2.30.

Fig. 2.31.

Fig. 2.32.

8. *Sensitivity of a bridge.* (QMC 1984) In the bridge circuit shown in figure 2.30, the bridge wire AB is 1 m long and has a resistance of 200 Ω; the galvanometer has a resistance of 37.5 Ω and gives a noticeable reading for a current of 10 μA. Construct an equivalent circuit for the voltage or current applied to the galvanometer. Hence find (i) the position of the sliding contact for balance; (ii) the distance the contact may be moved before the galvanometer gives a noticeable reading. (Ans: 62.5 cm from A; 0.625 mm.)

9. *Thevenin, Norton and nodes.* (QMC 1968) Find the Thevenin and Norton equivalents of figure 2.31. (Ans: $V_{EQ} = 4/11$ V; $R_{EQ} = 10/11$ Ω; $I_{EQ} = 2/5$ A.)

10*. *Thevenin.* (QMC 1970) What is the Thevenin equivalent of the circuit in figure 2.32 to the left of AB? (Ans: $V_{EQ} = 0$; $R_{EQ} = 4R/3$.)

11**. *Reciprocity.* (QMC 1987) Use the reciprocity theorem to calculate the value of R_1 in figures 2.33(a) and (b). (Hint: Relate the voltages V_{AB} and V_{CD} to currents I_1 and I_4 by

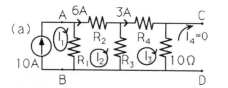
(a)

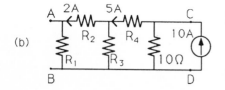
(b)

Fig. 2.33.

$I_1 = Y_{11}V_{AB} + Y_{14}V_{CD}$ and $I_4 = Y_{41}V_{AB} + Y_{44}V_{CD}$.) As a check, use mesh currents to solve for all of R_{1-4}. (Ans: $R_1 = 15\ \Omega$; $R_2 = 15/4\ \Omega$, $R_3 = 12.5\ \Omega$, $R_4 = 2.5\ \Omega$.)

12*. *Thevenin and Reciprocity.* (After UC 1974) Show that the Wheatstone bridge of figure 2.34 has an output voltage

$$V_{out} = \frac{E(R_2 R_3 - R_1 R_4)}{(R_1 + R_2)(R_3 + R_4)}.$$

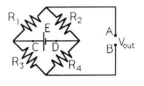

Fig. 2.34.

If R_2 is a platinum resistance thermometer of resistance $R_t = R_0(1 + \alpha t)$, where R_0 and R_t are the resistances at $0\ °C$ and $t\ °C$, respectively, and R_1, R_3 and R_4 are all equal to R_0, show that V_{out} is proportional to t if $\alpha t \ll 1$. What is then the output impedance of the circuit, assuming the battery has negligible internal resistance? What must be the input impedance of a DC amplifier connected across AB to ensure the maximum power transfer to the DC amplifier from the thermometer circuit?

What is the current through AB if the amplifier has input impedance R_{in}? If the battery E is moved from CD to the arm AB, leaving R_{in} in the arm AB, solve for I_{CD} (lengthy algebra) and hence verify the reciprocity theorem. (Ans: $V_{out} = \alpha Et/4$; output resistance $= R_0 =$ input impedance of the amplifier; $I_{CD} = \alpha Et/[4(R_0 + R_{in})]$.)

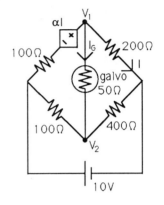

Fig. 2.35.

13. *Dependent voltage.* In figure 2.35, the dependent voltage source is of magnitude αI V, where I is in mA. What is the value of α when the bridge is balanced, i.e. $I_G = 0$? If α is instead 0.6 of this value, what are V_1, V_2 and I_G? (Hint: write three expressions for I_G.) (Ans: 0.05; $V_1 = 7.62$ V, $V_2 = 7.77$ V, $I_G = -3$ mA.)

14. *Dependent current.* Find I and V_X in figure 2.36. What power is supplied by the battery and by the dependent current source? What powers are dissipated in the 3 kΩ, 2 kΩ and 6 kΩ resistors? (Ans: 0.4 mA, 4.8 V; 2.4 mW, 33.6 mW, 0.48 mW, 11.52 mW, 24 mW.)

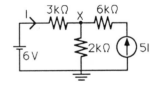

Fig. 2.36.

15. *Thevenin and dependent voltage.* In figure 2.37, terminals XY are open circuit. Find the Thevenin and Norton equivalents of the circuits to the left of AB. Note that in (b) method (iii) of section 2.1 is not applicable. (Ans: (a) $V_{EQ} = \alpha V_1 R_2/(R_1 + R_2)$, $I_{EQ} = \alpha V_1/R_1$, $R_{EQ} = R_1 R_2/(R_1 + R_2)$; (b) $V_{EQ} = \alpha V_1 R_2/[R_1 + R_2(1 + \alpha)]$, $I_{EQ} = \alpha V_1/R_1$, $R_{EQ} = R_1 R_2/[R_1 + R_2(1 + \alpha)]$.)

16. *Thevenin and dependent current.* Find the Thevenin equivalent resistance of the circuit feeding terminals AB of figure 2.38. For what value of α is $R_{EQ} \leq 0$? What value of α

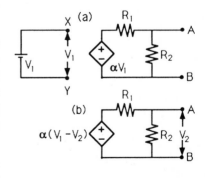

Fig. 2.37.

would you choose for (a) an ideal voltage source, (b) an ideal current source? (Ans: $42/(10 - \alpha)$ kΩ; $\alpha \geq 10$; (a) $\alpha = -\infty$, (b) $\alpha = 10$.)

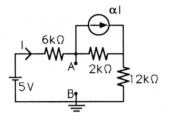

Fig. 2.38.

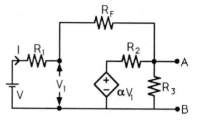

Fig. 2.39.

17**. *Thevenin and Norton.* Find the Thevenin and Norton equivalents of the circuit to the left of terminals AB in figure 2.39. (Ans: $V_{\text{EQ}} = V R_3 (R_2 + \alpha R_\text{F}) / [(R_1 + R_\text{F}) R_2 + R_3 (R_1 + R_\text{F} + R_2 - \alpha R_1)]$, $I_{\text{EQ}} = V(1 + \alpha R_\text{F}/R_2)/(R_1 + R_\text{F})$; $R_{\text{EQ}} = V_{\text{EQ}}/I_{\text{EQ}}$.)

3

Capacitance

3.1 Charge and Capacitance

Capacitance plays a crucial role in understanding the speed at which electronics can switch. This is a fundamental issue in high-speed computers and telecommunications. In a typical computer today, transistors switch in a few nanoseconds (10^{-9} s). Oscilloscopes capable of sampling and displaying waveforms at 10^{-10} s intervals are available (at a price). Oscillators working up to a frequency of 100 GHz (10^{11} cycles/s) are now appearing, and faster devices and lower operating currents are always being looked for. It turns out that capacitance is the feature which sets these limits. We shall first introduce the concept with an idealised example and then go on to discuss practical everyday capacitors.

Suppose an isolated plate is suspended in space. If charge is to be assembled on the plate, work must be done against the electrostatic repulsion between the charges. The necessary potential energy may be supplied by a battery, as in figure 3.1. If a second plate is connected to earth and brought up close to the first one, as in figure 3.2, charges will flow from earth, producing a negative charge on the lower plate. The charges get as close to neutralising one another as the geometry allows, and ideally the charge on the lower plate exactly balances that on the first. In this case, the result is a system having net zero charge, with no external electric field. Viewed from a distance, it looks neutral (or more precisely, like a dipole). If the separation of the plates is s, there is an electric field $\boldsymbol{E} = V/s$ running directly from one plate to the other (figure 3.3).

This pair of plates is the simplest version of a **capacitor**: a system having no net charge, but balancing positive and negative charges stored on two separate plates. The charge Q stored on each plate is related to the applied voltage V by

$$\boxed{Q = CV} \tag{3.1}$$

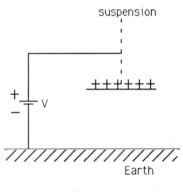

Fig. 3.1. Charging a plate.

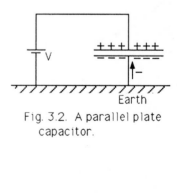

Fig. 3.2. A parallel plate capacitor.

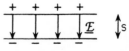

Fig. 3.3. Electric field within a capacitor.

39

where C is the **capacitance**, measured in **farads** (F). Capacitors can be made from plates of any shape. However, the linear relation between Q and V continues to hold, and can be proved in general from Coulomb's law. The magnitude of the capacitance depends on the geometry. For parallel plates of area A and separation s,

$$C = \epsilon_0 A/s \qquad (3.2)$$

where ϵ_0 is a fundamental constant, $10^7/4\pi c^2 = 8.854 \times 10^{-12}$ F m^{-1}, and c is the velocity of light. The units are such that the capacitance C is usually very small. For example, if $A = 1$ cm^2 and $s = 0.1$ cm, $C = 0.885 \times 10^{-12}$ F. Thus capacitances are commonly of order picofarads (10^{-12} F, written pF or $\mu\mu$F), or at most microfarads 10^{-6} F, written μF.

Switching speed

Capacitances between the terminals of a transistor are typically 0.5–50 pF. When the transistor switches, the voltages between its terminals change and this requires movement of charges. From equation (3.1),

$$I = \frac{dQ}{dt} = C\frac{dV}{dt} \qquad (3.1a)$$

so the rate at which the voltage can change is governed by C and the available current I. Fast switching requires large I (hence small circuit resistance) or small C or both. One of the virtues of miniaturising transistors in integrated circuits is the reduction in capacitance: in such circuits, capacitances of 1.5×10^{-15} F μm^{-2} are achieved.

The linear relation (3.1) between Q and V may be checked experimentally with the circuit of figure 3.4. When the switch is closed, the oscilloscope records a signal proportional to current I like that shown in figure 3.5. It is easy to demonstrate on the oscilloscope that the magnitude of I scales with V, but that otherwise the shape does not change. A formula for the curve will be derived later. The area under it measures the charge Q which has accumulated on each plate:

$$Q = \int_0^t I \, dt. \qquad (3.3)$$

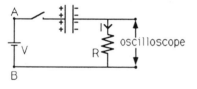

Fig. 3.4. Circuit to charge a capacitor.

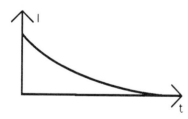

Fig. 3.5. Current v. time.

When current flows 'through' the capacitor, positive and negative charges accumulate at the same rate on the two plates, so the current flowing into one plate equals that leaving the other, and there appears to be a continuous current through the capacitor.

The reality of the charge stored on the plates may be demonstrated by removing the battery and then shorting AB: the oscilloscope records a current in the reverse sense while the capacitor discharges. Whenever large DC voltages are applied, capacitors should be discharged after use, since the stored charge can be hazardous to anyone touching the terminals.

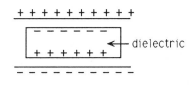

Fig. 3.6. Effect of a dielectric.

3.2 Energy Stored in a Capacitor

Energy must be supplied in order to assemble charge on the plates of a capacitor. This energy is given by

$$E = \int IV \, \mathrm{d}t = \int \frac{\mathrm{d}Q'}{\mathrm{d}t} \frac{Q'}{C} \, \mathrm{d}t = \frac{1}{C} \int_0^Q Q' \, \mathrm{d}Q'$$
$$= \tfrac{1}{2}Q^2/C = \tfrac{1}{2}QV = \tfrac{1}{2}CV^2. \qquad (3.4)$$

It is stored as potential energy and can be returned when the capacitor is discharged.

3.3 The Effect of a Dielectric

Suppose an isolated capacitor carries charge Q and voltage V and suppose an insulator is inserted between the plates. The electric field polarises the dielectric (figure 3.6) and the net field within the material is now equal to the original field less that due to the induced charges. Because the field drops, the potential difference between the plates likewise drops: remember $\boldsymbol{E} = V/s$. The charge Q has not changed. Hence $C = Q/V$ has increased. It goes up by a factor ϵ which is called the **permittivity** or **dielectric constant** of the material:

$$C \rightarrow \epsilon\epsilon_0 A/s. \qquad (3.2a)$$

Alternatively, if the plates are connected to an external battery so as to keep V constant, then Q increases: more charge is stored.

3.4 Practical Capacitors

Values of ϵ for some materials are shown in table 3.1. To make a large capacitance of order 1 μF requires big A and small s.

    Paper capacitors. One way of achieving this is to roll up two sheets of aluminium foil, separated by two sheets of insulating paper, like a Swiss roll, as sketched in figure 3.7. Such capacitors cover the range 0.001 to 1 μF and are readily available with working voltages of 100 to 1500 V. The colour coding scheme, illustrated in figure 3.8, is similar to that for resistors.

    Mica capacitors. A second form of capacitor, generally using mica as dielectric, is illustrated in figure 3.9. Values in the range 5 to 5000 pF are available, with working voltages up to 2 kV.

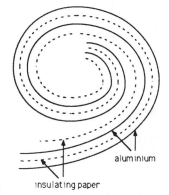

Fig. 3.7. How to make a paper capacitor.

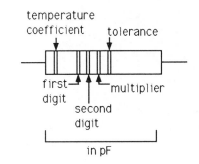

Fig. 3.8. Coding of paper capacitors.

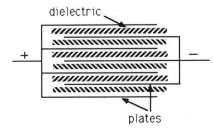

Fig. 3.9. The mica capacitor.

Table 3.1   Dielectric constants ϵ and dielectric strength.

Dielectric	ϵ	Dielectric strength (V mm^{-1})
Free space	1	∞
Air	1.0006	3 000
Paper	2.5	20 000
Transformer oil	4	16 000
Mica	5	200 000
Glass	6	12 000
Barium strontium titanate (ceramic)	7500	3 000

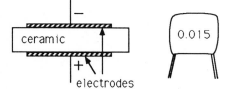

Fig. 3.10. The ceramic capacitor.

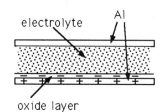

Fig. 3.11. The electrolytic capacitor.

Ceramic capacitors. Barium strontium titanate has a very large dielectric constant, and simple, cheap capacitors are made, as in figure 3.10, by depositing electrodes on to the ceramic and then encapsulating the material to provide electrical insulation. Capacitors from 1 pF to 0.1 μF are available with working voltages up to 20 kV.

Electrolytic capacitors. Capacitors in the range 1 to several hundred μF are often electrolytic (figure 3.11). Two strips of aluminium are separated by gauze soaked in an electrolyte of aluminium hydroxide. If the proper polarity is applied, an oxide layer a few nanometres thick forms on the positive strip. This layer is a good insulator and forms a capacitor with the positive plate. However, if the wrong polarity is applied, the oxide layer is destroyed and the electrolyte conducts; gas is liberated by the electrolysis and may cause the capacitor to explode! Electrolytic capacitors must not be used with alternating current. The negative terminal is indicated by a black band or negative sign. It is better, where possible, to use tantalum capacitors.

3.5   Capacitors in Parallel

In the parallel arrangement shown in figure 3.12, the same voltage V exists across both capacitors. Thus

$$Q_1 = C_1 V \qquad Q_2 = C_2 V$$

and the total charge stored Q is

$$Q = Q_1 + Q_2 = (C_1 + C_2)V.$$

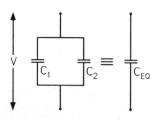

Fig. 3.12. Capacitors in parallel

The arrangement behaves like an equivalent capacitor with capacitance

$$\boxed{C_{EQ} = C_1 + C_2.}$$ (3.5)

It is physically obvious that parallel capacitances will add in this way, because of the increase in the area on which charge is stored.

3.6 Capacitors in Series

In the series arrangement (figure 3.13)

$$V = V_1 + V_2 = Q_1/C_1 + Q_2/C_2.$$

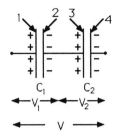

Because the plates labelled 2 and 3 are isolated electrically from the outside world, the net charge stored on them must be zero, so $Q_1 = Q_2$. Hence this arrangement is equivalent to a capacitance C_{EQ} where

Fig. 3.13. Capacitors in series.

$$V = Q/C_{\mathrm{EQ}} = Q/C_1 + Q/C_2$$

and

$$\boxed{\frac{1}{C_{\mathrm{EQ}}} = \frac{1}{C_1} + \frac{1}{C_2}.} \tag{3.6}$$

Can you make sense of equations (3.5) and (3.6) in terms of the total energy stored, using equations (3.4)?

3.7 The CR Transient

The remainder of the chapter will be concerned with the question of switching speed. If, as in figure 3.14, a pulse generator applies a square pulse to a resistor and capacitor in series, the waveforms observed across C and R are as sketched in the figure. They are called **transients**. The voltage V_R across the resistor decays exponentially from an initial value V_0 according to

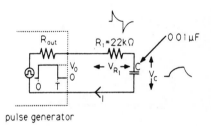

Fig. 3.14. Charging a capacitor.

$$V_R \propto V_0 \mathrm{e}^{-t/CR}.$$

First we derive the exponential shape of V_R, then the result will be extended to more elaborate situations where voltage steps follow one another faster than V_R can settle to zero.

In figure 3.14, suppose the capacitor is initially uncharged and the applied voltage is zero. At time $t = 0$, the generator applies a single square voltage pulse of height V_0 for a time T and then returns to zero. Physically, what happens is that the capacitor charges towards the DC voltage V_0. After the capacitor has charged fully, it settles at this voltage. However, initially the capacitor is uncharged and all of V_0 is across $(R_{\mathrm{out}} + R_1)$; current $I_0 = V_0/R$ flows, where R is the total resistance $(R_{\mathrm{out}} + R_1)$. This current begins to charge the capacitor; the voltage across R then falls, and with it the current. Thus $I(t)$ falls steadily until either the capacitor is fully charged or the input pulse ends. When the latter happens, the sequence is reversed and the capacitor discharges through R_{out} and R_1.

Algebraically, we need to apply Kirchhoff's voltage law to the circuit. The voltage across the resistors is $V_R = IR$ and that across the capacitor is Q/C, so

$$V_0 = RI + \frac{Q}{C} = RI + \frac{1}{C}\int_0^t I(t')\,dt'. \tag{3.7}$$

Differentiating with respect to time,

$$0 = R\frac{dI}{dt} + I/C. \tag{3.8}$$

This is a simple **differential equation**. It involves differentials only as high as dI/dt, and is therefore called a first-order differential equation. A circuit obeying such an equation is called a **first-order system**.

Exponential decay
From (3.8),

$$dI/dt = -I/CR \tag{3.8a}$$

i.e., current decreases at a rate proportional to its magnitude. The curve of figure 3.5 clearly has this property. This type of equation arises very frequently and its solution is named the **exponential function**. Its properties are reviewed systematically in appendix B. It occurs so commonly (like trigonometric functions) that it is available on most hand calculators.

The equation is solved by writing

$$dI/I = -dt/CR$$
$$\ln I = -t/CR + \text{constant}$$

$$\boxed{I = I_0\exp(-t/CR).}$$

The solution may be checked by differentiating it and substituting back into equation (3.8a):

$$dI/dt = -(1/CR)I_0\exp(-t/CR) = -I/CR.$$

The differential equation does not determine I_0, which arises from the integration constant. Any first-order differential equation involves one such constant. Any first-order differential equation involves one such constant. To find it, the **initial conditions** must be considered. At time $t = 0$, $I = I_0 = V_0/R$, so

$$I = (V_0/R)\exp(-t/\tau).$$

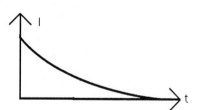

Fig. 3.5 again.

Here

$$\boxed{\tau = CR} \tag{3.9}$$

is called the **time constant** (or relaxation time). It is the time in which I falls from I_0 to $I_0/e \simeq 0.37 I_0$. If C is in farads and R in ohms, τ is in seconds. As a numerical example, suppose $C = 0.01~\mu\text{F}$ and $R = 22~\text{k}\Omega$; then $\tau = 0.22$ ms. Finally, the algebraic expressions for all waveforms are:

$$V_R = V_0 e^{-t/\tau}$$
$$V_{R1} = (R_1/R) V_R$$
$$V_C = V_0(1 - e^{-t/\tau}) \tag{3.10}$$
$$Q = CV_0(1 - e^{-t/\tau}).$$

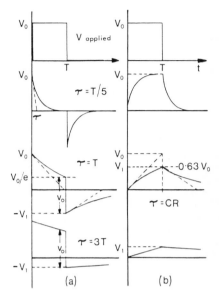

The form of these results is displayed in figure 3.15 for several values of τ. This is known as the **transient response** of the circuit, i.e. its response to a change in DC conditions.

Show that the tangent to $V_R(t)$ at $t = 0$ goes through $V = 0$ at $t = \tau$.

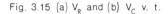

Fig. 3.15 (a) V_R and (b) V_C v. t.

It is well worth experimenting yourself with these waveforms using an oscilloscope and square waves from a pulse generator. Suppose you want to determine τ experimentally. A little trick is to make the height of the square wave 2.7 cm on the oscilloscope trace. Then τ is given by the time at which V_C deviates from its asymptotic value by 1 cm; at this time, V_R will read 1 cm on the trace.

In the circuit of figure 3.14, the source supplies charge Q at voltage V_0 and hence delivers energy QV_0; half of this is stored in the capacitor, but where does the other half go?

All practical devices have some capacitance. An oscilloscope has an input capacitance of typically 10–25 pF and so does a transistor. The coaxial cable used to connect to an oscilloscope has a capacitance of typically 150 pF per metre.If they are to follow an input signal rapidly, this capacitance must be charged quickly, i.e. V_C must follow a step in voltage as closely as possible. This requires small τ, i.e. small CR.

In a digital circuit, a capacitance of 10 pF is to be charged to 5 V through a 50 Ω resistor. What is τ and what is the initial charging current?

If the input pulse is long enough, V_C eventually reaches V_0 and $V_R \to 0$. Suppose, however, that the input ends before the capacitor is fully charged, as in the lower half of figure 3.15. The subsequent discharge of the capacitor is governed by algebra very similar to the charging process, but with different initial conditions. We now find a set of expressions analogous to equation (3.10) describing the resulting waveforms.

The charge on the capacitor cannot change instantly when the input voltage drops to zero. The change V_0 in the input pulse must therefore appear in full in the waveform V_R. The new voltage across the resistor is $-V_1$ and current $-V_1/R$ flows initially. The differential equation (3.8) has not changed, so the capacitor discharges from voltage V_1 towards zero with the same time constant τ as before. The previous expressions for I, V_R, etc, can be taken over with t replaced by $(t-T)$:

$$I = -(V_1/R)e^{-(t-T)/\tau}$$
$$V_R = -V_1 e^{-(t-T)/\tau}$$
$$V_{R1} = (R_1/R)V_R$$
$$V_C = V_1 e^{-(t-T)/\tau}$$
$$Q = CV_C.$$

These results are used in figure 3.15. It is recommended that readers familiarise themselves with exponentials by plotting out a waveform for one value of T (e.g. $T = \tau$) using a hand calculator.

After a capacitor has charged up, the full voltage V_0 from the source appears across it and the voltage across the resistors is zero. The capacitor is said to block direct current, because no current flows; this is one of its common uses. The input to an oscilloscope labelled 'AC' has such a capacitor, as shown in figure 3.16. The AC input is useful for eliminating unwanted DC levels which may be present in a circuit being studied. For example, you may want to look at a very small AC signal superimposed on a large DC background and it would be inconvenient if the scale of the display were dictated by the DC signal. The time constant of the circuit in figure 3.16 is $C(R + R_{in})$. The value of C is made large so that AC current flows as if the capacitor were transparent. In practice, CR_{in} is chosen to be about 0.1 s. The capacitor is called a **coupling capacitor** and the voltage source is said to be **AC coupled** to the oscilloscope.

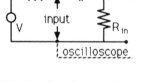

Fig. 3.16. AC input to an oscilloscope.

Leakage resistance

No insulator is perfect, so in practice charge in an isolated capacitor will gradually leak through the dielectric and the capacitor will discharge. An equivalent circuit describing this is shown in figure 3.17. The time constant for the discharge is CR_{leak}.

Fig. 3.17. Leakage resistance.

3.8   AC Coupling and Baseline Shift

Suppose the input signal to a CR circuit figure 3.18(a) has the irregular shape shown in figure 3.18(b). The voltage V_C across the capacitor will follow with time constant CR, as shown in figure 3.18(c).

Consider next what happens if the time constant CR is large and a train of square waves, as shown in figure 3.19(a), is applied from a pulse generator to the circuit of figure 3.16. The ratio p/q of the times for which the input is high and low respectively is called the **mark-to-space ratio**. Because of the large time constant, the circuit responds sluggishly and V_C tends towards the mean value of the input voltage,

$$\bar{V}_C = V_0 p/(p+q).$$

At the arrival of the first pulse, the capacitor begins to charge towards V_0, but it reaches only $V_1 \simeq V_0 t/\tau$ before the end of the pulse. The capacitor then begins discharging according to $V_C = V_1 \exp(-t/\tau)$. The discharge is incomplete before the second pulse arrives, so the capacitor acquires some net charge during the first cycle. This process continues during subsequent pulses, as shown in figure 3.19(b). Eventually equilibrium is reached when the rates of charging and discharging are equal.

It is easy to treat this quantitatively. The charging current over the duration of the square pulse is $(V_0 - V_C)/R'$, where $R' = R_{\text{in}} + R$; the discharging current in the absence of the input pulse is V_C/R'. When these two currents are equal on average:

$$(V_0 - \bar{V}_C)p/R' = \bar{V}_C q/R'$$

$$\bar{V}_C = V_0 p/(p+q).$$

The voltage V_X across R_{in} drifts negative by $\bar{V}_C R_{\text{in}}/R'$, due to the discharge current flowing through the resistances (figure 3.19(c)). This is called **baseline shift**. The discharge current may be put to use to make a simple **ratemeter**, producing an output signal proportional to the rate of the input pulses. However, in many situations it is unwelcome, since the voltage at X does not reproduce faithfully the input voltage, but has a DC shift superimposed. If, for example, the height of the pulses above zero plays a crucial role in subsequent circuitry, the DC shift can upset the intended operation. For high-speed circuitry, where pulses follow one another in quick succession, it is desirable to eliminate coupling capacitors by using DC coupling. Alternatively, the pulses may, as in figure 3.20, be shaped deliberately to have equal positive and negative areas; such pulses are called **bipolar**.

3.9 Stray Capacitance

In an electrical circuit, most of the electric field is along the wires and through the components. However, a very small part of the

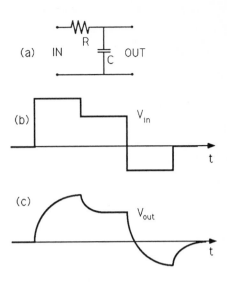

Fig. 3.18. Response of a CR circuit to an irregular waveform.

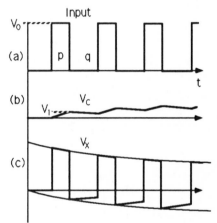

Fig. 3.19. Baseline shift generated by charging a capacitor.

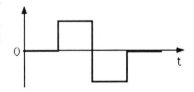

Fig. 3.20. Bipolar pulse.

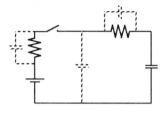

Fig. 3.21. Stray capacitance.

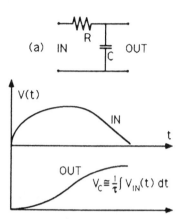

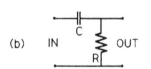

INTEGRATION (large CR)

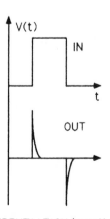

DIFFERENTIATION (small CR)

Fig. 3.22 (a) Integration,
(b) Differentiation.

electric field is through the air. For example, in figure 3.1, lines of electric field run from charges on the plate to induced charges in the nearby earth and elsewhere. Associated with them is **stray capacitance**. Thus, to figure 3.14 one should strictly add the capacitances shown by the broken lines in figure 3.21. Such stray capacitance is very difficult to calculate, but is of the order of $\epsilon_0 A/s$, where A is the area of an element and s is the separation of two elements, i.e. of order picofarads or less. Despite its small magnitude, stray capacitance ultimately limits high-frequency performance.

Electrical circuits are usually mounted inside a chassis which is earthed; this eliminates stray capacitance to distant sources and hence eliminates capacitative pick-up. Likewise, signal cables are protected by an earth braid. Shielding and earthing is a complicated art, and the reader is referred to specialist discussions (for example, Horowitz P and Hill W, 1989, *The Art of Electronics*, Cambridge University Press).

3.10 Integration and Differentiation

The algebra of the previous sections applies specifically to the case when the applied voltage is constant or a square wave. More generally, for an applied voltage $V(t)$ with arbitrary time dependence,

$$V_C = \frac{1}{C} \int I \, dt' = \frac{1}{CR} \int (V - V_C) \, dt'$$

where R is the total series resistance in the circuit. For large values of CR, the capacitor charges slowly and V_C is small. In this case, if the capacitor is uncharged initially,

$$V_C \simeq \frac{1}{CR} \int_0^t V(t') \, dt'. \tag{3.11}$$

For this reason, the circuit of figure 3.22(a) with large CR is called an **integrating circuit** or **integrator**. If a pulse is applied to the circuit, the capacitor is said to **integrate** the input pulse. The snag is that the output signal is small, so CR should not be *too* large. The approximation in equation (3.11) is accurate only for $t \ll CR$. To see this, the exponential in equation (3.10) is expanded in a power series with V equal to a constant V_0. The result is

$$V_C = V_0 \left[\frac{t}{CR} - \frac{1}{2} \left(\frac{t}{CR} \right)^2 + \dots \right]$$

Equation (3.11) gives only the first term in the bracket.

If $V =$ constant, after what time does $V_C(t)$ differ by 10% from its initial straight line?

The other extreme arises when CR is small. Figure 3.22(b) shows that in this case V_R displays a narrow spike at the leading and trailing edges of the input pulse. The output V_R is large where the slope of the input voltage is large. For this reason the circuit is known as a **differentiating** circuit and the process of generating the spike is called **differentiation**. It is often convenient to use this arrangement to generate from a square pulse a spike which can be used as a trigger in subsequent circuits; using the negative spike on the trailing edge is a simple way of generating a **time delay** equal to the width of the square pulse.

3.11* Thevenin's Theorem Again

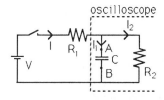

Fig. 3.23. Illustrating Thevenin's theorem.

Suppose you want to measure the input capacitance of your oscilloscope (and its leads). This can be done by measuring the time constant for charging it. However, it is necessary to allow for the input resistance R_2 in parallel with C (figure 3.23).

The algebra goes much the same as before, except that it is necessary to account separately for I_1 and I_2:

$$I_2 R_2 = V - I R_1 = V - (I_1 + I_2) R_1$$

or

$$I_2 = (V - R_1 I_1)/(R_1 + R_2).$$

Then

$$V_c = \frac{1}{C} \int I_1 \, \mathrm{d}t' = I_2 R_2 = \frac{V R_2}{R_1 + R_2} - \frac{R_1 R_2 I_1}{R_1 + R_2}.$$

This is an equation for I_1 of precisely the same form as equation (3.7) except that V_0 has been replaced by $V R_2/(R_1 + R_2)$ and R has been replaced by $R_1 R_2/(R_1 + R_2)$. Obviously, we have constructed a Thevenin equivalent for the circuit across AB, i.e. across the capacitor. With AB open circuit, the voltage across AB is reduced from V to $V R_2/(R_1 + R_2)$, i.e. that given by the potential divider made by R_1 and R_2; and the resistance of the Thevenin equivalent is given by R_1 and R_2 in parallel from A to B (with the battery short circuited).

3.12 Summary

$$Q = CV$$
$$I = C\frac{\mathrm{d}V}{\mathrm{d}t}.$$

Parallel capacitors have capacitance $C_1 + C_2$; in series, $C = C_1 C_2/(C_1 + C_2)$. The time constant for charging through resistance R is $\tau = CR$. For a square-wave input, $V_C = V_0(1 - \mathrm{e}^{-t/CR})$

and $V_R = V_0 e^{-t/CR}$. Asymptotically, $V_C \to V_0$ and $V_R \to 0$. The voltage across a capacitor cannot change instantaneously.

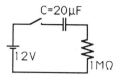

Fig. 3.24.

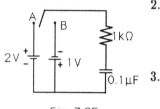

Fig. 3.25

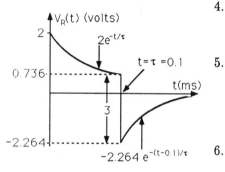

Fig. 3.26

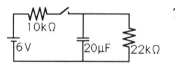

Fig. 3.27

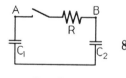

Fig. 3.28

3.13   Exercises

1. A capacitor of 2 μF is connected across a 12 V battery. What charge and what energy are stored in the capacitor? (Ans: 24 μC, 0.144 mJ.)

2. The current through a 10 μF capacitor is $I = (10t - 4t^2)$ μA from $t = 0$ to 2.5 s. If the capacitor is initially uncharged, what is the voltage across it as a function of time? (Ans: $0.1(5t^2 - 4t^3/3)$ V.)

3. The voltage across a 1.5 μF capacitor is $V = (10t - 4t^2)$ V from $t = 0$ to 2.5 s. What current flows through it as a function of time? (Ans: $1.5(10 - 8t)$ μA.)

4. Initially the switch in figure 3.24 is open and at time $t = 0$ it is closed. Find the voltage $V_C(t)$ across the capacitor against time t. (Ans: $12(1 - e^{-t/20})$ V, with t in s.)

5. At time $t = 0$, the capacitor of figure 3.25 carries no charge and the switch is moved to position A. After one time constant, it is moved to position B. Sketch the voltage V_R across the resistor as a function of time and label your sketch with quantitative values. (Ans: figure 3.26.)

6. Initially the switch in figure 3.27 is open and the capacitor is uncharged. At time $t = 0$ the switch is closed. Find the voltage $V_C(t)$ across the capacitor as a function of time. (Ans: $4.125[1 - \exp(-t/0.1375)]$ V, with t in s.)

7*. (QMC 1985) In the circuit of figure 3.28, C_1 initially carries charge q. At time $t = 0$, the switch is closed. What is the current through R as a function of time? Show that equilibrium is reached eventually with voltage $q/(C_1 + C_2)$ across each capacitor and with signs such that no current flows through the resistor. How much energy is dissipated in the resistor after closing the switch? (Ans: $I = q(1 - e^{-t/C'R})/C_1 R$ where $C' = C_1 C_2/(C_1 + C_2)$; $E = \frac{1}{2}q^2 C_2/[C_1(C_1 + C_2)]$.)

8**. (UC 1974) Figure 3.29 shows a circuit used for pulsing spark chambers. Bearing in mind that $R_1 \gg R_2 \gg R_3$ (i.e. you can make approximations), answer the following:

(i)   With what time constant does the capacitor C_1 charge up, and what are the potentials at A and B when C_1 is fully charged?

(ii) What happens to the potentials at A and B when the spark gap S breaks down, assuming it then has zero resistance.

(iii) With what time constant would you expect C_2 to charge, and what is the final value of the potential at A if the spark chamber does not break down?

(iv) With what time constant does the voltage at A decay to zero if the spark gap stays conducting and the chamber breaks down? Suggest a method by which you could observe the voltage change at A on an oscilloscope. If the chamber does *not* break down, what is the time constant?

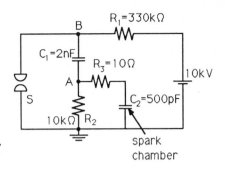

Fig. 3.29

(Ans: (i) $C_1(R_1 + R_2) = 0.68$ ms, neglecting current through R_3; $V_A = 0$, $V_B = 10$ kV; (ii) $B \to 0$ and $A \to -10$ kV; then (iii) $\tau = C_1 C_2 R_3/(C_1 + C_2) = 4$ ns; $V_A = -8$ kV if the spark chamber C_2 does not break down; (iv) C_1 charges through R_3 with $\tau = R_3 C_1 = 20$ ns; or $\tau = R_2(C_1 + C_2) = 25$ μs if the chamber does *not* break down.)

9. (Chelsea College 1974) In figure 3.30, the capacitors C_1 and C_2 are both initially uncharged and keys S_1 and S_2 are open. S_1 is now closed, and after C_1 has charged fully, S_2 is also closed. Describe without detailed mathematical analysis what happens in the circuit subsequently, assuming the influence of any resistance in the closed loop formed by C_1 and C_2 may be neglected. Obtain an expression for the voltage across C_2 at a time t after the closure of S_2. Explain *qualitatively* in what way the behaviour is modified if a small resistance is in series with S_2. (Ans: Initially $V_X = V_0$. After S_2 is closed, $V_X = C_1 V_0/(C_1 + C_2) + (V_0 C_2/C')[1 - \exp(-t/RC')]$, where $C' = C_1 + C_2$; providing $R(S_2) \ll R$, C_1 and C_2 come to equilibrium with $\tau = R(S_2)C_1 C_2/(C_1 + C_2)$; then as before.)

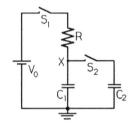

Fig. 3.30

10* Predict the output voltage waveform from the two circuits in figure 3.31 when the input voltage is a sawtooth pulse with duration T. (Ans: figure 3.32.)

11. What are the differences between a battery and a capacitor?

12. (Chelsea College 1974) Why is the Wheatstone bridge not used for the measurement of very high resistances? Describe how you would measure a resistance of the order of 10 MΩ by allowing a capacitor to discharge through it. Indicate suitable values for the components to be used. How can a check be made in order to discover whether the capacitor itself is 'leaky'? If it is found to be so, explain how the procedure can be modified to eliminate the resulting error without using another capacitor.

13. (QMC 1978) The train of positive square pulses shown in figure 3.33(a) is applied to the network of (b) and the out-

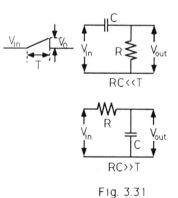

Fig. 3.31

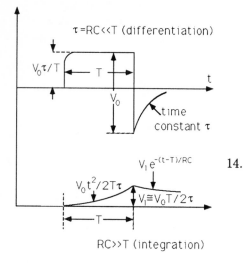

τ =RC≪T (differentiation)

$V_0\tau/T$

T

V_0

time constant τ

$V_1 e^{-(t-T)/RC}$

$V_0 t^2/2T\tau$

$V_1 \cong V_0 T/2\tau$

T

RC≫T (integration)

Fig. 3.32

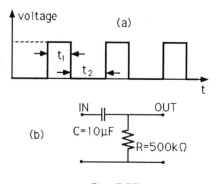

voltage

(a)

t_1

t_2

t

IN ——||—— OUT

(b)   C=10μF   R=500kΩ

Fig. 3.33

put is displayed on an oscilloscope. Before the pulse train is switched on, the vertical position of the trace is adjusted to be central. When the pulse train is switched on, the display initially reproduces figure 3.33(a), but gradually the display drifts downwards until the top of the waveform appears at $V_1 < V$ and the bottom at $V_2 < 0$, with $V_1 - V_2 = V$. Explain what is happening and relate V_1 and V_2 to t_1 and t_2. If the input is switched off at some instant when it is at zero voltage, what will one see on the oscilloscope?

14. (QMC 1990) In the circuit of figure 3.34, V_1 is a pulse generator giving square pulses which rise from zero to 6 V and last for 1 ms. Sketch the waveforms across AB when (a) a 5 kΩ resistor is applied there, and (b) when a 0.1μF capacitor is connected there. Label your sketches with magnitudes describing the waveforms quantitatively. (Ans: figure 3.35.)

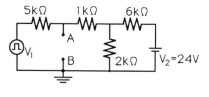

Fig. 3.34.

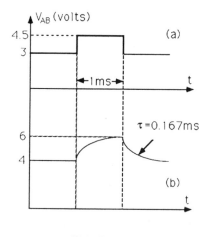

Fig. 3.35.

4

Alternating Current (AC); Bandwidth

4.1 Introduction

In the previous chapter we studied the charging of a capacitor when a voltage step was applied to a circuit. Now we turn to the way it responds to a voltage $V_0 \cos \omega t$ or $V_0 \sin \omega t$. This is an **alternating voltage**. An example might be a radio signal. The voltage supplied by the mains is also close to a sine wave. Later, in Chapter 15, we will show that *any* voltage waveform $V(t)$ can be expressed by Fourier's theorem as a sum of sines and cosines (or an integral involving them). So if we can understand how a circuit responds to a sine-wave input we can, with some effort, reconstruct its response to an input of absolutely any time dependence. This is the rationale for discussing alternating currents and voltages.

In Britain the mains voltage has a peak value $V_0 \simeq 240\sqrt{2}$ volts; in the rest of Europe it is $220\sqrt{2}$ volts and in the USA $110\sqrt{2}$ volts. The factor $\sqrt{2}$ will be explained shortly. One cycle $V_0 \cos \omega t$ takes a time T such that $\omega T = 2\pi$ (figure 4.1(a)); so the time period $T = 2\pi/\omega$. The frequency is $f = 1/T = \omega/2\pi$. In Europe, $f = 50$ Hz (or cycles per second); in the USA, $f = 60$ Hz.

4.2 Power in a Resistor: RMS Quantities

The origin of the factor $\sqrt{2}$ will now be examined. The power dissipated in a resistor is $P = VI$. If a voltage $V = V_0 \sin \omega t$ (figure 4.1(a)) is applied directly across resistance R, the current $I = V_0 \sin \omega t / R$ and the instantaneous power is

$$P = VI = (V_0^2/R) \sin^2 \omega t = (V_0^2/2R)(1 - \cos 2\omega t).$$

Integrated over a complete cycle, the term $\cos 2\omega t$ is zero, so the mean power dissipated is $\bar{P} = V_0^2/2R$. The factor 2 in the denominator accounts for the fact that the mean value of V^2 over the cycle

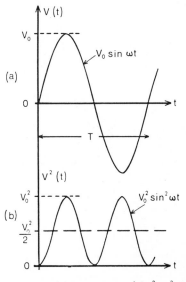

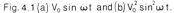

Fig. 4.1 (a) $V_0 \sin \omega t$ and (b) $V_0^2 \sin^2 \omega t$.

53

is $\frac{1}{2}V_0^2$. Figure 4.1(b) demonstrates that V^2 is symmetrical about $\frac{1}{2}V_0^2$ if you turn the page upside down. To bring the expression for $\bar{P}$ into line with the DC result $P = V^2/R$, it is conventional to rewrite this expression as $\bar{P} = V_{\mathrm{RMS}}^2/R$, where

$$V_{\mathrm{RMS}} = V_0/\sqrt{2}. \qquad (4.1)$$

RMS stands for root mean square and indicates that the square of V_{RMS} gives the mean value of $V^2(t)$. You may verify the factor $\sqrt{2}$ for yourself by measuring V_{RMS} of an AC signal on the AC range of a multimeter and determining V_0 by displaying the signal on an oscilloscope.

The mains are used largely to provide heating and lighting and it is conventional to refer to V_{RMS} rather than V_0; in Britain, the 240 volt mains provide $V_{\mathrm{RMS}} = 240$ V or $V_0 = 240\sqrt{2}$ V. Likewise,

$$I_{\mathrm{RMS}} = I_0/\sqrt{2}$$

where $I = I_0 \cos \omega t$, so

$$\bar{P} = V_{\mathrm{RMS}} I_{\mathrm{RMS}} = V_{\mathrm{RMS}}^2/R = I_{\mathrm{RMS}}^2 R. \qquad (4.2)$$

4.3 Phase Relations

For a resistance, $V_R = IR$; so an alternating current produces an alternating voltage and vice versa. For a capacitance, $V_C = (I/C) \int I \; \mathrm{d}t$. The integral of a sine is a cosine, so an alternating current again leads to an alternating voltage and vice versa. However, the change from sine to cosine implies a shift in time dependence between current and voltage. Later, we shall find the same to be true for an inductor too.

The quantity ωt is called the **phase** of the voltage $V_0 \sin \omega t$. We could equally well have written $V = V_0 \cos \omega t$, or more generally $V = V_0 \cos(\omega t + \phi)$ by choosing the origin of time, $t = 0$, differently. Results are independent of which form is chosen. However, the phase *relations* amongst different quantities will be important, and it will be necessary to manipulate expressions like $\cos(\omega t + \phi)$. Figure 4.2 provides a brief reminder of some of the simple properties of sines and cosines:

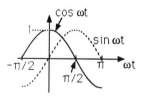

Fig 4.2. Relations between sines and cosines.

$$\cos 0 = 1 \qquad \sin 0 = 0$$
$$\cos \pi/2 = 0 \qquad \sin \pi/2 = 1$$
$$\cos(-\phi) = \cos \phi$$
$$\sin(-\phi) = -\sin \phi$$
$$\cos(\omega t - \pi/2) = \sin \omega t$$
$$\cos(\omega t + \pi/2) = -\sin \omega t.$$

The last two are special cases of the important general formulae

$$\cos(\omega t - \phi) = \cos\omega t \cos\phi + \sin\omega t \sin\phi$$
$$\cos(\omega t + \phi) = \cos\omega t \cos\phi - \sin\omega t \sin\phi.$$

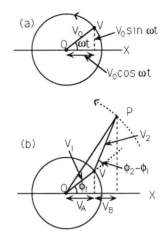

If you have trouble remembering these expressions (particularly the signs), try trial substitutions with $\omega t = 0$ and $\pi/2$.

It is often convenient to represent an AC voltage V graphically as a point (figure 4.3(a)) which describes a circle with angular frequency ω. The line OV makes an angle ωt with OX and the length of OV is V_0. Thus V is described by a rotating vector like the beam of a lighthouse. Such a diagram is called a **phasor diagram** and the vectors are called **phasors**. Later it will emerge that they are in fact representations of complex numbers. The projection of OV on to the horizontal axis is $V_0 \cos\omega t$ and that on to the vertical axis is $V_0 \sin\omega t$, so this diagram can describe either form, sines or cosines. The angular frequency ω is measured in radians per second. The time period T is the time required for the point to move once round the circle.

Fig 4.3(a) Vector representation of an AC voltage, (b) addition of two alternating voltages.

An an illustration of the use of this representation, consider the addition of two voltages:

$$V_A = V_1 \cos(\omega t + \phi_1) \qquad V_B = V_2 \cos(\omega t + \phi_2).$$

Their sum is represented by the point P in figure 4.3(b), which has projection $V_A + V_B$ on to OX. The angle PVO is $\pi - (\phi_2 - \phi_1)$, so

$$OP = [V_1^2 + V_2^2 + 2V_1V_2 \cos(\phi_2 - \phi_1)]^{1/2}.$$

This result is independent of t, so the sum is also a voltage of angular frequency ω. The sum P is said to *lead* V_A because its phase angle is larger than ϕ_1 and the vector OP is describing its circle in advance of OV. Conversely, V_A is said to *lag* P and V_B.

Does V_B lead or lag the sum P?

4.4 Response of a Capacitor to AC

Now consider the relation between AC voltage and current for a capacitor alone (figure 4.4). If a current $I_0 \cos\omega t$ flows, the charge on the capacitor is

$$Q = \int I \, dt$$

and

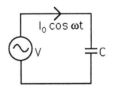

Fig. 4.4 Alternating voltage applied to a capacitor.

$$V = \frac{1}{C} \int I_0 \cos\omega t \, dt = \frac{I_0}{\omega C} \sin\omega t$$

$$= \frac{I_0}{\omega C} \cos(\omega t - \pi/2). \qquad (4.3)$$

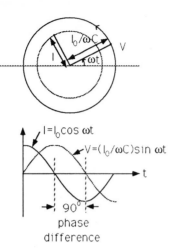

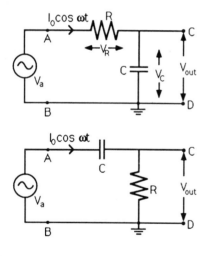

Fig. 4.5 The phase relation between voltage and current for a capacitor.

The magnitude of the voltage is given by $I_0/(\omega C)$. The quantity $1/\omega C$ plays a role similar to resistance in Ohm's law. It is called the **reactance** of the capacitor. The reactance drops to zero as $\omega \to \infty$; at high frequency, a capacitor acts like a short circuit. Conversely, as $\omega \to 0$, the current drops to zero: a capacitor blocks direct current.

However, despite the similarity to Ohm's law, there is an essential difference: I leads V in phase by 90° or $\pi/2$, as shown in figure 4.5. The physical reason for this is that current must flow first if charge, and hence voltage, is to become established on the capacitor.

In a capacitor, voltage follows current as charge reaches the plates. Does this suggest a physical explanation of why the reactance $\to 0$ as $\omega \to \infty$?

An AC voltage is applied to capacitors C_1 and C_2 in parallel. Show that they behave like a single capacitor $(C_1 + C_2)$.

4.5   Simple Filter Circuits

The remainder of this chapter extends the relation between current and voltage for a capacitor to more elaborate circuits; this introduces the fundamental notion of **bandwidth**. Two circuits which occur very commonly are shown in figure 4.6. Their qualitative behaviour can be appreciated easily from the fact that the reactance of the capacitor drops to zero as $\omega \to \infty$. A voltage will appear across the capacitor only at low frequencies; at high frequencies it is entirely across the resistor. Thus figure 4.6(a) transmits an output only at low frequencies and is called a **low-pass filter**. Figure 4.6(b) does the converse; it is called a **high-pass filter**.

In order to analyse both circuits quantitatively, suppose the AC voltage is switched on and left running for sufficient time for transients to die away. This gives the **steady-state** behaviour. Also assume no load is applied externally to terminals CD. If the current flowing round the circuit is $I_0 \cos \omega t$, the applied voltage is

Fig. 4.6 (a) Low-pass filter, (b) high-pass filter.

$$V_a = V_R + V_C$$
$$= IR + (1/C)\int I \, dt$$
$$= RI_0 \cos \omega t + (1/\omega C)I_0 \sin \omega t. \tag{4.4}$$

We anticipate that the voltage V_a lags behind current by an angle between $\pi/2$ (the result for a capacitor alone) and 0 (the result for a resistor). So we write

$$V_a = V_0 \cos(\omega t - \phi) = V_0(\cos \omega t \cos \phi + \sin \omega t \sin \phi).$$

In order for the coefficients of $\cos\omega t$ and $\sin\omega t$ to agree with (4.4) we must have

$$V_0 \cos\phi = RI_0 \tag{4.5}$$
$$V_0 \sin\phi = I_0/\omega C. \tag{4.6}$$

Squaring (4.5) and (4.6) and adding,

$$V_0^2 = [R^2 + (1/\omega C)^2]I_0^2 \tag{4.7}$$

and dividing (4.6) by (4.5)

$$\tan\phi = 1/\omega CR. \tag{4.8}$$

These results justify the trial solution $V_a = V_0 \cos(\omega t - \phi)$.

Phasors

It is convenient to draw a snapshot at $t = 0$. Any other value of t will, however, give the same relations. At $t = 0$, the vector describing the current is along the horizontal axis and so is the vector describing V_R. The voltage V_C lags I in phase by $90°$. The vectors OA and OB add to give OC, representing the applied voltage V_a. We could have derived V_a graphically like this, bypassing the algebra.

The magnitude of V_a is given by the hypotenuse of the triangle, and hence by equation (4.7). The output of the high-pass filter, V_R, leads V_a by angle ϕ, figure 4.7(b). When R dominates (at high frequencies), $\phi \to 0$; when C dominates (at low frequencies), $\phi \to 90°$. The output of the low-pass filter, V_C, lags V_a by angle θ, where

$$\tan\theta = \omega CR. \tag{4.9}$$

The magnitudes of V_R and V_C are given by the reciprocal of equation (4.7):

$$|V_R| = I_0 R = V_0 R/[R^2 + (1/\omega C)^2]^{1/2}$$
$$= V_0/[1 + (1/\omega CR)^2]^{1/2} \tag{4.10}$$
$$|V_C| = I_0/\omega C = V_0/[1 + (\omega CR)^2]^{1/2}. \tag{4.11}$$

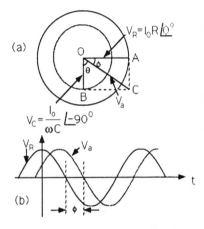

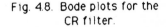

Fig. 4.7 (a) Phasor diagram for the CR filter circuit, (b) waveforms.

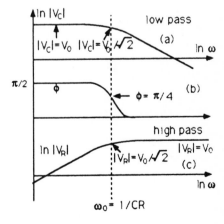

Fig. 4.8. Bode plots for the CR filter.

Bode plots

Figure 4.8 shows the dependence of $|V_C|$, ϕ and $|V_R|$ on ω. It is convenient to use logarithmic scales for both $|V|$ and ω, so as to compress a large range of values on to a single graph. The resulting diagram is called a Bode plot. At low frequency, $|V_C| \simeq V_0$ from equation (4.11), so the low-pass filter transmits the input voltage to the output across C; at low frequencies $\theta \to 0$, so the output of the low-pass filter is in phase with the input. This simply reflects the fact that at low frequencies $V_R \ll V_C$ and $V_C \simeq V_a$. Conversely, the high-pass filter transmits at high frequency, from equation (4.10); at high frequencies $\phi \to 0$, so the output of the high-pass filter, V_R, is again in phase with the applied voltage V_a. Here $V_C \ll V_R$ and $V_R \simeq V_a$.

Off the top of either of curves (a) or (c), a phase difference develops between input and output of the filter. As ω rises from 0, $|V_C|$ falls to $V_0/\sqrt{2}$ when $\omega = \omega_0 = 1/CR$; also $\phi \to \pi/4$ (45°) at this frequency. It is no accident that this value of ω is equal to $1/\tau$, where $\tau = CR$ is the time constant of the CR transient (equation (3.9)). Later, Fourier's theorem will reveal an intimate relation between the response of any circuit to alternating current and to a step function.

It is well worth measuring the Bode plot for yourself on the oscilloscope, using say $R = 1$ kΩ and $C = 0.1$ μF. Vary ω and observe $|V_C|$ and $|V_R|$. The phase difference ϕ between the applied voltage and V_R may be observed by displaying V_{AB} and V_{CD} of figure 4.6(b) on the two traces of an oscilloscope.

Decibels

The vertical scale of the Bode plot is usually expressed in **decibels** (a notation derived historically from loudspeakers). This notation may at first sight look complicated, but is nonetheless useful. The ratio of signal power at the input and at the output of the filter circuit is V_0^2/V_{CD}^2. The quantity

$$r = \log_{10}(V_0^2/V_{CD}^2)$$

is expressed in bels (after Alexander Graham Bell); alternatively it can be expressed in decibels (db),

$$r = 10 \log_{10}(V_0^2/V_{CD}^2) \text{ db} = 20 \log_{10}(V_0/V_{CD}) \text{ db}.$$

A factor 2 in V_0/V_{CD} corresponds to $r = 20 \log_{10} 2 = 20 \times 0.3010 \simeq$ 6 db. Table 4.1 illustrates the conversion from V_0/V_{CD} to decibels. Every factor 2 in V_0/V_{CD} corresponds to 6 db and every factor 10 in V_0/V_{CD} to 20 db.

Table 4.1   Conversion of attenuations to decibels.

V_0/V_{CD}	1	$\sqrt{2}$	2	3	4	8	10	100	1000
Attenuation (db)	0	3	6	9.5	12	18	20	40	60

Suppose a signal is multiplied by factors f_i in several successive pieces of electronics and $V_{out} = f_1 f_2 f_3 \ldots V_{in}$. The attenuations or amplifications *add* if expressed in decibels:

$$r = 20 \log_{10}(V_{out}/V_{in}) = 20 \log_{10}(f_1 f_2 f_3 \ldots) \text{ db}$$
$$= r_1 + r_2 + r_3 + \ldots \qquad \text{where} \qquad r_i = 20 \log_{10} f_i \text{ db}.$$

For example, if there are attenuations by factors 2, 2 and 4 in successive circuits, the net attenuation is $(6 + 6 + 12)$ db $= 24$ db. This is a valuable property of Bode plots: overall attenuation or amplification is obtained by adding them. This addition rule is also true for the phase angles ϕ, as demonstrated later using complex numbers.

The horizontal scale of figure 4.8 is expressed in **decades** or **octaves**. A factor 10 in frequency is one decade and $\log_{10} \omega$ increases by 1; a factor 2 in frequency is one octave and $\log_{10} \omega$ increases by 0.3. The jargon used to describe the fall-off of V_C at high frequencies is **roll-off**. For frequencies well above ω_0, equation (4.11) reads

$$V_C \simeq V_0/\omega C R$$

so

$$\log_{10}(V_C/V_0) \simeq -\log_{10}(CR) - \log_{10} \omega.$$

This is a straight-line relation on the Bode plot, figure 4.9. The slope of the roll-off is -6 db/octave or -20 db/decade. It is often adequate to approximate the behaviour of the filter by the two straight lines of figure 4.9 meeting at the point X at the cut-off frequency ω_0. This is called the **corner frequency**. At this frequency the true curve is actually a factor $1/\sqrt{2}$ or 3 db below X. Exercise 8 shows that two cascaded filters give a roll-off at high frequencies of -12 db/octave or -40 db/decade. For n filters in sequence, the roll-off is $-6n$ db/octave or $-20n$ db/decade.

Worked examples on phasors

(1) Suppose it is known that the voltage V_C across the 20 μF capacitor in figure 4.10(a) is $2 \cos(100t + 20°)$ V. What is the applied voltage V_a? To find V_R, the current must be determined. This leads V_C by 90° and is of magnitude $\omega C V_C = 2 \times 100 \times 2 \times 10^{-5} = 4 \times 10^{-3}$ A; it has phase angle 110°. The voltage across the 1 kΩ resistor is 4 V at the same angle. The phasor diagram is shown in figure 4.10(b). The applied voltage V_a is the vector sum of V_C and V_R; it has magnitude $\sqrt{20}$ V and leads V_C by angle $\theta = \tan^{-1} 2 = 63.4°$. So V_a is $\sqrt{20} \cos(100t + 83.4°)$ V.

(2) Suppose the same voltage V_a is applied to the same components in parallel (figure 4.10(c)). What total current flows? Firstly,

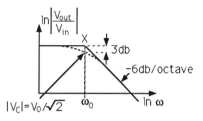

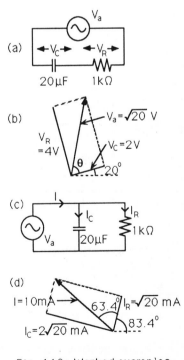

Fig. 4.9. An approximation to the Bode plot of a CR filter circuit.

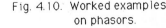

Fig. 4.10. Worked examples on phasors.

$I_R = V_a/R = \sqrt{20}\angle 83.4°$ mA. Next, the magnitude of I_C is $\omega C V_a = 2\sqrt{20}$ mA and it leads the voltage by 90°. The phasor diagram for these currents is shown in figure 4.10(d). The total current is $10\cos(100t + 146.8°)$ mA.

In Chapter 6, an algebra will be developed which avoids drawing these diagrams; nonetheless, they are useful for understanding simple circuits, and you are recommended to try examples from the exercises at the end of the chapter.

4.6   Power Factor

The instantaneous power P dissipated in the CR circuit is

$$P = V_a I = V_0 \cos(\omega t - \phi) I_0 \cos \omega t$$
$$= \tfrac{1}{2} V_0 I_0 [\cos(2\omega t - \phi) + \cos \phi].$$

Averaged over a complete cycle, this becomes

$$\boxed{\bar{P} = \tfrac{1}{2} V_0 I_0 \cos \phi = V_{\text{RMS}} I_{\text{RMS}} \cos \phi.} \tag{4.12}$$

The factor $\cos \phi$ is called the **power factor**.

An important result is that $\cos \phi = 0$ for a capacitor alone and $\bar{P}$ is then zero. Energy is held in the capacitor twice per cycle as it charges first one way then the other, but no power is dissipated. Another way of expressing $\bar{P}$ is, using $\cos \phi = V_R/V_0$ (figure 4.7(a)),

$$\bar{P} = \tfrac{1}{2} V_R I_0 = \tfrac{1}{2} V_R I_R. \tag{4.12a}$$

This demonstrates that power is dissipated purely by the resistor.

4.7   Amplifiers

Most amplifiers have an input which behaves simply as a resistance at low frequencies. (To be strictly accurate, it is often a slope resistance, as discussed in section 1.11). Ideally, they can be DC coupled to a source. However, in practice internal operation of the amplifier may involve DC voltages which would upset the source, or vice versa, and it is quite common to insert a large decoupling capacitor C_1 between source and amplifier, as in figure 4.11(a). The DC levels in the source and the amplifier may then be different. The signal reaching the amplifier is cut off at low frequencies by C_1, making a high-pass filter as in figure 4.6(b). It drops by a factor $\sqrt{2}$ at a frequency $\omega_1 = 1/C_1(R_1 + R_2)$. To get a little feeling for the numbers, suppose the source has an output impedance $R_1 = 2$ kΩ and the amplifier has input impedance $R_2 = 1$ kΩ. These are typical of a simple transistor amplifier (the common emitter

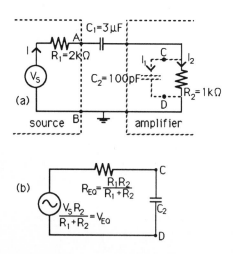

Fig. 4.11 (a) An amplifier AC coupled to a source, (b) its high-frequency equivalent circuit.

amplifier of Chapter 18). If the circuit is to amplify audio signals down to say $\omega_1 = 100$ rad/s, we require $C \geq 1/(R_1 + R_2)\omega_1$ or $3.3 \ \mu\text{F}$. For frequencies well above ω_1, the reactance of C_1 can be neglected completely and the circuit behaves as if C_1 were absent.

Now consider high frequencies. The amplifier always has some input capacitance C_2, and eventually at high frequencies the reactance of this capacitance will become small compared to R_2 and the input to the amplifier will drop. The high-frequency behaviour of the circuit will be analysed, first using Thevenin's theorem, and secondly algebraically. In both cases C_1 is neglected.

Suppose the circuit is broken at points CD and the circuit across the capacitor C_2 is replaced by its Thevenin equivalent. When CD is open circuit, $V_{EQ} = V_S R_2/(R_1 + R_2)$. The resistance of the equivalent circuit is obtained by shorting V_S, with the result R_1 in parallel with R_2; so $R_{EQ} = R_1 R_2/(R_1 + R_2)$. The Thevenin equivalent, figure 4.11(b), is a low-pass filter, and the overall behaviour of the amplifier will be as shown in figure 4.12. At high frequencies, the voltage at CD falls by $1/\sqrt{2}$ when $\omega_2 = 1/C_2 R_{EQ}$. Again, to get a rough idea of numbers, suppose $C_2 = 100$ pF, a value typical of an ordinary transistor; and $R_{EQ} = \frac{2}{3}$ kΩ. Then $\omega_2 = 15 \times 10^6$ rad/s.

As a further worked example on phasor diagrams, the high-frequency behaviour of the amplifier will be repeated algebraically. This example will illustrate how to treat the parallel combination of C_2 and R_2 using phasors; it is a little awkward and in Chapter 6 a superior algebraic method will be introduced, so do not labour too long over the present example.

Let the output voltage V_{CD} across this parallel combination be $V_2 \cos \omega t = I_2 R_2$, and let the source voltage be $V_S \cos(\omega t + \theta)$. Then, from figure 4.11(a),

$$V_S \cos(\omega t + \theta) = R_1 I + R_2 I_2 = R_1 I_1 + (R_1 + R_2)I_2$$

$$V_{CD} = I_2 R_2 = (1/C_2) \int I_1 \, dt.$$

The current I_1 leads I_2 by $90°$, so the vector diagram for V_S is as shown in figure 4.13. The horizontal component is due to I_2 and the vertical component is due to I_1. The output voltage V_{CD} is in phase with I_2 and lags V_S by an angle θ where

$$\tan \theta = \frac{R_1 \omega C_2}{(R_1 + R_2)/R_2} = \frac{R_1 R_2}{R_1 + R_2}\omega C_2$$

and the magnitudes of V_S and V_2 are related by

$$|V_S|^2/|V_2|^2 = (\omega R_1 C_2)^2 + [(R_1 + R_2)/R_2]^2 \qquad (4.13)$$

or

$$|V_2| = [R_2/(R_1 + R_2)]|V_S| \left[1 + \left(\frac{R_1 R_2}{R_1 + R_2}\omega C_2\right)^2\right]^{-1/2}$$

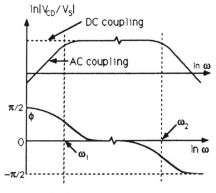

Fig. 4.12. Bode plot for the amplifier.

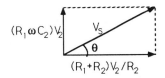

Fig. 4.13. Phasor diagram for V_S and V_2

These results are similar to equations (4.9) and (4.11), except that R has been replaced by $R_{\mathrm{EQ}} = R_1 R_2/(R_1 + R_2)$ and the voltage at frequencies below ω_2 is $V_2 = V_{\mathrm{EQ}} = R_2 V_{\mathrm{S}}/(R_1 + R_2)$. This gives an algebraic *derivation* of the Thevenin equivalent circuit of figure 4.11(b). In this example, the simpler derivation comes from *assuming* Thevenin's theorem, as in figure 4.11(b).

4.8   Bandwidth

The gain of the amplifier on the flat part of figure 4.12 at intermediate frequencies is called the **mid-band gain**. The frequencies ω_1 and ω_2 at which the output drops to $1\sqrt{2}$ of its maximum are called the **lower** and **upper cut-off frequencies**. The range of ω from ω_1 to ω_2 over which the response curves are approximately flat is known as the **bandwidth** of the circuit. This is a very important quantity. It is the range of frequencies over which the performance is constant, or nearly so. An oscilloscope, for example, has a large bandwidth, so as to be able to examine signals over a wide frequency range. Generally speaking, it is hard to achieve large bandwidth, but easy for the user to restrict it for particular applications (for example, when selecting a radio station). Manufacturers try to produce circuits and transistors with large bandwidth for multipurpose use.

With the numerical values given for the amplifier of figure 4.11, $\omega_1 = 10^2$ and $\omega_2 = 1.5 \times 10^7$ rad/s. Suppose an audio amplifier is to be optimised instead for the range $\omega = 10^2$ to 10^5 rad/s. The bandwidth may be moved down to centre on this range by using values of R_1 and R_2 a factor 10 or so larger. On the other hand, suppose an amplifier is required for TV frequencies (100 MHz) or the even higher frequencies used in computers or radio astronomy. To achieve a bandwidth reaching such frequencies, it is essential to keep C_2 small and also R_1 and R_2. Much of the present development of transistors is aimed at reducing capacitance to achieve higher speeds, and gallium arsenide devices now work up to 10^{11} Hz. Standards for high-frequency work are to make $R_1 = R_2 = 50 \ \Omega$ (or less often 70, 100 or 125 Ω).

4.9   Noise and Bandwidth

No electrical circuit is ever completely quiescent. Even if there is no voltage applied to it, thermal fluctuations give rise to fluctuating voltages. This is called **Johnson noise** or **thermal noise**. If you turn the sensitivity of an oscilloscope up to the maximum, you will usually see such noise on the trace. In a resistance R, it may be shown that these fluctuations give rise to a mean power

$$V_{\mathrm{RMS}}^2/R = 4kT \, \Delta f \qquad (4.14)$$

in a frequency interval Δf. Here, k is Boltzmann's constant and at room temperature $kT \simeq 1.4 \times 10^{-23} \times 300$ J. If $R = 10^5$ Ω and $\Delta f = 10^6$ Hz, for example, $V_{\text{RMS}} = 40$ μV. The power is uniformly spread over all frequencies, and this type of noise is therefore called **white noise**. In detecting weak signals, it is necessary to restrict the bandwidth to a small range, to reduce noise.

A second type of noise is **shot noise**. This arises from the fact that current is not continuous, but is carried by electrons whose charge is quantised in units of $e = 1.6 \times 10^{-19}$ C. Imagine, for example, an electron crossing the gap of a capacitor. As it approaches one plate, it repels electrons in the plate and induces a surge of current. If mean current I flows, the RMS fluctuation in that current is given by

$$I^2_{\text{RMS}} = 2eI \, \Delta f \qquad (4.15)$$

and again the effect is uniformly distributed over all frequencies. If $I = 1$ mA and $\Delta f = 10^6$ Hz, $I_{\text{RMS}} = 1.8 \times 10^{-8}$ A.

At what current is $I_{\text{RMS}} = I$ if $\Delta f = 10^6$ Hz? To what does this correspond in terms of electrons/unit bandwidth?

These two sources of noise are inescapable, though they can be minimised by suitable choice of circuit parameters (R, I and temperature T). In addition, there may be stray pick-up from the mains, corona discharge from power lines, car ignition, earth loops and so on. These can be minimised by shielding. There is also a further source of noise, called **contact** or **$1/f$ noise**. It originates from poor contacts within conductors; it is much worse in a carbon-composition resistor, for example, than in a metal-film or wire-wound resistor. Empirically, it is given by

$$I^2_{\text{RMS}} = KI \, \Delta f/f \qquad (4.16)$$

where K depends on the material. Because its frequency spectrum is proportional to $1/f$, it dominates at low frequency, typically below 1 kHz in practice.

How do noise signals of different frequencies and from different sources combine? Now

$$V_{\text{total}} = V_1 + V_2 + V_3 + \ldots$$
$$V^2_{\text{total}} = V_1^2 + V_2^2 + V_3^2 + \ldots + 2V_1V_2 + 2V_1V_3 + 2V_2V_3 + \ldots.$$

Because the signals are random, there is no correlation between V_1 and V_2 and so on. The average value of V_1V_2 is zero and likewise for other cross-terms. So values of V_i^2 add, i.e. noise powers add.

4.10   Summary

A capacitor has reactance $1/\omega C$; it blocks direct current and transmits high-frequency signals. Alternating current through a capacitor leads the voltage across it by 90°. The lower cut-off frequency of a CR filter is $\omega_1 = 1/CR_s$, where R_s is the *series* resistance; the upper cut-off frequency of a filter or amplifier is $\omega_2 = 1/CR_p$, where R_p is the resistance in *parallel* with the capacitor. The bandwidth of the circuit, from ω_1 to ω_2 is the range of ω over which its output is roughly constant and within a factor $1/\sqrt{2}$ of its maximum value. The phase of the applied voltage is 45° ahead of the output of the amplifier at ω_2 and 45° behind it at ω_1 (figure 4.12). The power dissipated in a CR filter circuit is $V_{\text{RMS}} I_{\text{RMS}} \cos\phi$ and is dissipated purely by the resistor; capacitors store energy temporarily, but do not dissipate it. The magnitudes of thermal noise and shot noise are proportional to bandwidth and are uniformly distributed with ω; noise powers add incoherently.

4.11   Exercises

1. A resistance $R = 120\ \Omega$ and a capacitance $C = 12.5\ \mu F$ are connected in series; current $I = 1.5\cos(5000t+25°)$ mA flows through them. What is the voltage across each component and the total applied voltage? What is the power factor of the circuit and what power is dissipated in each component? (Ans: $V_R = 180\cos(5000t + 25°)$ mV, $V_C = 24\cos(5000t - 65°)$ mV, $V_a = 181.6\cos(5000t + 17.4°)$ mV; 0.9912; $\bar{P}_R = 135\ \mu W$, $\bar{P}_C = 0$.)

2. A resistor $R = 1\ k\Omega$ is in series with an unknown capacitor C. The voltage V_R across the resistor is $12\cos 10^4 t$ V, and V_R leads the applied voltage by 33.7°. What is C? (Ans: 0.15 μF.)

3. What is the current in each of the three circuits of figures 4.14(a)–(c) and what are the voltages across each of the six components after transients have died away, supposing that the applied voltage is $12\cos\omega t$ V in each case, with $\omega = 10^3$ rad/s? (Ans: (a) $150\cos\omega t\ \mu A$, $4.95\cos\omega t$ V, $7.05\cos\omega t$ V; (b) $-9\sin\omega t$ mA, $9\cos\omega t$ V, $3\cos\omega t$ V; (c) $7.68\cos(\omega t+39.8°)$ mA, $7.68\cos(\omega t - 50.2°)$ V, $9.22\cos(\omega t + 39.8°)$ V.)

4. If in each of figures 4.14(a)–(c) the components are rearranged in parallel, what is the total current in each circuit and the current through each component? (Ans: (a) $619\cos\omega t\ \mu A$, $364\cos\omega t\ \mu A$, $255\cos\omega t\ \mu A$, (b) $-48\sin\omega t$ mA, $-12\sin\omega t$ mA, $-36\sin\omega t$ mA; (c) $15.6\cos(\omega t + 50.2°)$ mA, $-12\sin\omega t$ mA, $10\cos\omega t$ mA.)

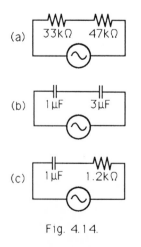

(a)   33kΩ   47kΩ

(b)   1 μF   3 μF

(c)   1 μF   1.2 kΩ

Fig. 4.14.

5*.   (QMC 1973) Design an AC high-pass filter to give a half-power frequency $f = 5$ kHz and an input impedance $|V|/|I|$ of 2 kΩ at that frequency. Determine the frequency f' for half-voltage output, and draw the vector diagram for the voltages in the circuit at that frequency, indicating the scale on your diagram. (b) If R is replaced by $2R$, redesign your circuit for the same half-power frequency and draw the new vector diagram at frequency f' to the same scale; what is the new input impedance at $f = 5$ kHz? (c) What rearrangement of components is necessary in (a) to give a low-pass filter, and how does the magnitude of the output voltage vary with frequency? (Ans: (a) $R = 1.4$ kΩ, $C = 0.023$ μF, $f = (5/\sqrt{3})$ kHz, voltage diagram figure 4.15(a), (b) 0.0115 μF, voltage diagram unchanged, 4 kΩ, (c) figure 4.15(b).)

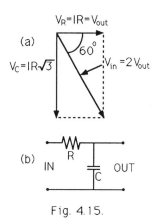

(a)

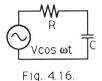

(b)

Fig. 4.15.

6.   The voltage source in figure 4.16 is $V \cos \omega t$. Find the power dissipation in the circuit and show that it is the same for two different values of R. What is the value of R for maximum power dissipation? The magnitude of R is adjusted to this value and an equal resistance is placed in parallel with C. Find the Thevenin equivalent across C and hence the current flowing in C. (Ans: $\bar{P} = \frac{1}{2}V^2 R/[R^2 + (1/\omega C)^2]$; $1/\omega C$; $(\omega C V/\sqrt{5}) \cos(\omega t + 63.4°)$.)

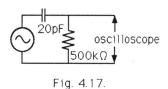

Fig. 4.16.

7**.   A voltage $20 \sin \omega t$ V is applied to the circuit in figure 4.17 with $\omega = 10^5$ rad/s; the voltage across the 500 kΩ resistor is measured using an oscilloscope with input resistance 1 MΩ and input capacitance 1 pF. What is the observed voltage? By how much is the observed phase angle wrong due to the loading by the oscilloscope? (Ans: $10.9 \sin(\omega t + 55.0°)$ V; $+10.0°$.)

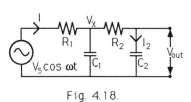

Fig. 4.17.

8.   Find the output voltage from the two-stage filter in figure 4.18 for frequencies large compared with $1/C_1 R_1$ and $1/C_2 R_2$. Hint: find the current I through R_1 ignoring the reactance of C_1; for this current find V_X, then I_2 ignoring the reactance of C_2, and hence an approximate value of the output voltage. (Ans: $V_{out}/V_S \simeq -1/\omega^2 C_1 C_2 R_1 R_2$.)

9.   (Westfield 1974) Explain the meaning of the decibel scale and show how it can be expressed in terms of voltage ratios. A switched attenuator has two controls calibrated in decibels, in tens and units respectively. To what values must these controls be set if a reduction in voltage by a factor of twenty is required? (Ans: 26 db.)

Fig. 4.18.

5

Inductance

5.1 Faraday's Law

Suppose a coil is wound around a cylindrical support or 'former' as in figure 5.1. This arrangement is called a **solenoid**. If a current is passed through the coil, a magnetic field H is generated along the length of the solenoid. Its direction and magnitude may be investigated with the aid of a compass needle, using the Earth's magnetic field as reference for comparison. If the solenoid has 100 turns/cm, a suitable current is $I = 20$ mA. The fundamental experiments of Ampére, who established the laws governing the magnetic field, are described in textbooks of electromagnetism. The points of concern here are that H is parallel to the axis of the solenoid and is proportional to current and to the number of turns m:

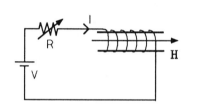

Fig. 5.1. The field of a solenoid.

$$H \propto mI. \tag{5.1}$$

Mutual inductance

Next suppose two separate coils are brought close together, as in figure 5.2. A current I_1 is fed through coil 1 and the second is connected to an oscilloscope. When I_1 is steady no signal is observed on the oscilloscope. But if I_1 changes (for example, by breaking the first circuit), a signal is observed from the second coil. This observation was first made by Faraday, using a sensitive galvanometer. He studied the dependence on I_1, on the orientation of the two coils and on movement of the second coil with respect to the first. He concluded that a *voltage* (rather than a current) is generated with a magnitude given by the rate of change of magnetic flux F in the second coil. This flux is defined as $F = H_n A$, where A is the area of the coil and H_n is the component of magnetic field normal to the plane of the coil; more exactly, if H_n varies over the area, $F = \int H_n \, dA$. Faraday's law of electromagnetic induction is one of

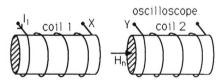

Fig. 5.2. Mutual Inductance.

the five basic laws of electromagnetism and states that the induced voltage V is

$$V = -\frac{dF}{dt} = -\frac{d}{dt}\int H_n \, dA.$$ (5.2)

The minus sign indicates that the induced voltage has a sign such as to produce a current which opposes the change of field.

From equation (5.1), $H_1 \propto I_1$ and thus the voltage in coil 2, $V_2 \propto dI_1/dt$. The proportionality constant M is called the **mutual inductance**:

$$V_2 = M\frac{dI_1}{dt}.$$ (5.3)

It is measured in **henries** (H). With M in these units,

$$V(\text{volts}) = M(\text{henries})\frac{dI(\text{amps})}{dt(\text{seconds})}.$$ (5.4)

Experiment and theory both demonstrate that the behaviour of the two coils is reciprocal: if a current I_2 varies in the second coil, a voltage V_1 is induced in the first with

$$V_1 = M\frac{dI_2}{dt}$$ (5.5)

with precisely the same value of M.

Magnetic fields are greatly enhanced in some ferromagnetic materials, namely iron, cobalt and nickel. If a cylindrical iron core is inserted down the axes of the two coils of figure 5.2, the observed voltages are enhanced by a large factor for small currents. For large currents, the iron saturates and the enhancement factor is reduced.

5.2 Self-inductance

The magnetic field produced by one coil can induce a voltage in a second. Why not in itself too? If you join coil 1 and coil 2 of figure 5.2 at points X and Y, what difference does it make?

Observations described in the next section demonstrate that a changing current I in a coil indeed generates a voltage V across this same coil where

$$V = L\frac{dI}{dt}$$ (5.6)

and L is called **self-inductance**. The sign of the voltage is such as to produce a current resisting the change. (If it had the opposite sign, there would be a runaway situation where any change of current in a coil would produce a continuously increasing current.)

The remainder of this chapter explores the response of a coil or **inductor** first to a voltage step and then to an AC voltage. There is a close parallel to the behaviour of a capacitor, studied in the previous two chapters. Mutual inductance plays little role in most electronic circuits but is important in power supplies. The discussion of these is postponed to Chapter 22.

5.3   *LR* Transient

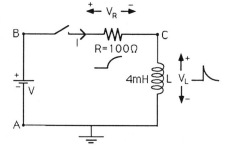

Fig. 5.3. Series LR circuit.

An inductance resists a change of current. If a voltage is applied across an inductor by closing the switch in figure 5.3, the current will grow only gradually and a transient is observed as current climbs to the value V/R which flows in the absence of the inductor. This transient turns out to be very similar to the CR transient for charging a capacitor. In the present case, the voltage V_L across the inductor falls exponentially with time:

$$V_L = V_0 e^{-t/\tau}$$

where

$$\tau = L/R.$$

This is sketched in figure 5.3.

When the switch is closed at $t = 0$, the initial current is zero, and the full applied voltage V appears across the inductor. As the current grows, V_L falls as V_R grows: $V_L = V - IR$. Setting $V_L = L\,dI/dt$,

$$V = L\frac{dI}{dt} + IR. \qquad (5.7)$$

This is a first-order differential equation similar to the one in Chapter 3, except for the appearance of the constant V on the left-hand side. It may be solved by the trial substitution

$$I = A\exp(-t/\tau) + B$$

where A and B are constants. The constant term B is required to give Ohm's law $I = V/R$ as $t \to \infty$. Differentiating I with respect to t,

$$dI/dt = -(A/\tau)\exp(-t/\tau).$$

Substituting in (5.7)

$$V = (-LA/\tau)\exp(-t/\tau) + AR\exp(-t/\tau) + BR.$$

This equation is satisfied for any A and any t if

$$\boxed{\tau = L/R} \qquad (5.8)$$

and $B = V/R$. With L in henries and R in ohms, τ is in seconds.

To find the constant A, the initial conditions must be used. The condition $I = 0$ at $t = 0$ requires $A = -B$, so finally

$$I = (V/R)[1 - \exp(-t/\tau)].\tag{5.9}$$

Figure 5.4 shows the voltage against time across the resistor and across the inductor. A large value of L hinders the creation of the current for a long time and for this reason an inductor is often called a **choke**.

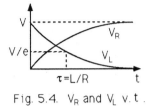

Fig. 5.4. V_R and V_L v. t .

If the initial current is I_0 instead of zero, how are equation (5.9) and figure 5.4 affected?

From electromagnetic theory, the self-inductance of a long solenoid of volume V and n turns/metre is $L = 4\pi \times 10^{-7}n^2V$ henries, so you may observe these phenomena using a coil of about 1000 turns wound on a former of 2 cm diameter and 10 cm long. This produces a self-inductance of about 4 mH$= 4 \times 10^{-3}$ H. The battery of figure 5.3 may be replaced by a square-wave generator, as in figure 5.5. The coil will have a resistance R_L of a few ohms, which can be measured with a multimeter. The voltage V_L is observed with an oscilloscope connected across CD. It is necessary to allow for the voltage drop IR_L if this is significant. It is instructive to insert a cylinder of soft iron inside the former of the inductor and observe the change in L.

In interpreting the observations, it is necessary to remember that R of equation (5.8) is replaced by the total series resistance $(R + R_{\text{out}} + R_L)$ of figure 5.5. Providing the output resistance of your pulse generator is small ($< 100\ \Omega$), τ will be $\geq 4 \times 10^{-5}$ s and the current rise can be observed readily. Suppose, however, your generator has an output impedance which is large ($> 1000\ \Omega$). In this case, τ is inconveniently small. The remedy lies in figure 5.6(a), where the variable resistor R is put in parallel with the coil. The Thevenin equivalent circuit across CD, figure 5.6(b), has R_{EQ} given by the parallel combination R and R_{out}, and can be made arbitrarily small by reducing R. The time constant becomes $\tau = L/(R_L+R_{\text{EQ}})$. The penalty is that the voltage of the Thevenin equivalent circuit is $V_{\text{EQ}} = RV_S/(R+R_{\text{out}})$ and this decreases with decreasing R. However, by turning up the pulse generator to give an output of, say, 10 V, you can still observe a reasonable size of signal across CD.

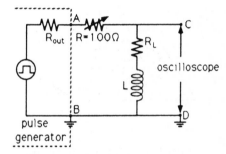

Fig. 5.5. Circuit to observe V_L.

5.4   Energy Stored in an Inductor

If a current I is established in an inductor, the work done is given by

$$E = \int VI\ \mathrm{d}t = \int L\frac{\mathrm{d}I}{\mathrm{d}t}I\ \mathrm{d}t = \int LI\ \mathrm{d}I$$

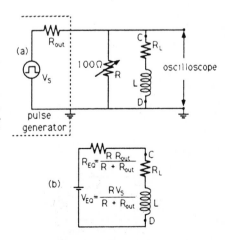

Fig. 5.6 (a) Variant if R_{out} is large, (b) its equivalent circuit.

or

$$E = \tfrac{1}{2}LI^2.$$ (5.10)

For large currents in a transformer coil, this stored energy can be quite large. If a sudden break occurs in the circuit, the voltage induced in insulation and the air can be sufficient to cause a spark. Ignition coils of cars work on this principle. In circuits where you do *not* wish this to happen, you should put a parallel resistor across the inductor to provide a discharge path (see exercise 11).

Can you understand in terms of equation (5.10) why a small time constant τ goes with large R?

5.5   Stray Inductance

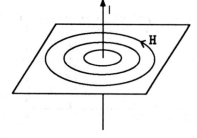

Fig. 5.7. Field of a straight wire.

Inductive effects are not restricted to the geometry of a coil. They arise for a conductor of any shape, even a straight wire. Figure 5.7 shows the field around a straight wire. It is at right angles to the current and in the direction of a right-handed screw pointing along the direction of the current. Some of this magnetic field lies within the wire itself. If the current changes, an EMF is generated within the wire causing currents which oppose the change. If the current is uniformly distributed in the wire, it can be shown that the self-inductance of length ℓ (metres) is

$$L = \mu_0 \ell / 8\pi$$

where $\mu_0 = 10^{-7}$. Although numerically small, it is not always negligible. At high frequencies, there is also the complication that the current is concentrated near the surface of the wire (the skin effect). Resistors, and even wires, have such **stray inductance** and at high frequencies leads must be kept as short as is practicable.

5.6   Response of an Inductor to Alternating Current

When an alternating voltage is applied to an inductor (figure 5.8),

$$V = L\frac{\mathrm{d}I}{\mathrm{d}t}$$

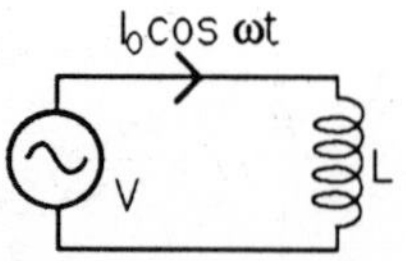

and if

$$I = I_0 \cos \omega t$$

$$\boxed{V = -\omega L I_0 \sin \omega t = \omega L I_0 \cos(\omega t + \pi/2).}$$

Fig. 5.8. Alternating voltage applied to an inductor.

The magnitude of the voltage is $\omega L I_0$ and the reactance of the inductor is ωL. The voltage leads current by $\pi/2$ (figure 5.9). The physical reason is that an inductor hinders current flow and current follows voltage. Averaged over a cycle, no power is dissipated in an inductor, because of the phase difference of $\pi/2$.

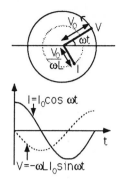

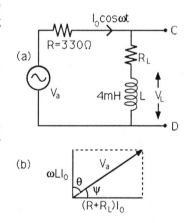

Fig. 5.9. Current in an inductor lags V by 90°.

5.7   Phasors

We can extend to inductors quite straightforwardly the phasor diagrams introduced for capacitors in the previous chapter. Suppose, for example, an AC voltage is applied to a series combination of a resistance R and an inductor (figure 5.10). Let the current in the circuit be $I = I_0 \cos \omega t$. Then, from Kirchhoff's voltage law,

$$V_a = RI + R_L I + L\frac{dI}{dt}$$
$$= (R + R_L)I_0 \cos \omega t + \omega L I_0 \cos(\omega t + \pi/2).$$

The corresponding phasor diagram is shown in figure 5.10(b). The inductor has reactance ωL and gives rise to a voltage $\omega L I_0$ leading the current by 90°. Then

$$V_a^2 = [(R + R_L)^2 + (\omega L)^2]^{1/2} I_0^2 \qquad (5.11)$$
$$\tan \psi = \omega L/(R + R_L).$$

The applied voltage leads the current by angle ψ, but it lags V_L, the voltage across the inductor, by angle $\theta = (\pi/2 - \psi)$. From equation (5.11), the circuit operates as a high-pass filter:

$$\frac{|V_L|}{|V_a|} = \frac{\omega L}{[(R + R_L)^2 + (\omega L)^2]^{1/2}} = \frac{1}{[1 + (1/\omega L)^2(R + R_L)^2]^{1/2}}.$$

The cut-off frequency is $\omega_0 = (R + R_L)/L$. It is straightforward to find L by using an oscilloscope to observe V_{CD} and sweeping the frequency of the generator from high to low values to find ω_0; however, the voltage V_{CD} includes IR_L which must be included in the arithmetic if it is significant compared to $\omega L I_0$.

Fig. 5.10 (a) RL series circuit and (b) its phasor diagram.

5.8   Summary

Just as it takes time for charge to build up on a capacitor, so it takes time for a current to become established in an inductor: current through an inductor cannot change instantaneously. The time constant is L/R. The reactance is ωL. An inductor transmits direct current but hinders alternating current. The current through an inductor lags the voltage across it by 90°. Averaged over one cycle of alternating current, no power is dissipated by an inductor. Energy $\frac{1}{2}I^2L$ is stored in it. If the current is suddenly broken, the voltage $L\,dI/dt$ creates a surge which can damage other components in the circuit (for example, transistors) unless a low-resistance discharge path is provided.

5.9   Exercises

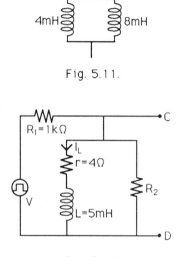

Fig. 5.11.

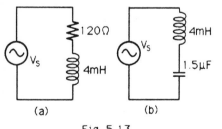

Fig. 5.12.

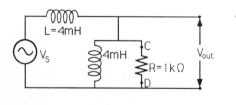

(a)   (b)

Fig. 5.13.

Fig. 5.14.

1.   What is the equivalent inductance of figure 5.11? (Ans: 8/3 mH.)

2.   An inductance of 0.1 H carries a current $25\cos 100t$ A. What is the voltage across it? (Ans: $-250\sin 100t$ V.)

3.   The voltage across a 4 mH inductor is $12\sin(1000t+20°)$ mV; what current flows through it? (Ans: $3\sin(1000t-70°)$ mA.)

4.   A constant voltage of 12 V is applied to a series combination $R = 5\,\Omega$ and $L = 20$ H by closing a switch at time $t = 0$. Find the current against time and the voltages V_R and V_L across R and L. At what time is $V_R = V_L$? (Ans: $I = 2.4(1-e^{-t/4})$ A, $V_R = 12(1 - e^{-t/4})$ V, $V_L = 12e^{-t/4}$ V, 2.77 s.)

5**. Show algebraically that in figure 5.12 the current I_L through the inductor is given by

$$\frac{V R_2}{R_1 + R_2} = \frac{R_1 R_2 I_L}{R_1 + R_2} + L\frac{dI_L}{dt} + rI_L \quad \text{(Thevenin equivalent).}$$

If the voltage source is a generator of 4 V square pulses, what value of R_2 is required to observe pulses at CD with a time constant of 0.1 ms, and what is then the height of the pulses at CD? (Ans: $R_2 = 48\,\Omega$, 184 mV.)

6.   What currents flow in the circuits of figure 5.13(a) and (b) after transients have died away if the voltage source is $2\cos(2\times 10^4 t + 30°)$ V? What are the voltages across the individual components? (Ans: (a) $13.9\cos(2\times 10^4 t - 3.7°)$ mA; $V_R = 1.67\cos(2\times 10^4 t - 3.7°)$ V; $V_L = 1.11\cos(2\times 10^4 t + 86.3°)$ V; (b) $42.9\cos(2\times 10^4 t - 60°)$ mA; $V_L = 3.43\cos(2\times 10^4 t + 30°)$ V; $V_C = -1.43\cos(2\times 10^4 t + 30°)$ V.)

7.   If the components in each of figures 5.13(a) and (b) are rearranged in parallel, what current flows through each component and what is the total current in each circuit? (Ans: (a) $I_R = 16.7\cos(2\times 10^4 t + 30°)$ mA, $I_L = 25\cos(2\times 10^4 t - 60°)$ mA, $30\cos(2\times 10^4 t - 26.3°)$ mA, (b) $I_L = 25\cos(2\times 10^4 t - 60°)$ mA, $I_C = 60\cos(2\times 10^4 t - 60°)$ mA, $I = 35\cos(2\times 10^4 t + 120°)$ mA.)

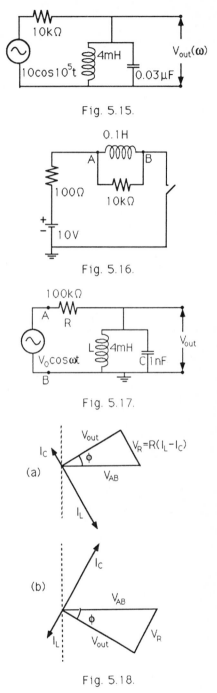

8. A wire-wound resistor of 20 Ω has a stray inductance of 5×10^{-7} H. Will this appear in series or in parallel with R? At what frequency does this stray inductance become significant and what is then its effect? (Ans: in series, for a sensible result at low frequency; $\omega_0 = 4 \times 10^7$ rad/s; makes a low-pass filter and for $\omega > \omega_0$, L dominates.)

Fig. 5.15.

9*. In the circuit of figure 5.14, what is the output voltage $V_{\text{out}}(\omega)$ if $V_S = 10 \cos \omega t$ V? Does the circuit behave as a low- or high-pass filter? What is the cut-off frequency? Hint: consider the Thevenin equivalent across CD. (Ans: $V_{\text{out}} = \frac{1}{2} V_S R \cos(\omega t - \psi)/[(\frac{1}{2}\omega L)^2 + R^2]^{\frac{1}{2}}$ with $\tan \psi = \omega L/2R$; low-pass filter; $\omega_0 = 5 \times 10^5$ rad/s.)

10*. In figure 5.15, what is the output voltage $V_{\text{out}}(\omega)$? Hint: let $V_{\text{out}} = V_0 \cos(\omega t - \psi)$ and find I_L and I_C. (Ans: $V_{\text{out}} = (4/2.04) \cos(\omega t - \psi)$ with $\tan \psi = 5$.)

Fig. 5.16.

11**. (QMC 1981) The switch in figure 5.16 has been closed for a long time before it is suddenly opened. Sketch voltage waveforms for points A and B with respect to earth and the waveform of current in the inductor, marking values immediately before and after the opening of the switch and the final values attained after the decay of the transient voltages and currents. What is the time constant of these transients? Assume the inductor is loss free. (Ans: $V_A = V_B = 0$ before; $V_A = 10$ V, $V_B = 1010$ V after; V_B falls exponentially to 10 V with $\tau = 10^{-5}$ s.)

Fig. 5.17.

12. (QMW 1990) What are the phase relations between AC voltage and current in (a) a capacitor, (b) a resistor and (c) an inductor? Show that, for the circuit shown in figure 5.17,

$$|V_{\text{out}}| = V_0 \left[1 + \left(\frac{CR}{L}(\omega L - 1/\omega C) \right)^2 \right]^{-1/2}$$

and find the relation between V_{out} and V_0. Find expressions for the currents through the capacitor and inductor and display these currents as phasor diagrams relative to V_{AB} at (a) angular frequency $\omega = 1/\sqrt{LC}$, (b) somewhat lower frequency, and (c) somewhat higher frequency. Sketch the Bode plots for V_{out} as a function of frequency, indicating the essential qualitative features. (Ans: (a) $\phi = 0$, $I_R = 0$, (b) and (c) figures 5.18(a) and (b).)

Fig. 5.18.

6

Complex Numbers and Impedance

6.1 Complex Numbers

Complex numbers are very helpful in many areas of physics and engineering. The reason is that they express information about both magnitude and phase in a very simple way. They play an absolutely central role in the treatment of AC circuits, since they handle capacitances and inductances in a neat, compact way. The phasor diagrams of the previous two chapters are in fact graphical representations of complex numbers. In many cases, it is easier to handle them algebraically than with diagrams, so this chapter will develop and illustrate this algebra. First, we review the pure mathematics; then we apply it to AC circuits.

As a preliminary, consider the properties of 'ordinary' or **real** numbers. Examples are -4, -3, -2, ..., 1, e, π, 10, and so forth. They can be represented by points plotted along a single axis (figure 6.1), called the 'real axis'. The sum or difference or product of any two real numbers is another real number: for example, $2+3 = 5$ and $2 \times 3 = 6$. The square of any real number is a real number: $3^2 = 9$. However, the converse is not always true: $\sqrt{9} = +3$ or -3, but there is no real number whose square is -9.

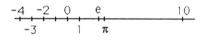

Fig. 6.1. Real numbers.

The notion of the square root of a negative number was discussed as far back as 1575 by Bombelli, but it was Euler who in 1777 introduced a specific notation. Mathematicians use the symbol i for the square root of -1. In electrical engineering, it is conventional to use instead the symbol j, to avoid confusion with currents:

$$\boxed{j^2 = -1} \tag{6.1}$$

or

$$\boxed{j = \sqrt{-1}.} \tag{6.2}$$

By this formal device, algebraic expressions may be written for the square roots of negative numbers, e.g. $\sqrt{-9} = 3j$ or $-3j$. These

are called **imaginary** numbers. Examples are −4j, −3j, −2j, ...,
j, πj, and so on. They are distinct from real numbers, but differ
only by the appearance of the factor j. They may be plotted along
a new axis called the 'imaginary axis'. It is conventional to plot
the real and imaginary axes at right angles to one another, as in
figure 6.2. This plot is called the **Argand diagram** or **complex
plane**.

Complex numbers have both real and imaginary parts. An example is the point $Z = 4 + 5j$ shown in figure 6.2. It may also
be written $Z = (4,5)$, i.e. as an ordered pair of real and imaginary numbers. Real numbers, for example (4,0), are a special
case of complex numbers, as are imaginary numbers, for example
$(0,5) \equiv 5j$.

Complex numbers obey precisely the same rules of addition,
subtraction, multiplication and division as real numbers. For example:

$$\text{addition}: (2 + 3j) + (5 + 9j) = 7 + 12j$$
$$\text{subtraction}: (7 + 12j) − (5 + 9j) = 2 + 3j$$
$$\text{multiplication}: (A + jB)(C + jD) = (AC + j^2 BD) + j(BC + AD)$$
$$= (AC − BD) + j(BC + AD).$$
$$(6.3)$$

(Division will be dealt with below.) These examples demonstrate
that the sums and products of complex numbers are themselves
complex numbers. They are the simplest possible generalisation
of the notion of a real number. The only fresh ingredient is the
definition $j^2 = −1$.

Modulus and phase

The point Z of figure 6.2 may alternatively be described in terms
of R and θ (figure 6.3). The distance R of the point from the
origin is called the **modulus** of Z and is written $|Z|$. The angle
θ is called its **phase** or sometimes the **argument** of Z, arg. By
convention, θ is measured from the real axis and is positive towards
the imaginary axis; thus θ_1 is positive for Z_1 in figure 6.3 and θ_2
is negative for the point Z_2.

In terms of these quantities,

$$Z = R(\cos \theta + j \sin \theta). \qquad (6.4)$$

This is a very useful form for expressing complex numbers. We say
that 'the real part of Z is $R \cos \theta$' or

$$\text{Re } Z = R \cos \theta \qquad (6.5)$$

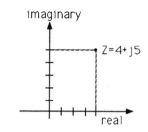

Fig. 6.2. A complex number.

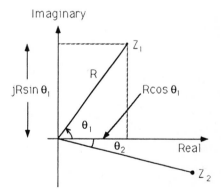

Fig. 6.3. The R, θ representation.

and 'the imaginary part of Z is $R \sin \theta$' or

$$\text{Im } Z = R \sin \theta. \qquad (6.6)$$

Note that θ is not unique: adding 2π (or multiples) to θ gives the same complex number.

Complex exponentials

There is yet a third way of expressing a complex number. This is as a complex exponential. It is very important for AC circuits. The properties of exponentials are reviewed in appendix B. The property of interest here is that

$$e^{\alpha} e^{\beta} = e^{(\alpha + \beta)}. \qquad (6.7)$$

Suppose

$$Z_1 = R_1(\cos \theta_1 + j \sin \theta_1)$$

and

$$Z_2 = R_2(\cos \theta_2 + j \sin \theta_2).$$

Then

$$\begin{aligned} Z_1 Z_2 &= R_1 R_2[(\cos \theta_1 \cos \theta_2 - \sin \theta_1 \sin \theta_2) \\ &\quad + j(\sin \theta_1 \cos \theta_2 + \cos \theta_1 \sin \theta_2)] \\ &= R_1 R_2[\cos(\theta_1 + \theta_2) + j \sin(\theta_1 + \theta_2)]. \end{aligned}$$

This is just the result obtained by writing

$$Z = R e^{j\theta} \equiv R \exp(j\theta) \qquad (6.8)$$

and using (6.7) to multiply Z_1 by Z_2:

$$Z_1 Z_2 = R_1 R_2 e^{j(\theta_1 + \theta_2)}. \qquad (6.9)$$

In words, when two complex numbers Z_1 and Z_2 are multiplied, their phases add and their moduli multiply. The form (6.8) is very useful in relating AC voltage and current, as discussed below.

A particularly important complex number is j itself. It lies on the unit circle, and can be written

$$j = e^{j\pi/2}$$

i.e. its phase angle is $\pi/2$. Multiplying any complex number by j therefore has the effect of adding $\pi/2$ to its phase:

$$jZ_1 = R_1 e^{j(\theta_1 + \pi/2)}.$$

As an example, multiplying any real number by j turns it into an imaginary number and rotates its phase by $\pi/2$. Likewise,

$$1/j = -j^2/j = -j = e^{-j\pi/2}$$

so dividing any complex number by j has the effect of subtracting $\pi/2$ from its phase:

$$Z_1/j = R_1 e^{j(\theta_1 - \pi/2)}.$$

Complex numbers are a generalisation of real numbers. Can you see the possibility of further generalisations? For example, does the cube root of -1 or the square root of j introduce any new possibility or can they be expressed in terms of complex numbers?

A final piece of terminology is the **complex conjugate** of a complex number. If $Z = X + jY = Re^{j\theta}$, its complex conjugate is $Z^* = X - jY = Re^{-j\theta}$. It is useful because

$$ZZ^* = R^2 = |Z|^2.$$

The complex conjugate of j is $-j$, and $j \times (-j) = 1$.

In terms of phase angles, why is $(A+jB)(A-jB)$ real and what is its magnitude?

In summary, complex numbers may be expressed in three forms:

$$\boxed{\begin{aligned} Z &= X + jY \\ Z &= R(\cos\theta + j\sin\theta) \\ Z &= Re^{j\theta} \end{aligned}}$$

and follow just the same rules of addition and multiplication as real numbers, except for the extra rule that $j^2 = -1$. Division is simply the converse of multiplication, so that

$$Z_3/Z_2 = Z_1 \quad \text{where} \quad R_1 = R_3/R_2 \quad \text{and} \quad \theta_1 = \theta_3 - \theta_2. \tag{6.10}$$

Worked examples on complex numbers
Suppose $Z_1 = 3 + 4j$ and $Z_2 = 5 - 12j$. They can be rewritten

$$Z_1 = 5e^{j\theta_1} \quad \text{with} \quad \theta_1 = \tan^{-1}(4/3) = 53.13°$$
$$Z_2 = 13e^{j\theta_2} \quad \text{with} \quad \theta_2 = \tan^{-1}(-12/5) = -67.38°.$$

Then

$$Z_3 = Z_2 - Z_1 = (5-3) - (12+4)j = 2 - 16j$$
$$Z_4 = Z_1 Z_2 = (15+48) + (20-36)j = 63 - 16j$$
$$= 65e^{j\theta_4} \quad \text{with} \quad \theta_4 = \tan^{-1}(-16/63) = -14.25°$$
$$\text{and} \quad Z_5 = Z_2/Z_1 = R_5 e^{j\theta_5}$$

where $R_5 = 13/5$ and $\theta_5 = \theta_2 - \theta_1 = -120.51°$. Further examples are given at the end of the chapter.

6.2   AC Voltages and Currents

So far, AC problems have been tackled using $V_0 \cos \omega t$ and occasionally using $V_0 \sin \omega t$. Of course, both give the same answers, since one may be transformed to the other just by redefining $t = 0$. Now

$$e^{j\omega t} = \cos \omega t + j \sin \omega t.$$

It is plausible that the same problems can be solved using $V = V_0 e^{j\omega t}$. Indeed this is so. Then the earlier results may be recovered using

$$\text{Re } V = V_0 \cos \omega t$$

or

$$\text{Im } V = V_0 \sin \omega t.$$

However, there are two great virtues of complex numbers. They actually *simplify* the algebra, and they lead to further insight.

Consider as an example an AC voltage applied simply to a capacitor (figure 6.4). Voltage and current differ in phase, so we take

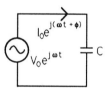

Fig. 6.4. AC applied to a capacitor.

$$V = V_0 e^{j\omega t} \qquad \text{and} \qquad I = I_0 e^{j(\omega t + \phi)} = I_0 e^{j\omega t} e^{j\phi}. \qquad (6.11)$$

Then

$$\begin{aligned}
V &= Q/C \\
Q &= CV_0 e^{j\omega t} \\
I &= dQ/dt = j\omega CV_0 e^{j\omega t} = (j\omega C)V. \qquad (6.12)
\end{aligned}$$

In summary,

$$V = IZ \qquad (6.13)$$

with

$$Z = 1/j\omega C = -j/\omega C = (1/\omega C)e^{-j\pi/2}. \qquad (6.14)$$

How is this algebra to be interpreted?

From the multiplicative law (6.9) for complex numbers, we can immediately read off from equations (6.13) and (6.14) that

$$|V| = |Z||I|$$

or

$$V_0 = (1/\omega C)I_0$$

and

$$\text{Phase}(V) = \text{Phase}(Z) + \text{Phase}(I)$$
$$= -\pi/2 + \text{Phase}(I).$$

These relations summarise what is already known from Chapter 4, but they do so in a very simple form: the reactance of the capacitor is $1/\omega C$ and the current leads voltage by 90° in phase. We can recover the same results as were obtained with sines and cosines by taking the real part of equations (6.11) and (6.12):

$$V = \text{Re}(V_0 e^{j\omega t}) = V_0 \cos \omega t$$
$$I = \text{Re}(\omega C V_0 e^{j(\omega t + \pi/2)}) = \omega C V_0 \cos(\omega t + \pi/2).$$

However, the form (6.13) using complex numbers is actually simpler because the complex quantity Z expresses the phase difference; Z is called the **impedance** of the capacitor.

The important virtues of (6.13) and (6.14) are that

(i) the relation between V and I is linear; this permits us to bring to bear on AC circuits all the pure mathematics of linear algebra; it is the basis of Thevenin's theorem for capacitors (and inductors, see below); and

(ii) using complex numbers, the equations express *both magnitude and phase*.

Amplitudes in optics and quantum mechanics are likewise expressed using complex numbers.

6.3   Inductance

For an inductance carrying current $I_0 \cos \omega t$,

$$V_L = L\frac{dI}{dt} = -\omega L I_0 \sin \omega t = \omega L I_0 \cos(\omega t + \pi/2). \qquad (6.15)$$

Again, the relation may be simplified by writing $I = I_0 e^{j\omega t}$, when

$$V_L = L\frac{dI}{dt} = j\omega L I_0 e^{j\omega t} = (j\omega L)I.$$

There is once more a linear relation $V_L = Z_L I$ with

$$Z_L = j\omega L = (\omega L)e^{j\pi/2}. \qquad (6.16)$$

This time

$$|V_L| = |Z_L||I| \qquad \text{or} \qquad V_0 = \omega L I_0 \qquad (6.17)$$

$$\text{Phase}(V_L) = \text{Phase}(Z_L) + \text{Phase}(I) = \pi/2 + \text{Phase}(I). \qquad (6.18)$$

The voltage leads current by 90° in phase. Combining (6.17) and (6.18)

$$V_L = \omega L I_0 \cos(\omega t + \pi/2)$$

which reproduces (6.15).

6.4   Summary on Impedance

Once you have grasped the idea of using complex numbers, there is very little to memorise. Resistance, capacitance and inductance may be put on the same footing by using complex numbers to write

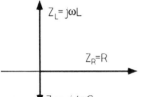

Fig. 6.5. Impedances Z_R, Z_C and Z_L.

$$\boxed{\begin{aligned} V(\text{complex}) &= Z(\text{complex})\, I(\text{complex}) \\ Z(\text{resistance}) &= R \\ Z(\text{capacitance}) &= 1/\text{j}\omega C \\ Z(\text{inductance}) &= \text{j}\omega L. \end{aligned}}$$

In all three cases

$$\boxed{\begin{aligned} |V| &= |Z||I| \\ \text{phase}(V) &= \text{phase}(Z) + \text{phase}(I). \end{aligned}}$$

On the Argand diagram, the complex impedances may be represented by the points shown in figure 6.5. This diagram allows us to read off immediately that for an inductance V leads I by a phase angle of 90° and for a capacitance it lags by 90°.

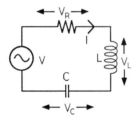

Fig. 6.6. Series combination of R, C and L.

6.5   Impedances in Series

For the series combination shown in figure 6.6:

$$\begin{aligned} V &= V_R + V_L + V_C \\ &= (Z_R + Z_L + Z_C)I \\ &= (R + \text{j}\omega L - \text{j}/\omega C)I. \end{aligned}$$

The combined impedance is shown in figure 6.7. Its magnitude is

$$|Z| = [R^2 + (\omega L - 1/\omega C)^2]^{1/2}$$

and its phase ψ is given by

$$\tan \psi = (\omega L - 1/\omega C)/R.$$

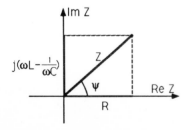

Fig. 6.7. Impedance diagram for the RCL series combination.

Voltage leads current by angle ψ and $|V| = |Z||I|$. The impedance is a minimum when $\omega L = 1/\omega C$, i.e. $\omega = 1/\sqrt{LC}$. At this frequency there is a peak in the current of $I = V/R$ and the impedance is purely resistive: $\psi = 0$. This circuit will be studied more extensively in Chapter 14.

In a particular problem, we might know I or V or the voltage across one of the elements R, C or L. Getting from one to another involves simple manipulations of magnitudes and phases. Suppose, for the sake of argument, we know that $V = V_0 \cos(\omega t + 30°)$. Then

$$I = I_0 \cos(\omega t + 30° - \psi) \qquad \text{where} \qquad I_0 = V_0/|Z|$$
$$V_R = RI = RI_0 \cos(\omega t + 30° - \psi)$$
$$V_C = (1/j\omega C)I = (I_0/\omega C)\cos(\omega t - 60° - \psi)$$
$$V_L = (j\omega L)I = \omega L I_0 \cos(\omega t + 120° - \psi)$$
$$= \omega L I_0 \sin(\omega t + 30° - \psi).$$

In the last relation, it is not necessary to remember the trigonometrical relationship between cos and sin. A useful trick is to make use of the fact that

$$e^{j\omega t} = \cos \omega t + j \sin \omega t.$$

Then if

$$I \propto \cos(\omega t + 30° - \psi) = \mathrm{Re}\{\exp[j(\omega t + 30° - \psi)]\}$$
$$V_L \propto jI \propto \mathrm{Im}\{\exp[j(\omega t + 30° - \psi)]\} \propto \sin(\omega t + 30° - \psi).$$

A second example is given in figure 6.8. For this circuit

$$V = (R - j/\omega C_1 - j/\omega C_2)I.$$

In this case, $V = |Z|I_0 \cos(\omega t - \phi)$, where

$$Z = [R^2 + (1/\omega C_1 + 1/\omega C_2)^2]^{1/2}$$
$$\tan \phi = (1/\omega C_1 + 1/\omega C_2)/R.$$

This gives a simple derivation of the impedance of two capacitors in series.

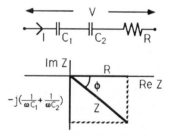

Fig. 6.8 (a) series combination of C_1, C_2 and R, (b) the corresponding impedance diagram.

6.6 Impedances in Parallel

Consider the parallel combination of elements shown in figure 6.9. The same voltage V exists across all three elements, so

$$I = I_R + I_C + I_L = \frac{V}{R} + \frac{V}{Z_C} + \frac{V}{Z_L} = V\left(\frac{1}{R} + j\omega C + \frac{1}{j\omega L}\right) = V/Z \tag{6.19}$$

$$\frac{1}{Z} = \frac{1}{R} + j\left(\omega C - \frac{1}{\omega L}\right). \tag{6.20}$$

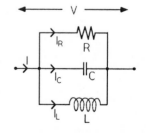

Fig. 6.9. Parallel combination of R, C and L.

Because of the linear relation betwen voltage and current, impedances may be added in series and parallel combinations in just the same way as resistors in Chapter 1. When $\omega C = 1/\omega L$, the second term is zero and $Z \to R$. At this frequency there is a maximum in the impedance and zero phase difference between voltage and current.

There is, however, a final algebraic step required to interpret equations (6.19) and (6.20). They can be used to write

$$|I| = |V||1/Z|$$
$$I_0 = V_0[(1/R)^2 + (\omega C - 1/\omega L)^2]^{1/2}$$
$$\text{Phase}(I) = \text{Phase}(V) + \text{Phase}(1/Z)$$
$$= \text{Phase}(V) + \tan^{-1}[(\omega C - 1/\omega L)R].$$

A resistor has stray capacitance and stray inductance. As $\omega \to \infty$, which dominates: capacitance or inductance?

There may be circumstances where an explicit algebraic form is required for Z itself. From (6.20)

$$Z = \frac{R}{1 + j(\omega C - 1/\omega L)R}.$$

This expression contains j in the denominator, and if we want to know its real and imaginary parts this may be inconvenient. There is a standard way to rearrange it, by multiplying top and bottom by the complex conjugate of the denominator:

$$Z = \frac{R[1 - j(\omega C - 1/\omega L)R]}{[1 + j(\omega C - 1/\omega L)R][1 - j(\omega C - 1/\omega L)R]}$$
$$= \frac{R - jR^2(\omega C - 1/\omega L)}{1 + R^2(\omega C - 1/\omega L)^2}.$$

As a final example, consider the cascade of two CR filters shown in figure 6.10(a). This would be a difficult problem using phasor diagrams. In order to find V_{out}, the elements are rearranged as shown in figures 6.10(b)–(d). The impedances form a potential divider, with the result

$$V_{\text{out}} = \frac{V}{1 + j\omega C_1 R_1} \times \frac{1}{j\omega C_2} \times \left(\frac{1}{j\omega C_2} + R_2 + \frac{R_1}{1 + j\omega C_1 R_1}\right)^{-1}$$
$$= \frac{V}{1 + j\omega(C_1 R_1 + C_2 R_2 + C_2 R_1) - \omega^2 C_1 C_2 R_1 R_2}$$
$$\equiv \frac{V}{(1 + j\omega\tau_1)(1 + j\omega\tau_2)}.$$

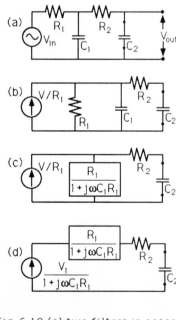

Fig. 6.10 (a) two filters in cascade, (b)–(d) rearrangements of the circuit.

where

$$\tau_1\tau_2 = C_1C_2R_1R_2 = B \qquad \text{say}$$

and

$$\tau_1 + \tau_2 = \tau_1 + B/\tau_1 = (C_1 + C_2)R_1 + C_2R_2 = 2A \qquad \text{say}.$$

Solving this quadratic equation,

$$\tau_1 = A - (A^2 - B)^{1/2}$$

and

$$\tau_2 = A + (A^2 - B)^{1/2}.$$

The Bode plot for this circuit is shown in figures 6.11(a) and (b) with corner frequencies at $\omega_1 = 1/\tau_1$ and $\omega_2 = 1/\tau_2$. At high frequencies, the roll-off is -12 db/octave and the phase lag of the output is π. It should be clear by extrapolation that if n filters are cascaded, the roll-off at high frequencies is $-6n$ db/octave and the phase lag is $n\pi/2$.

You are advised to practice plenty of examples from the exercises at the end of the chapter, since fluent manipulation of impedances is important throughout electronics.

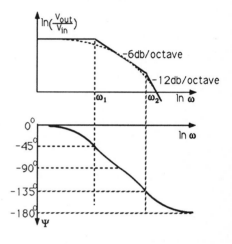

Fig. 6.11. Bode plot for the circuit of Fig. 6.10.

Admittance

Sometimes, it is convenient to work with $Y = 1/Z$ instead of with Z itself. The quantity Y is called admittance. Then, instead of $V = ZI$, we have

$$I = YV. \tag{6.21}$$

A clear example where this is convenient is in dealing with parallel combinations of elements, as in figure 6.9. In that example,

$$Y = Y_R + Y_C + Y_L$$
$$= \frac{1}{R} + j\omega C + \frac{1}{j\omega L} = \frac{1}{R} + j\left(\omega C - \frac{1}{\omega L}\right).$$

The inverse of resistance is called **conductance**, and the official SI unit is the **Siemen**. A common alternative is to call this unit the **mho** (ohm written backwards).

6.7 Power

Expressions will now be derived for power in terms of complex numbers. Repeating part of an earlier discussion, the power dissipated in an electrical network is given by

$$P = IV = I_0 \cos\omega t \, V_0 \cos(\omega t + \psi) = \tfrac{1}{2}I_0V_0[\cos(2\omega t + \psi) + \cos\psi].$$

Averaged over a complete cycle, the first term contributes zero, so the mean power dissipated is

$$\bar{P} = \tfrac{1}{2}I_0 V_0 \cos\psi.$$

If $V = V_0 e^{j(\omega t + \psi)}$ and $I = I_0 e^{j\omega t}$, then

$$\bar{P} = \tfrac{1}{2}\mathrm{Re}(VI^*) = \tfrac{1}{2}\mathrm{Re}(V^*I) \qquad (6.22)$$

where the star denotes the complex conjugate. Alternatively (figure 6.12),

$$\cos\psi = \frac{\mathrm{Re}(Z)}{|Z|}$$

Fig. 6.12. Re Z.

and

$$\bar{P} = \tfrac{1}{2}I_0^2 |Z|\frac{\mathrm{Re}(Z)}{|Z|} = \tfrac{1}{2}I_0^2 \mathrm{Re}(Z). \qquad (6.22a)$$

Alternatively,

$$\frac{1}{Z} = \left|\frac{1}{Z}\right| e^{-j\psi}$$

$$\mathrm{Re}(1/Z) = (1/|Z|)\cos\psi$$

and

$$\bar{P} = \tfrac{1}{2}V_0^2 \mathrm{Re}(1/Z) = \tfrac{1}{2}V_0^2 \mathrm{Re}(Y). \qquad (6.22b)$$

All of equations (6.22) are simple and obvious generalisations of the corresponding results for resistors.

6.8   Bridges

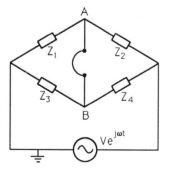

Fig. 6.13. AC Wheatstone bridge.

The familiar Wheatstone bridge is generalised in the arrangement shown in figure 6.13, where impedances $Z_1, \ldots, Z_4$ replace resistances. The out-of-balance current between A and B may be measured by headphones or by a multimeter on its AC range. The condition for zero current in this arm is that $V_A = V_B$. If the current in the arm AB is zero,

$$V_A = V e^{j\omega t}\frac{Z_1}{Z_1 + Z_2}$$

$$V_B = V e^{j\omega t}\frac{Z_3}{Z_3 + Z_4}$$

so the condition for balance is that

$$\frac{Z_3 + Z_4}{Z_3} = \frac{Z_1 + Z_2}{Z_1}$$

$$Z_4/Z_3 = Z_2/Z_1 \qquad (6.23)$$

$$Z_1 Z_4 = Z_2 Z_3. \qquad (6.23a)$$

These equations require that both real and imaginary parts of (6.23) balance.

Many types of bridge have been devised to measure capacitances and inductances. One worked example will be given here and several more are given in the exercises. A warning to the student is that in practice stray capacitance between points in the circuit and stray inductance in the resistors can present problems, particularly at high frequency.

If the generator does not give pure sine waves but contains harmonics, does it alter the balance point of the bridge? If so, what can you do about it?

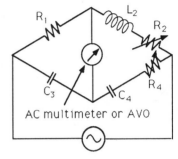

Fig. 6.14. The Owen bridge.

Worked example

For the Owen bridge, illustrated in figure 6.14, equation (6.23a) gives

$$R_1 \left(R_4 + \frac{1}{j\omega C_4} \right) = (R_2 + j\omega L_2) \frac{1}{j\omega C_3}.$$

This requires

$$R_1 R_4 = L_2/C_3$$
$$R_1/C_4 = R_2/C_3.$$

The variable resistors R_2 and R_4 may be adjusted independently to satisfy these two conditions. The balance in this particular bridge is independent of ω; this is not always the case.

This bridge provides a simple way of measuring L. If L is a few mH, suitable component values are $C_3 = C_4 = 0.01 \ \mu F$ and $R_1 = 1 \ k\Omega$. It is necessary to adjust R_2 and R_4 successively to achieve balance.

6.9   Exercises

1. *Representation of complex numbers.* Draw real and imaginary axes on graph paper and plot the following complex numbers. Convert each one to polar form $re^{j\phi}$. (a) $3+j4$, (b) $-3+j3$, (c) $-8-6j$, (d) -1, (e) $-2j$. (Ans: $5e^{j53.13°}$, $3\sqrt{2}e^{j135°}$, $10e^{j216.87°}$, $1e^{j180°}$, $2e^{270°}$.)

2. *Products of complex numbers.* Calculate the products of the following pairs of complex numbers. As a check, convert them into polar form, and again compute their product. (a) $(3+j4)(-3+j3)$; (b) $(-3+j3)(-8-6j)$; (c) $(-8-6j)(-2j)$; (d) $(x-jy)(x+jy)$. (Ans: $-21-j3 = 15\sqrt{2}e^{j188.13°}$; $42-6j = 30\sqrt{2}e^{-j8.13°}$; $-12+16j = 20e^{j126.87°}$; x^2+y^2.)

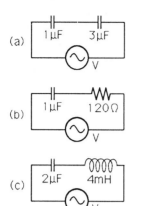

(a)

(b)

(c)

Fig. 6.15.

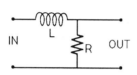

IN   OUT

Fig. 6.16.

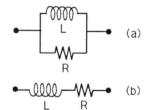

(a)

(b)

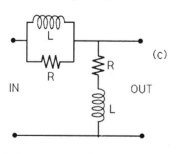

(c)

IN   OUT

Fig. 6.17.

3. *Quotients of complex numbers.* Find the quotient in expressions (a)–(d) by multiplying both numerator and denominator by the complex conjugate of the denominator. As a check, convert the numbers to polar form, and evaluate the quotient in this form. (a) $j7/(3-j3)$; (b) $(3+j4)/(-3+j3)$; (c) $(-3+j3)/(3+j4)$; (d) $(3+j3)/(1-j1)$. (Ans: $7(j-1)/6 = 7e^{j135°}/3\sqrt{2}$; $(1-7j)/6 = 5e^{-j81.87°}/3\sqrt{2}$; $3(1+7j)/25 = 3\sqrt{2}e^{j81.87°}/5$; $3j = 3e^{j90°}$.)

4. $e^{j\phi}$. Express each complex number in cartesian form and plot graphically. (a) $3e^{-j\pi/4}$, (b) $4e^{j\pi/2}$, (c) $5e^{-j2\pi/3}$, (d) $6e^{j4\pi/3}$. (Ans: $3(1-j)/\sqrt{2}$, $4j$, $-5(1+\sqrt{3}j)/2$, $3(-1-\sqrt{3}j)$.)

5. $e^{j(\omega t+\phi)}$. A point moves in the complex plane with time as $e^{jt/2}$. Plot its movement at $t = 0, 1, 2, \ldots, 10$. What happens after $t = 4\pi$? Plot a graph against t of $Re(e^{jt/2})$ and $Re(e^{j(t/2+\pi/2)})$. Satisfy yourself that $d(e^{jt/2})/dt = (j/2)e^{jt/2}$. (Ans: repeats after $t = 4\pi$.)

6. An inductance of 4 mH passes a current $I = 5\cos 500t$ mA. What is the voltage across it? Show that $I = 5Re(e^{j500t})$ mA and find the voltage in terms of complex exponentials. Repeat this problem for a capacitance of 20 μF. (Ans: $-10\sin 500t$ mV$= Re(10je^{j500t})$; $0.5\sin 500t$ V $= Re(0.5e^{j500t}/j)$.)

7. *Addition and subtraction of complex amplitudes.* (a) Represent the two currents $I_1 = 2\sin(\omega t - \pi/4)$ and $I_2 = \sin(\omega t + \pi/2)$ on the complex plane at $t = 0$ and hence find $I_1 + I_2$ for any t; (b) find the difference $V_1 - V_2$ where $V_1 = 5\cos(\omega t + 50°)$ and $V_2 = 3\cos(\omega t + 150°)$. (Ans: $1.47\sin(\omega t - 16.32°)$; $6.26\cos(\omega t + 21.85°)$.)

8*. *Complex impedance.* If in figures 6.15(a)–(c) $V = 10e^{j\omega t}$ mV, where $\omega = 10^4$ rad/s, find the magnitude and phase of the current and the voltage across each of the six components. (Ans: (a) $0.075e^{j(\omega t+\pi/2)}$ mA, $7.5e^{j\omega t}$ mV, $2.5e^{j\omega t}$ mV; (b) $64e^{j(\omega t+39.81°)}$ μA, $6.4e^{j(\omega t-50.19°)}$ mV, $7.68e^{j(\omega t+39.81°)}$ mV; (c) $e^{j(\omega t+\pi/2)}$ mA, $-40e^{j\omega t}$ mV, $50e^{j\omega t}$ mV.)

9. If the components in each of figures 6.15(a)–(c) are rearranged in parallel, what current flows in each component, and what is the total current in each circuit? (Ans: (a) $100e^{j(\omega t+\pi/2)}$ μA, $300e^{j(\omega t+\pi/2)}$ μA, $400e^{j(\omega t+\pi/2)}$ μA; (b) $100e^{j(\omega t+\pi/2)}$ μA, $83.3e^{j\omega t}$ μA, $130.17 \times e^{j(\omega t+50.19°)}$ μA; (c) $0.25e^{j(\omega t-\pi/2)}$ mA, $-0.2e^{j(\omega t-\pi/2)}$ mA, $0.05e^{j(\omega t-\pi/2)}$ mA.)

10. A low-pass filter consists of a capacitance C and resistance R in series. Show that the output voltage and input voltage are related by $V_{out}/V_{in} = 1/(1 + j\omega CR)$. The same resistance is used in series with an inductance L to make a low-pass filter

of identical frequency dependence. Draw the circuit and find the value of L. (Ans: figure 6.16, $L = CR^2$.)

11. (QMC 1975) A series LCR circuit with $L = 15$ mH has an instantaneous voltage $V = 45\sin(4000t + 30°)$ mV and an instantaneous current $I = 0.5\sin(4000t)$ mA. Find the values of R and C. At what frequency would the voltage and current always be in phase? (Ans: $45\sqrt{3}\ \Omega$, $16.7\ \mu\text{F}$, $\omega = 2000$ rad/s, $f = 1000/\pi$ Hz.)

12. Find expressions for the impedances of the networks in figures 6.17(a) and (b). At what frequency does the circuit of figure 6.17(c) produce zero phase shift? (Ans: (a) $Z = j\omega LR/(R+j\omega L)$, (b) $Z = R+j\omega L$; $\omega = R/L$.)

13. Find the impedances of the networks in figures 6.18(a), (b) and (c). (Ans: $R_2(1+j\omega CR_1)/[1+j\omega C(R_1+R_2)]$, $R_2(R_1 + j\omega L)/(R_1 + R_2 + j\omega L)$, $(R+j\omega L)/[(1-\omega^2 CL)+j\omega CR]$.)

14*. (QMC 1972). What is the power factor in circuits used at angular frequency ω consisting of (a) an ideal capacitor, (b) an inductance L in series with a resistance $R = \omega L$, and (c) the elements (a) and (b) in parallel? (Ans: (a) 0, (b) $1/\sqrt{2}$, (c) $(2 - 4\omega^2 LC + 4\omega^4 L^2 C^2)^{-1/2}$.)

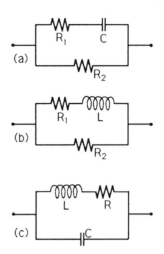

Fig. 6.18.

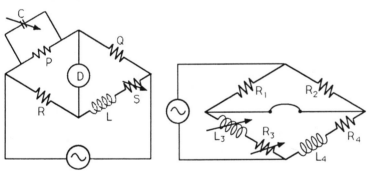

Fig. 6.19.   Fig. 6.20.

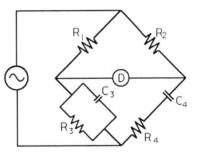

Fig. 6.21.

15. Find the conditions for zero signal in the detector D of Maxwell's L/C bridge (figure 6.19). (Ans: $S/R = Q/P$ and $L = CQR$.)

16. The Maxwell self-inductance bridge of figure 6.20 is used to find the resistance R_4 and self-inductance L_4 of an inductor by varying L_3 and R_3. Find the balance conditions. (Ans: $R_4 = R_2 R_3/R_1$ and $L_4 = R_2 L_3/R_1$.)

17. Find the balance conditions for the Wien bridge of figure 6.21. (Ans: $\omega^2 = 1/C_3 C_4 R_3 R_4$ and $(C_3/C_4)+(R_4/R_3) = R_2/R_1$.)

18**. Anderson's bridge, figure 6.22, is used to calibrate an inductance L against a standard capacitor C. Find the two balance

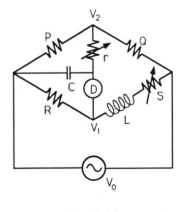

Fig. 6.22.

conditions for S and L. Hint: use node voltages V_2 and V_1. (Ans: $S = QR/P$ and $L = CR[Q + r(1 + Q/P)]$.)

19. (IC 1974) Show that the impedance between the terminals AB of figure 6.23 for a signal of frequency ω is $Z_1 = [R(1 - \omega^2 LC) + j\omega L]/(1 + j\omega CR)$ and that between A'B' is $Z_2 = [R(1 - \omega^2 LC) + j\omega L]/[(R + j\omega L)j\omega C]$. The circuits are now joined at AA' and BB'. Calculate the impedance of the complete circuit between AA' and BB', and show that if $L = 2CR^2$ the circuit acts as a pure resistance of magnitude R at all frequencies.

20*. (QEC 1974) Find the conditions for zero output from the network of figure 6.24. (Ans: $\omega^2 CLR = 2R_0$ and $RR_0 + 2L/C = (1/\omega^2 C^2)$.)

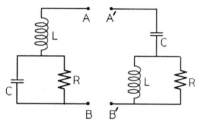

Fig. 6.23.

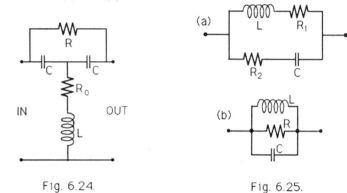

Fig. 6.24.   Fig. 6.25.

21. Find the admittance of the networks in figures 6.25(a) and (b). (Ans: $[1 - \omega^2 CL + j\omega C(R_1 + R_2)]/[(R_1 + j\omega L)(1 + j\omega CR_2)]$ and $(1/R) + j(\omega^2 CL - 1)/\omega L$.)

22*. *Node voltage and Thevenin's theorem.* Find V_X in figures 6.26(a) and (b) by using Kirchhoff's current law for that node. What is the Thevenin equivalent circuit across the points XY in both cases? Use this to check your evaluation of V_X. (Ans: (a) $\frac{1}{2}V_S/[1 + \frac{1}{2}R_1/(R_2 + j\omega L)]$, $V_{EQ} = \frac{1}{2}V_S$, $Z_{EQ} = \frac{1}{2}R_1$; (b) $V_S/(1 - \omega^2 CL + j\omega L/R)$, $V_{EQ} = V_S/(1 - \omega^2 CL)$, $Z_{EQ} = j\omega L/(1 - \omega^2 CL)$.)

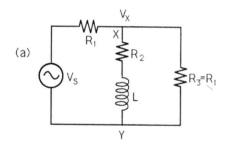

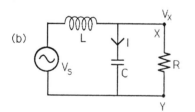

Fig. 6.26.

23**. *Superposition and Thevenin's theorem.* Find the current through the capacitor of figure 6.27 originating separately from V_1 and V_2. Check by finding the Thevenin equivalent circuit across terminals XY and hence the current I. (Ans: $I_1 = -V_1\omega^2 CL/[R(1 - \omega^2 CL) + j\omega L]$, $I_2 = V_2 j\omega CR/[R(1 - \omega^2 CL) + j\omega L]$, $V_{EQ} = (V_2 R + V_1 j\omega L)/(R + j\omega L)$, $Z_{EQ} = j\omega LR/(R + j\omega L)$.)

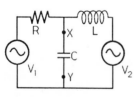

Fig. 6.27.

24. A generator with output impedance $R_1 + jX_1$ is connected directly across a load $R_2 + jX_2$. Show that maximum power is transferred to the load when $R_1 = R_2$ and $X_1 = -X_2$.

7

Operational Amplifiers and Negative Feedback

7.1 Introduction

So far, we have mostly considered passive elements (resistors, capacitors and inductors), which simply transmit currents and voltages. Amplifiers involve active devices—transistors. The amplifier is a central feature of all electronic circuits. Sometimes it may be hidden from view inside the circuit and may not be immediately apparent in the relation between input and output, but it is always there. Even in digital circuitry where the output voltage is often the same as at the input, the circuit will generally offer current amplification.

There are certain general principles which allow circuits to be designed so that they are insensitive to the precise details of the amplifier. As a consequence, circuits designed 40 or 60 years ago for valves can be redesigned with changes of detail to work with bipolar transistors or FETs or any other amplifier. As each new device is invented, the principles of circuit design do not change in an essential way, although details do.

This chapter will be concerned with those principles. For the moment the properties of individual transistors can be sidestepped. It is fortunate that manufacturers have done much of the detailed hard work by marketing cleverly designed integrated circuits (containing many transistors) which act as nearly ideal amplifiers. They are conservatively designed to avoid common pitfalls, and as long as we do not want to stretch performance to the limit they are a great convenience. Those readers who aspire to the ultimate in performance (or to design actual integrated circuits) will need to understand the idiosyncracies of individual transistors in addition to the principles described here.

Consider initially a voltage amplifier for which

$$\boxed{\text{voltage gain, } G = \frac{\text{output voltage}}{\text{input voltage}}.} \qquad (7.1)$$

(Gain G should not be confused with conductance g.) Ideally, G is large and independent of input voltage, current and frequency. The ideal voltage amplifier has large input resistance, so that it puts a negligible load on the previous circuit. It has a low output resistance, so that most of the output voltage is delivered to the subsequent circuit.

Opamps

Nowadays, a wide variety of **operational amplifiers** or 'opamps' are available very cheaply (35p upwards) satisfying most of these criteria. They look similar to the sketch of figure 7.1(a). These are called 'dual-in-line' (DIL) packages. Connections are made via the legs, which can be pushed into a breadboard or soldered to a printed circuit board (PC). The legs are easily bent, and you need to ease the chip out of a circuit board carefully using pliers.

The μA741C will be adopted for illustrative purposes. It is used widely for applications that are not too demanding. The layout of its connections is shown in figure 7.1(b). Power supplies, typically of $+15$ and -15 V, are connected at pins 7 and 4, labelled $+V$ and $-V$. Ultimately, the gain of the amplifier arises by drawing currents from these supplies. For reasons which will soon become apparent, all operational amplifiers are actually differential amplifiers, having two inputs V_+ and V_- (at pins 3 and 2 in this case) and with

$$V_{\text{out}} = G(\omega)(V_+ - V_-). \tag{7.2}$$

Pin 3 is called the non-inverting input and pin 2 the inverting input: if $V_+ = 0$, $V_{\text{out}} = -G(\omega)V_-$. The remaining pins are used for fine tuning of the performance.

Figure 7.2 shows a schematic representation of the amplifier. Its essential characteristics are:

(a) input impedance $r_{\text{in}} \sim 1$ MΩ; (for some other types of operational amplifier this can be as high as 10^{12} Ω);

(b) output impedance $r_{\text{out}} \sim 75$ Ω; small letters r_{in} and r_{out} are used since these are usually slope resistances of active devices (see Chapter 1) rather than actual resistors; this detail will be elaborated in Chapters 17 and 18;

(c) temperature stability $dG/G \sim 10^{-5}$ °C^{-1};

(d) supply voltages $+V$ and $-V$ which may be varied from 3 to 18 V; they need not be the same magnitude;

(e) input and output protected against short circuits and against voltages up to ± 15 V;

(f) output currents up to 20 mA;

(g) a gain G up to 2×10^5 at low frequencies; however, the output voltage limits when it gets within about 1 V of the power supply voltages. The gain varies with frequency in the manner shown in figure 7.3. There are good reasons for this, which

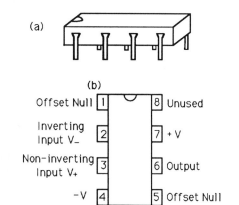

(a)

(b)

Offset Null [1]   [8] Unused

Inverting Input V_- [2]   [7] $+V$

Non-inverting Input V_+ [3]   [6] Output

$-V$ [4]   [5] Offset Null

Fig. 7.1 (a) an operational amplifier as a "chip", (b) the pin layout of the μA741C

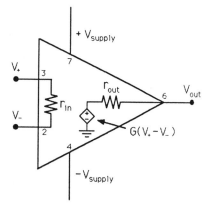

Fig. 7.2. Schematic of the amplifier.

will be explained in Chapter 19. It is desirable as a safeguard against simple amplifiers breaking into oscillation, a problem all too familiar to those who have tried to build their own high-gain amplifier from transistors.

7.2   Series Voltage Feedback

Such an amplifier is very rarely used in isolation. Almost invariably it is built into a **feedback loop**. Figure 7.4(a) shows the general case schematically where a feedback voltage $B_v V_{out}$ is derived from the output voltage and fed back to the input of the amplifier. There it is summed with the input voltage V_{in}. Then

$$V_{out} = G_v(V_{in} + V_F) = G_v(V_{in} + B_v V_{out})$$

providing the output voltage does not reach the saturation limit available from the power supply; the amplification A of the whole circuit is

$$A = \frac{V_{out}}{V_{in}} = \frac{G_v}{1 - G_v B_v}. \qquad (7.3)$$

This formula plays a fundamental role in the design of electronic circuits. Note the distinction between G_v, the voltage gain of the operational amplifier itself, and A, the amplification of the whole circuit between input and output voltages. The quantity G_v is called the **open-loop gain**, i.e. the gain with no feedback; A is called the **closed-loop gain**.

The feedback fraction B_v may be positive or negative. **Positive feedback** $(G_v B_v > 0)$ is used in digital circuitry. **Negative feedback** $(G_v B_v < 0)$ will be discussed at length in this chapter. One simple way of achieving it is shown in figure 7.4(b), where the feedback is applied from a potential divider to the inverting input of a differential amplifier. In this case

$$G_v = G(\omega)$$
$$V_F = -[R_1/(R_1 + R_2)]V_{out}$$
$$B_v = -R_1/(R_1 + R_2) = -B \qquad \text{say.}$$

Then

$$A = \frac{G(\omega)}{1 + BG(\omega)} \qquad \text{for negative feedback.} \qquad (7.3a)$$

You will meet both forms (7.3) and (7.3a) in the literature and the difference in sign in the denominator is a source of confusion. The former covers the use of positive or negative feedback in general.

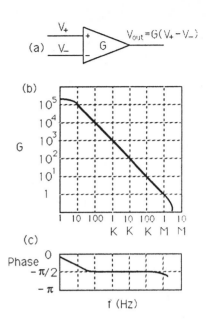

(a) V_+ / V_- / G / $V_{out} = G(V_+ - V_-)$

(b)

G 10^5 10^4 10^3 10^2 10^1 1

1 10 100   1 10 100   1 10
   K   K   K   M   M

(c) Phase   0 / $-\pi/2$ / $-\pi$

f (Hz)

Fig. 7.3 (a) Symbol for a differential amplifier, (b) G(f) of the μA741C, (c) phase difference between output and input.

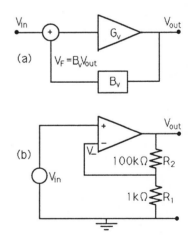

(a) V_{in} / G_v / V_{out} / $V_F = B_v V_{out}$ / B_v

(b) V_- / V_{out} / 100kΩ R_2 / 1kΩ R_1 / V_{in}

Fig. 7.4 (a) A general feedback loop, (b) negative feedback.

The latter conveniently takes account of the minus sign when B_v is negative.

For negative feedback, there is an important simplification if $BG \gg 1$. Then the 1 in the denominator can be neglected and

$$\boxed{A \simeq 1/B.} \tag{7.4}$$

(This result assumes that the input voltage is sufficiently small that the output does not saturate.) The great virtue of this arrangement is that the resulting amplification is independent of G, which can easily vary by a factor of two from one batch of operational amplifiers to another. If $G = 10^5$ and $B = 0.01$, then $A \simeq 100$. The exact solution (7.3a) gives $A = 99.90$ if $G = 10^5$ and 99.95 if $G = 2 \times 10^5$, showing that a change of G by a factor of two alters the circuit amplification A by only 0.05%. Since B is derived from a pair of resistors, it is very stable and is independent of frequency. Achieving this degree of stability is otherwise virtually impossible. The use of negative feedback makes circuit performance insensitive to the amplifier, which may be changed for another one of the same type or indeed one of a different type with minimal change in A. It is a general characteristic of negative feedback that it enhances stability. We shall see later that it has other virtues too.

Everyday examples

Negative feedback is commonplace in everyday life. If you argue a point with someone, he or she is likely to put the reverse point of view. Checks and balances in politics and the judiciary are important for stability. In the home, a thermostat measures room temperature and cuts off the heating when the temperature gets too high. When you drive a car you apply feedback to stabilise speed and direction. In Nature, feedback stabilises the population via the food supply. In every case, the signal fed back to the input depends on the output, making a closed loop.

Formulate the output and input signals algebraically in each of these cases, and the feedback.

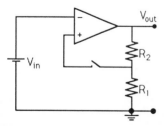

Fig. 7.5. Positive feedback.

Positive feedback

Positive feedback on the other hand leads to instability or saturation of the amplifier. You might think that we can still make the approximation $G_v B_v \gg 1$ and arrive at the result $A = -1/B_v$ again. However, what happens in this case is that the amplifier saturates. Suppose that V_{in} of figure 7.5 is a very small signal, so small that the output of the amplifier does not saturate. If the switch is closed and feedback is applied to the positive input of the amplifier, it reinforces V_{out} and drives it to the limit of what the amplifier will produce. When the amplifier saturates, G_v falls until $G_v B_v$ is just less than one; equation (7.3) is satisfied with V_{out} equal to whatever happens to be the voltage limit of the amplifier and G_v taking up an appropriate value, less than its unsaturated value. If V_{in} is reversed in sign, the same feedback loop drives the amplifier to saturation with the other polarity. In either case, the output of the amplifier swings to its extreme positive or negative limit and is insensitive to the precise value of the input. These two saturated states are useful for representing logical 0 and 1 in digital circuitry.

Positive feedback leads to saturation or in some cases oscillation (Chapter 19). In economics, positive feedback is dangerous because it leads to violent swings. War is also like this with the population united in a common aim. Dictators surround themselves with 'Yes Men' to achieve positive feedback within their empires; changes of dictator are frequently violent.

Let us return to negative feedback and equations (7.3*a*) and (7.4). You might be tempted to make B very small so as to achieve a large amplification $1/B$. However, remember the approximation $GB \gg 1$. The gain falls at high frequency according to figure 7.3, and the approximation will fail above frequencies where $GB = 1$. This limits the bandwidth of the circuit. On the straight part of the Bode plot, $G = G_0/\omega$, where $G_0 = 6 \times 10^6$ rad/s for the μA741C. Suppose an audio amplifier is required working up to $\omega = 10^5$ rad/s. If GB is to be 1 at this frequency, $B = \omega/G_0 = 1/60$, and the amplification of a single stage is limited to 60. If more amplification is required, the remedy is to cascade several stages of amplification. The upper limit of frequency at which amplification is possible with the μA741C is 6×10^6 rad/s where $G = 1$. If an amplifier is required working up to say $\omega = 10^8$ rad/s, this chip has insufficient bandwidth; either a much superior operational amplifier or individual transistors having this bandwidth must be used.

It is instructive to put together the circuit of figure 7.4(b) and examine the amplification with a DC input, and then with AC signals as a function of frequency. Input and output voltages may be displayed on two traces of an oscilloscope. Try varying the magnitude of V_{in} until the output saturates. Then vary the positive and negative supply voltages independently and observe their effect on the levels at which the output limits. Try varying the ratio of R_2 to R_1.

A practical point is that no pin of the operational amplifier is earthed; see figure 7.1(b). How does it know where earth potential is with respect to the supply voltages $+V$ and $-V$? The answer is via the load applied across V_{out}, for example, an oscilloscope input resistance if you are observing the waveform. It is important to earth the supply at the same point as the load (except on those rare occasions when you specifically wish to introduce a bias voltage so that the whole circuit is raised or lowered with respect to earth). If the supply voltages have a floating earth, the circuit will not function properly.

A second practical point is that the amplifier does actually draw small DC bias currents for transistor operation through the inverting and non-inverting input terminals V_+ and V_-. You must not cut off these bias currents by capacitors; if you do, the amplifier will not work: the intention is that it should be DC coupled.

7.3* Approximations in Voltage Feedback

The alert reader may have noticed that there were some approximations involved in reaching equation (7.3a). We digress now to see how good these are. With the component values chosen for figure 7.4(b) they are actually very good, but the unwary might choose much higher or lower values of the feedback resistors R_1 and R_2 and run into trouble.

Figure 7.6 shows the circuit of figure 7.4 drawn out in somewhat greater detail. It includes a load resistor R_L. There are two voltage generators: the source V_S which supplies the circuit and the generator $G(V_+ - V_-)$ at the output of the operational amplifier. The currents in the circuit are the superposition of those due to these two voltage sources, and the voltages in the circuit follow from the sum of these currents; deriving V_- from them is an alternative (and more precise) way of deriving the voltage gain. Try it.

The first approximation was that $V_{out} = G(V_+ - V_-)$, neglecting the voltage drop in r_{out}. This is a good approximation if $R_L \gg r_{out}$, but what happens if R_L is small or r_{out} is large? In this case, V_{out} is derived from $G(V_+ - V_-)$ by the potential divider formed from r_{out} and the parallel combination of R_L with $(R_1 + R_2)$. Let us call this latter combination R'_L. Then

$$V_{out} = G'(V_{in} - V_-)$$

where

$$G' = GR'_L/(R'_L + r_{out}).$$

The previous algebra can be repeated with G' replacing G. The voltage gain is insensitive to G, so this change has rather little effect on amplification. If, for example, $R'_L = 10\ \Omega$ and $r_{out} = 100\ \Omega$, $G' \simeq 10^4$ instead of 10^5 at low frequencies. The main effect is that the circuit runs out of gain at a lower frequency given by $G'(\omega)B = 1$.

Secondly, we took $B = R_1/(R_1 + R_2)$. This is not strictly correct. In parallel with R_1 is $(r_{in} + R_S)$. However, with the value chosen here ($r_{in} = 10^6\ \Omega$), the impedance across R_1 is affected

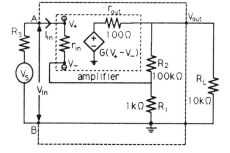

Fig. 7.6. Voltage feedback showing greater detail.

by at most one part in 10^3. If much larger values of R_1 and R_2 were chosen, for example, $R_1 = 10^5$ Ω and $R_2 = 10^6$ Ω, giving $B = 0.1$, this approximation would deteriorate. For a given operational amplifier, all that happens is that the voltage amplification is marginally different from $(R_1 + R_2)/R_1$. However, if the chip were changed for one with a quite different input impedance r_{in}, the voltage gain could be altered significantly.

Thirdly, V_{in} makes a contribution to V_-. It drives a current through r_{in} and then the parallel combination of R_1 and R_2 to earth (if r_{out} is neglected). The voltage drop across R_1 and R_2 due to this current makes a small contribution to V_{in}. Effectively, the voltage $(V_+ - V_-)$ applied to the operational amplifier is reduced, and with it the voltage gain. Again, this is small if $R_1 = 10^3$ Ω, but if R_1 and R_2 are $\simeq 10^5$ Ω, there is a noticeable correction.

If $R_1 = 100$ kΩ and $R_2 = 1$ MΩ, how big is it?

Finally, V_{in} makes a direct contribution to V_{out} via the current it drives through R_2, then the parallel combination of r_{out} and R_L. This is an example of **feedforward**. With the component values chosen here, it is entirely negligible.

The moral is that it is worth drawing out the full circuit diagram, figure 7.6, and making sure that the usual approximations are adequate; if not, they are easily corrected.

7.4   Shunt Feedback

Figure 7.7 shows an alternative feedback arrangement. The output is connected back to the negative input of the operational amplifier via the feedback resistor R_F. This supplies a feedback current to the amplifier. For this reason, the configuration is sometimes called **negative current feedback**. (However, this name allows the possibility of confusion with feedback which depends on output *current* rather than output *voltage*.) An alternative name, used here, is **shunt feedback**. This is because R_F appears across the operational amplifier, between input and output. Later, we find that R_F is effectively in parallel with the input signal.

There is an extremely simple but powerful way of deriving a close approximation to the amplification V_{out}/V_{in}. It will be demonstrated here and used in many exercises at the end of the chapter. From equation (7.2), $V_+ - V_- = V_{out}/G$. If G is large,

$$\boxed{V_+ \simeq V_-.} \qquad (7.5)$$

In figure 7.7, where V_+ is earthed, this ensures that the inverting input is very close to earth potential, despite the fact that there is no direct electrical connection to earth. The negative terminal is

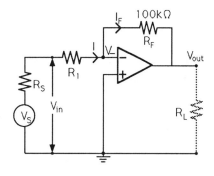

Fig. 7.7. Shunt feedback,
$V_{out} = -(R_F/R_1)V_{in}$

said to be a **virtual earth**. Then the current through resistor R_1 is $I \simeq V_{in}/R_1$. Very little of this current flows into the operational amplifier for two reasons. Firstly, $V_+ - V_-$ is very small; secondly, r_{in} is usually large compared with R_F. The current I_F through the feedback resistor R_F is therefore very close to I. Because $V_- \simeq 0$,

$$V_{out} \simeq -R_F I = -R_F V_{in}/R_1$$

so the amplification A is given by

$$\boxed{A = V_{out}/V_{in} \simeq -R_F/R_1.} \tag{7.6}$$

The arrangement in figure 7.7 acts rather like a lever pivoting about zero at the inverting input. It gives inversion and an amplification independent of G; the result looks like that for voltage feedback, equation (7.4), except for a sign change and $B = R_1/R_F$. The sign change arises because V_{in} is attached to the inverting input of the amplifier. Again, the formula for the amplification assumes that the input voltage is small enough that the output does not saturate.

Although the algebra given above is very simple, it does hide one vital point. There is actually a small difference between V_+ and V_- and it is this difference which drives the amplifier. It is instructive and simple to repeat the algebra keeping this small effect, but retaining the approximation that no current flows into the operational amplifier. Then

$$V_- = -V_{out}/G$$
$$I = (V_{in} - V_-)/R_1$$
$$V_{out} = V_- - IR_F = V_-\left(1 + \frac{R_F}{R_1}\right) - V_{in}\frac{R_F}{R_1}$$
$$= -\frac{V_{out}}{G}\left(1 + \frac{R_F}{R_1}\right) - V_{in}\frac{R_F}{R_1}.$$

Finally

$$V_{out}\left[1 + \frac{1}{G}\left(1 + \frac{R_F}{R_1}\right)\right] = -V_{in}\frac{R_F}{R_1}.$$

The effect of the small difference between V_+ and V_- is to introduce an extra term on the left-hand side of order $1/G$. At low frequencies this term is negligible because G is so large, and it simplifies the algebra greatly to omit it.

The current I flows towards the input of the amplifier if V_{in} is positive; I_F flows away from it. The tiny difference between these currents actually drives the sensitive amplifier. The input current I is almost exactly balanced by the feedback current I_F. This is the origin of the term **negative current feedback**.

Shunt feedback is used very widely. Suppose, for example, we want to amplify by 100 the voltage from a previous circuit having an output impedance of 1 kΩ. This output impedance can play the role of R_1 (providing its value is sufficiently stable) and then R_F is chosen to be 100 kΩ (figure 7.8(a)). The resulting amplification is independent of the choice of operational amplifier, providing that it has $G \gg 100$ and adequate bandwidth.

This simple arrangement is also a convenient way of transforming a current to a voltage. Suppose, for example, a current source such as a photodiode is available. It supplies an input current I, as in figure 7.8(b), and $V_{out} = -R_F I$.

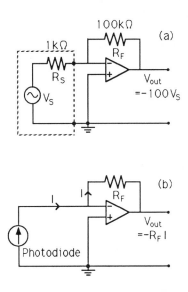

Fig. 7.8 (a) Amplifying by 100, (b) current to voltage conversion.

7.5   The Analogue Adder

The previous circuit produced an output voltage proportional to input voltage. Figure 7.9 shows a circuit which generalises this by adding and subtracting input signals; the output voltage is $\alpha V_3 - \beta V_1 - \gamma V_2$, where the coefficients α, β and γ depend only on resistor values.

To see in outline how the circuit works, remember that (a) $V_+ \simeq V_-$, and (b) the current flowing into the operational amplifier itself is negligibly small (at low frequencies). Then V_+, the voltage at the non-inverting input of the operational amplifier, is determined from V_3 by the potential divider R_3 and R_4. Next $V_- \simeq V_+$. So the currents through R_1 and R_2 are proportional to $(V_1 - V_+)$ and $(V_2 - V_+)$, respectively; these currents add and flow through R_F, so V_{out} depends linearly on V_1, V_2 and V_+ (hence V_3). This arrangement can be generalised to any number of inputs.

It is simple algebra to find the coefficients α, β and γ:

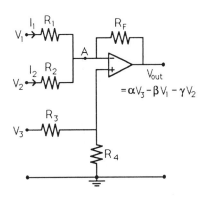

Fig. 7.9. The analogue adder.

$$V_+ = V_3 R_4 / (R_3 + R_4)$$
$$V_- \simeq V_+$$
$$V_{out} \simeq V_+ - R_F(I_1 + I_2)$$
$$\simeq V_+ - R_F \left(\frac{V_1 - V_+}{R_1}\right) - R_F \left(\frac{V_2 - V_+}{R_2}\right)$$
$$= V_3 \frac{R_4}{R_3 + R_4}\left(1 + \frac{R_F}{R_1} + \frac{R_F}{R_2}\right) - \frac{R_F}{R_1}V_1 - \frac{R_F}{R_2}V_2 \quad (7.7)$$
$$\alpha = \frac{R_4}{R_3 + R_4}\left(1 + \frac{R_F}{R_1} + \frac{R_F}{R_2}\right); \quad \beta = \frac{R_F}{R_1}; \quad \gamma = \frac{R_F}{R_2}.$$

The circuit is a convenient way of **mixing** or adding signals. This feature is one reason which makes shunt feedback a popular configuration, used in a wide variety of applications.

Sketch V_{out} if V_1 is a DC voltage, V_2 an AC voltage and $V_3 = 0$.

An entertaining demonstration is to supply any two inputs from two sine wave generators very close in frequency; V_{out} exhibits beats because the phase difference of the two inputs varies with time. If you do not have two generators, it is easy to generate one of them by inductive pick-up of a 50 Hz signal from the mains; tune your generator close to this frequency and adjust the two amplitudes to be similar in magnitude. Pick-up of mains signals is indeed all too common a problem.

7.6   The Differential Amplifier

A differential amplifier is the special case where $V_{out} \propto (V_3 - V_2)$. The input V_1 is eliminated and resistor values are adjusted so that $\alpha = \gamma$. This is achieved by setting $R_F/R_2 = R_4/R_3 = r$. The algebra simplifies to

$$V_{out} = r(V_3 - V_2). \tag{7.8}$$

This is a very valuable arrangement. Suppose V_3 and V_2 both contain an unwanted background of the same magnitude. It might be a mains signal or some similar interference. By taking the difference between V_2 and V_3, this background is removed. This trick is called **common mode rejection**. It is a feature built into many electronic instruments.

Two examples will illustrate its use. Firstly, suppose a weak radio signal (from a distant galaxy perhaps) is to be detected in the presence of local pick-up. One arranges two detectors, both receiving the local pick-up but only one detecting the distant source. The differential amplifier eliminates the background and selects out the distant source. It improves the signal/background ratio.

Secondly, suppose one wants to detect the small electrical signal from a patients's heartbeat. The body acts as a good aerial and picks up a large background signal, as you can verify by connecting yourself across the leads of a sensitive oscilloscope. To get rid of this background, one rigs up a dummy aerial of resistors and capacitors, whose values are adjusted by trial and error to simulate the body. Then a differential amplifier will reject the background and select the small signal from the heartbeat.

All operational amplifiers are designed to have a high common mode rejection ratio (CMRR). This figure is the ratio G_- for $(V_3 - V_2)$ compared with G_+ for a signal $(V_3 + V_2)$ common to both inputs:

$$CMRR = G_-/G_+.$$

For the μA741C, it is 90 db, or $2^{15} \simeq 33\,000$. What it implies is that the ideal performance of the operational amplifier expressed by equation (7.2) should be replaced by

$$V_{out} = G_-(V_+ - V_-) + G_+(V_+ + V_-) \tag{7.2a}$$

with $G_+/G_- < 1/33\,000$.

In order to achieve the best common mode rejection, it is important that the input impedance should be the same for V_2 and V_3; otherwise, different proportions of these voltages will be dropped in the source resistances. The input impedance for V_3 is (R_3+R_4), since negligible current flows into the amplifier. What about V_2? The voltage at A in figure 7.9 is $V_3 R_4/(R_3+R_4)$. So if $V_2 = V_3$, the voltage across R_2 is $V_3 R_3/(R_3 + R_4)$ and the input impedance for V_2 is $V_2/I_2 = R_2(R_3 + R_4)/R_3$. For this to be equal to $(R_3 + R_4)$ requires that $R_2 = R_3$, hence $R_4 = R_F$. This arrangement is shown in figure 7.10.

It is instructive to make up the circuit of figure 7.9 and apply the same AC signals to V_2 and V_3. If the common mode rejection is perfect, there is zero output. Using variable resistors for R_F and R_2 and viewing V_{out} with the oscilloscope, you quickly get a feeling for common mode rejection.

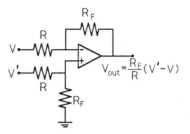

Fig. 7.10. Differential amplifier.

$$V_{\text{out}} = \frac{R_F}{R}(V'-V)$$

7.7 Gain–bandwidth Product

Shunt feedback gives an amplification

$$A \simeq -R_F/R_1 = -1/B.$$

As always, it is necessary to examine how good the approximations are leading to this result. Obviously it must break down at some high frequency where the gain of the amplifier falls. It will emerge that it breaks down above the frequency at which $GB \to 1$. This condition gives the bandwidth, i.e. the range of frequencies over which the amplification is constant (within a factor $1/\sqrt{2}$). To demonstrate this result, it is necessary to go through the algebra keeping terms which involve G. We shall take the opportunity to examine at the same time the effect of the input impedance of the operational amplifier, r_{in}; it turns out to have negligible effect for normal values.

In figure 7.11 the operational amplifier is replaced by its Thevenin equivalent circuit, but omitting its output impedance. (If the latter is included, the algebra gets rather messy, but exhibits no new features.) The following equations show a standard way of handling the algebra of shunt feedback. This procedure will be used with minor variants in subsequent examples. The voltage V_- at the inverting input is expressed in as many ways as possible, four in this case:

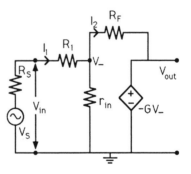

Fig. 7.11. Thevenin equivalent of Fig. 7.7.

$$V_- = V_S - I_1(R_1 + R_S) \qquad (7.9)$$
$$= V_{\text{out}} + I_2 R_F \qquad (7.10)$$
$$= (I_1 - I_2)r_{\text{in}} \qquad (7.11)$$
$$= -V_{\text{out}}/G. \qquad (7.12)$$

It is necessary to eliminate I_1 and I_2 and solve for V_{out}/V_{in}. From equations (7.9) and (7.12),

$$(R_1 + R_S)I_1 = V_S + V_{out}/G$$

and from equations (7.10) and (7.12),

$$R_F I_2 = -V_{out}(1 + 1/G).$$

Then equations (7.11) and (7.12) give

$$-\frac{V_{out}}{G} = \frac{r_{in}(V_S + V_{out}/G)}{(R_1 + R_S)} + \frac{r_{in} V_{out}}{R_F}\left(1 + \frac{1}{G}\right)$$

$$A = \frac{V_{out}}{V_S} = -R_F/(R_1 + R_S)$$

$$\times \left[1 + \frac{R_F}{G}\left(\frac{1}{R_F} + \frac{1}{(R_1 + R_S)} + \frac{1}{r_{in}}\right)\right]^{-1}. \quad (7.13)$$

For large G and small R_S, the voltage gain is $-R_F/R_1$, reproducing the elementary result, equation (7.6). The bracket in the denominator is clearly the admittance between the inverting input of the amplifier and earth. Writing $R_S + R_1 = R_1'$, the largest correction term in the denominator is R_F/GR_1', which produces a correction to the voltage gain of $(1 + R_F/GR_1')^{-1} = (1 + 1/GB)^{-1}$. If $G = 10^5$ and $B = 10^{-2}$, as in our example, figure 7.8(a), this correction term is one part in 10^3 at low frequencies. The remaining terms introduce even smaller corrections, since R_F and r_{in} are normally larger than R_1'. So if the dominant correction is retained, it is sufficient to take

$$A \simeq -\frac{R_F/R_1'}{1 + R_F/GR_1'} = -\frac{1/B}{1 + 1/GB} = -\frac{1}{B + 1/G} = -\frac{G}{1 + BG}.$$
$$(7.14)$$

This is the same formula as for voltage feedback, equation (7.3a), except for a sign change. This sign arises because the input is applied to the inverting input of the operational amplifier; in the general formula of equation (7.3),

$$G_V = -G(\omega)$$
$$B = R_1'/R_F$$

and

$$A = \frac{-G(\omega)}{1 + BG(\omega)}.$$

Now consider the frequency dependence. At low frequencies, $BG \gg 1$ and the amplification is $A_0 = -1/B$ on the flat part of figure 7.12. This is called the **mid-band gain**: the name arises because at low frequencies the amplification falls again if AC coupling is used (see figure 4.12). At high frequencies, $G(\omega)$ falls and eventually BG reaches 1. Around this point the amplification of the circuit begins to fall.

To examine this quantitatively, an expression is needed for $G(\omega)$. It is given by

$$G = \frac{G_0}{1 + j\omega/\omega_L} \qquad (7.15)$$

where $\omega_L \simeq 60$ rad/s for the μA741C. The factor j accounts for the 90° phase difference of figure 7.3(c) between input and output. The operational amplifier itself is behaving as a low-pass filter with lower cut-off frequency ω_L. (The reason is that a capacitor is deliberately built into it for reasons of stability, discussed in Chapter 19.) On the straight-line part of figure 7.3(b), $G \simeq G_0/j\omega$, where G_0 is a constant. So the bandwidth is given by

$$|BG| = |BG_0/j\omega_0| = 1$$
$$\omega_0 = BG_0 = G_0/A_0.$$

The larger the value of the mid-band gain A_0, the smaller is the bandwidth ω_0. Indeed the simplest way of determining how G depends on ω is by measuring figure 7.12 experimentally by varying B. At frequency ω_0, the amplification drops by a factor $1/\sqrt{2}$ and a phase difference of 45° develops between input and output.

The product

$$\boxed{A_0\omega_0 = G_0} \qquad (7.16)$$

is independent of B. It is called the **gain–bandwidth product** and is a fundamental parameter of the amplifier. A large bandwidth can be chosen or large amplification, but not both. Usually it is more difficult to achieve large bandwidth than large amplification. Often large bandwidth is important. In an oscilloscope, for example, it is vital that signals should be amplified by the same factor across the whole working range of frequencies. If necessary, large gain may be obtained by having more than one stage of amplification (within limits of noise and stability, to be discussed later). Thus it is commonplace to sacrifice amplification in order to achieve large bandwidth. These remarks apply equally well to voltage feedback, since the form of the voltage gain is precisely the same as equation (7.14), except for the sign.

The operational amplifier cannot provide any gain above the frequency where $G(\omega)$ falls to 1. This is called the **unity gain bandwidth**. For the μA741C it is $\omega \simeq 6 \times 10^6$ rad/s. More expensive operational amplifiers have a unity gain bandwidth as high as $\omega = 5 \times 10^9$ rad/s.

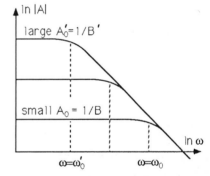

Fig. 7.12. Amplification v. ω .

7.8 Offset Voltage and Bias Current

So far the operational amplifier has been treated as ideal, except

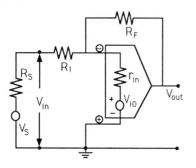

Fig. 7.13. Effect of V_{io} with shunt feedback.

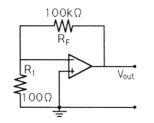

Fig. 7.14. Measurement of V_{io}.

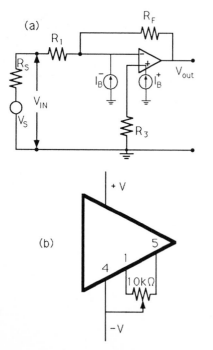

Fig. 7.15 (a) Input offset currents, (b) use of the offset null.

for the variation of G with frequency. Some practical limitations will now be discussed. Firstly, if there is no input to V_+ or V_-, the output voltage is not exactly zero; instead, there is a small DC offset:

$$V_{out} = G(V_+ - V_- + V_{io}) \qquad (7.17)$$

where V_{io} is called the **input offset voltage**. For the μA741C, $|V_{io}| < 1$ mV (and for some other operational amplifiers may be as low as 5 μV). Nonetheless, if $G = 10^5$, it may be sufficient to drive the amplifier to saturation when there is no input. An obvious question is what its effect will be including feedback.

For series voltage feedback (section 7.2), it is immediately obvious from equation (7.17) that the offset voltage simply adds to V_+, with the result that

$$V_{out} = \frac{(V_{in} + V_{io})G}{1 + GB}. \qquad (7.18)$$

The effect is to introduce a DC shift into the output of $\sim V_{io}/B$; AC signals are superposed on this level. If $B > 0.01$, the DC shift is < 0.1 V and is unimportant; for larger amplifications it may become serious.

For shunt feedback (figure 7.13), the effect of V_{io} is to move the virtual earth at the inverting input to voltage V_{io}. Then

$$\frac{V_{in} - V_{io}}{R_1} \simeq \frac{V_{io} - V_{out}}{R_F}$$

so

$$V_{out} \simeq \frac{-R_F}{R_1}V_{in} + V_{io}\left(1 + \frac{R_F}{R_1}\right). \qquad (7.19)$$

This introduces into the output a DC level of $V_{io}(1 + 1/B)$. If, for example, $V_{io} = 1$ mV and $B = 0.01$, the output DC shift is 100 mV. The value of V_{io} may be measured with the circuit shown in figure 7.14 (where $V_{in} = 0$) using R_F/R_1 in the range 100 to 1000.

Bias currents

A second correction to ideal performance is that both inputs of the operational amplifier draw small DC bias currents I_B^+ and I_B^-, indicated in figure 7.15(a) by the current generators. For the μA741C, $|I_B| < 200$ nA, but with $R_3 = 10^5$ Ω this could simulate an offset voltage of 20 mV and produce a corresponding DC output shift. The moral is that large resistors should not be used directly in series with the input terminals of the operational amplifier. It is a good practice to use a resistor R_3 between the positive input and earth such that I_B^- and I_B^+ produce cancelling effects. This resistor may be adjusted empirically to produce the cancellation. The currents produce only a DC shift at the output, as does V_{io}, so R_3 can usually be adjusted to cancel all three effects by adjusting the DC output of the circuit to be zero when the input is zero. From a detailed model of the amplifier, it will be shown later that R_3 should be made roughly equal to the parallel combination of R_F with $(R_1 + R_S)$ of figure 7.15(a). Figure 7.15(b) shows the alternative use of the offset null connections, pins 1 and 5, provided by the manufacturer.

The value of I_B^+ may be measured using the circuit shown in figure 7.16. The current I_B^- flows mostly through R_1, so

$$V_- \simeq \frac{V_{out} R_1}{R_1 + R_2} + I_B^- R_1 \simeq V_{io} + R_3 I_B^+ - V_{out}/G.$$

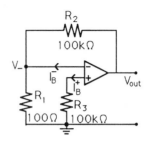

Fig. 7.16. Measurement of I_B^+.

Because R_1 is very small, I_B^- may be neglected, so

$$R_3 I_B^+ \simeq \frac{V_{out} R_1}{R_1 + R_2} - V_{io}. \tag{7.20}$$

Slew rate

A more significant limitation of the performance of the operational amplifier is that there is a maximum rate at which the output can change. This is called the **slew rate**. It is illustrated in figure 7.17. For the μA741C it is $0.5\ \mathrm{V}\,\mu s^{-1}$. If you feed in a sine wave, the output cannot rise faster than this and will be distorted at large amplitudes and large frequencies. It implies a limitation on the bandwidth for large signals. If $V_{out} = A\sin\omega t$, then $dV_{out}/dt = A\omega\cos\omega t$, and the slew rate S becomes a limitation when $A = S/\omega$. If A is equal to the full output range (15 V), the output of the μA741C is distorted above a frequency $\omega \simeq 3 \times 10^4$ rad/s called the **full power bandwidth**.

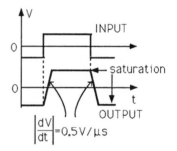

Fig. 7.17. Maximum rate of change of the output.

7.9 Other Opamps

The μA741C is a modest general purpose opamp. There are literally hundreds of others available, as a quick browse through manufacturers' catalogues will reveal. None will do everything. A brief selection is given in table 7.1 to illustrate the range of possibilities.

7.10 Complex Feedback Loops

Up to here the feedback has been via resistors. However, there is a multitude of useful circuits involving feedback via networks of resistors and capacitors. In figure 7.18, networks with impedances Z_1 and Z_F replace the resistors of figure 7.7. Then $V_{out}/V_{in} =$

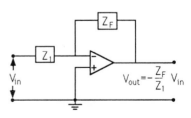

Fig 7.18. Feedback using impedances Z_1 and Z_F.

Table 7.1   A selection of specialised opamps.

Type	Supply voltage (V) Min. Max.		Max input offset voltage (mV)	Input bias current	Slew rate (V/μs)	Unity gain bandwidth (MHz)	CMRR (db)	Open loop gain (db)	Max output current (mA)	Comments
SL2541B	14	30	10	20 μA	900	800	47	45	10	Fast
LM675	16	60	10	2 μA	8	5.5	70	70	3000	High o/p current
AD707C	6	36	0.005	1 nA	0.3	0.9	130	138	12	High CMRR
ICL7641B	1	18	5	0.05 nA	1.6	1.4	60	80	5	Low voltage
3583	100	300	3	0.1 nA	30	5	110	94	75	High voltage
OP-43E	10	36	0.2	0.005 nA	6	2.4	100	120	15	Low noise

$-Z_F/Z_1$. This is called the **transfer function** of the circuit. Shaping the dependence of this transfer function on frequency can correct, to some degree, deficiencies in the frequency response of the source. For example, a record or a tape deck reproduces signals of different frequencies with different sensitivities. The signal needs correcting to reproduce the original faithfully. This can be achieved by making Z_F and Z_1 from networks of capacitors and resistors giving suitable frequency dependence. The technique is called **frequency-selective feedback**. You might well worry that such networks will introduce a frequency-dependent phase shift. This is indeed the case, but it turns out that the ear is not particularly sensitive to phase; in the first approximation it is sensitive only to intensity.

Logarithmic converters

Another important and amusing example arises when either Z_1 or Z_F is non-linear. Later it will be shown that the voltage V across a pn diode is related to the current I through it by

$$I \simeq I_0 \exp(\alpha V)$$

where I_0 and α are constants (except for some temperature dependence). Precisely the same relation holds also between the collector current of a bipolar transistor and the voltage V_{BE} between base and emitter. Suppose that Z_1 in figure 7.18 is a resistor R_1 and Z_F a diode or a transistor connected so that V_{out} is equal to $-V_{BE}$. The current through Z_1 and Z_F is V_{in}/Z_1 and

$$V_{out} = -\frac{1}{\alpha} \ln \left(\frac{V_{in}}{R_1 I_0} \right) = -\frac{1}{\alpha} \left(\ln V_{in} - \text{constant} \right).$$

There are tricks by which the constant may be cancelled using a second identical diode or transistor. Then the circuit takes the logarithm of V_{in} and is called a logarithmic converter. If Z_1 and Z_F are interchanged, the circuit forms the antilogarithm or exponential. Two voltages V_A and V_B can be multiplied by taking logs and adding, then taking the antilog: two logarithmic converters are used to take logs of V_A and V_B, followed by an analogue adder and then an antilog circuit:

$$\frac{1}{\alpha} \ln V_{out} = \frac{1}{\alpha} (\ln V_A + \ln V_B).$$

Multipliers

Such multipliers are available commercially from several manufacturers. They incorporate refined temperature stabilisation and

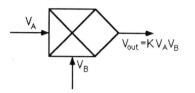

Fig 7.19. The symbol for a four-quadrant multiplier.

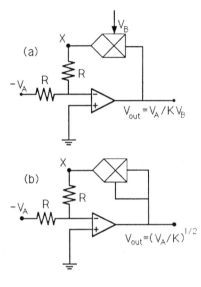

Fig. 7.20 (a) Using a multiplier to do division, (b) forming a square root.

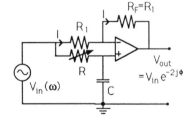

Fig. 7.21. Phase shifting circuit.

accurate cancellation of input bias currents (which would upset the result). One-quadrant multipliers accept positive inputs. By clever design, two-quadrant multipliers will multiply one positive input with a second of either polarity; four-quadrant multipliers accept two inputs of either polarity. As an example, the Analogue Devices AD534 is a four-quadrant multiplier with an accuracy of $\pm 0.25\%$ and a bandwidth of 1 MHz; input and output voltages are up to 10 V and $V_{\text{out}} = KV_A V_B$ where $K \simeq 0.1$. Other multipliers are available up to 25 MHz.

Figure 7.19 shows the symbol for such a multiplier. Figure 7.20(a) illustrates how to do division and figure 7.20(b) shows how to form a square root. In the former,

$$V_X = KV_{\text{out}} V_B = V_A$$

and in the latter,

$$V_X = V_A = KV_{\text{out}}^2.$$

Worked example

Complicated feedback loops will be discussed in a later chapter on active filters. At this point, plenty of practice is desirable in handling voltage and shunt feedback, and your attention is directed towards problems 1–8 at the end of the chapter. Here, one worked example will show the way. The circuit of figure 7.21 will be analysed assuming the input is an AC signal of angular frequency ω. The approach is to make use of the approximation $V_+ = V_-$, so the first step is to find V_+. If the impedance $1/j\omega C$ of the capacitor is written as Z_C,

$$V_+ = \frac{V_{\text{in}} Z_C}{Z_C + R} = \frac{V_{\text{in}}}{1 + R/Z_C} = \frac{V_{\text{in}}}{1 + j\omega CR}$$

then

$$R_1 I = V_{\text{in}} - V_- \simeq V_{\text{in}} - V_+.$$

If the current flowing into the operational amplifier is negligible, the current through R_F is equal to I, so

$$V_{\text{out}} = V_+ - R_1 I = 2V_+ - V_{\text{in}}$$

$$= \frac{V_{\text{in}}}{1 + j\omega CR}(2 - 1 - j\omega CR) = \frac{V_{\text{in}}(1 - j\omega CR)}{1 + j\omega CR}$$

$$= \frac{V_{\text{in}}(1 + \omega^2 C^2 R^2)^{1/2} e^{-j\phi}}{(1 + \omega^2 C^2 R^2)^{1/2} e^{j\phi}} = V_{\text{in}} e^{-2j\phi}$$

where $\phi = \tan^{-1} \omega CR$. This circuit retards the phase of V_{in} by an angle 2ϕ without any change in magnitude.

This example is typical of a host of tricks which can be played using feedback networks and operational amplifiers. Exercise 3 gives many more. If you look in hobby magazines or manuals you will find scores of examples. Derivation of the transfer function always follows lines similar to the one here. The essential steps are: (i) setting $V_+ \simeq V_-$, and (ii) neglecting the current into the operational amplifier.

7.11   Impedance Transformation

Figure 7.22 shows a circuit whose input impedance takes a curious and useful form, which can be used to manipulate impedances. The usual assumption will be made that the amplifier has large gain and high input impedance, so that the voltages V_1 and V_2 at the two input terminals are approximately the same. Then

$$I_1 = (V_1 - V_0)/R_1$$
$$I_2 = (V_0 - V_2)/R_2 = V_2/Z_L.$$

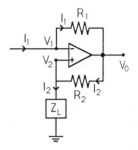

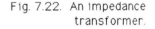

Fig. 7.22. An impedance transformer.

Setting $V_1 = V_2$ and dividing one equation by the other,

$$I_1 = -R_2 I_2/R_1 = -R_2 V_2/(R_1 Z_L) = -R_2 V_1/(R_1 Z_L).$$

The input impedance of the circuit is

$$R_{\text{in}} = V_1/I_1 = -R_1 Z_L/R_2. \qquad (7.21)$$

The curious feature is that the input impedance includes a minus sign. If the load Z_L is resistive, the circuit behaves as a **negative resistor**. Instead of dissipating power, it can provide controlled power to the circuit to which V_1 is connected. This is a consequence of the positive feedback provided by R_2 and Z_L.

If Z_L is a capacitor with impedance $-j/\omega C$, the circuit behaves like an inductor with impedance $j\omega L$ where $L = (R_1/R_2)/\omega^2 C$. In integrated circuits, it is impossible to form any significant inductance within the silicon. At one particular frequency, the circuit of figure 7.22 can be used to simulate an inductance; however, it has a different frequency dependence. A clever and more elaborate circuit, called the **gyrator** (exercise 12, figure 7.47) reproduces the frequency dependence of an inductor. Although it looks complicated, it is readily fabricated inside an integrated circuit.

7.12   Input and Output Impedances with Feedback

It is now time to examine in somewhat greater detail the subtleties of voltage and shunt feedback, and in particular input and output

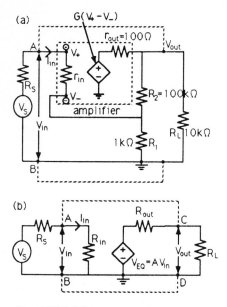

(a)

(b)

Fig. 7.23 (a) The circuit of the voltage
feedback amplifier of Fig. 7.6 and
(b) its equivalent circuit.

impedances. A general idea of their values is very helpful in planning how to link circuits together.

Figure 7.23(a) reproduces figure 7.6, showing in somewhat greater detail the circuit of figure 7.4(b) for voltage feedback. The operational amplifier itself is described as a 'black box' having input resistance r_{in} and an output drawn in Thevenin equivalent form with output impedance r_{out}. A load R_L is applied to the output and a source to the input.

For many purposes, it is useful to go further and represent everything inside the larger dashed rectangle by yet another black box, figure 7.23(b), having input impedance R_{in} and output impedance R_{out}. Once these two quantities are known, we can assess immediately the proportion of the source voltage V_S delivered to the circuit:

$$V_{in} = V_S R_{in}/(R_S + R_{in}) \qquad (7.22)$$

and the fraction of the output voltage V_{EQ} delivered to the load resistor R_L:

$$V_{out} = \frac{A V_{in} R_L}{R_L + R_{out}}. \qquad (7.23)$$

Input and output impedances R_{in} and R_{out} also play a second vital role. In association with series and parallel capacitances, they determine upper and lower cut-off frequencies, as discussed in Chapter 4. For example, R_{in} gives an idea of limitations to the bandwidth of the circuit from an educated guess about the capacitance across the input AB.

The input and output impedances R_{in} and R_{out} of the second black box are not the same as r_{in} and r_{out}, the values for the operational amplifier itself, because of the feedback. Summarising what we shall find in the rest of the chapter, there are four important rules.

(a) Series voltage feedback increases the input impedance to

$$R_{in} = r_{in}(1 + GB).$$

If for example $GB = 10^5 \times 10^{-2} = 10^3$, this has a dramatic effect on the qualitative behaviour of the circuit as a whole, and particularly on bandwidth.

(b) Shunt feedback decreases the input impedance; across the terminals of the operational amplifier itself,

$$R_{in} = r_{in}/(1 + GB).$$

This result, called the **Miller effect**, again has dramatic consequences.

(c) If the feedback stabilises the output *voltage* by deriving the feedback from a voltage in the output circuit, the output impedance is reduced to

$$R_{out} = r_{out}/(1 + GB).$$

(d) If the feedback is derived from an output *current*, the output impedance is increased by a factor $1 + GB$.

Input impedance

Let us begin with voltage feedback and figure 7.23. Consider R_{in} first. It is necessary first to solve for I_{in}, hence $R_{in} = V_{in}/I_{in}$. With feedback, the signal across r_{in} becomes

$$V_{in} - V_- = V_{in} - BV_{out} = V_{in} - \frac{BGV_{in}}{1 + GB} = \frac{V_{in}}{1 + GB}. \qquad (7.24)$$

This means that feedback reduces the signal applied to the input of the operational amplifier by the factor $(1 + GB)$ and I_{in} is correspondingly reduced:

$$I_{in} = V_{in}/[r_{in}(1 + GB)].$$

The input impedance of the circuit is

$$\boxed{R_{in} = V_{in}/I_{in} = r_{in}(1 + GB).} \qquad (7.25)$$

This demonstrates rule (a) given above.

Output impedance

The output impedance is deduced by applying Thevenin's theorem to the output CD of figure 7.23(b):

$$R_{out} = \frac{\text{open circuit voltage}}{\text{short circuit current}}.$$

The open-circuit voltage is just $GV_{in}/(1 + GB)$ from equation (7.3a). When the output is short circuited, there is no feedback voltage, so $I_{out} = GV_{in}/r_{out}$. Therefore

$$\boxed{R_{out} = \frac{GV_{in}}{1 + GB}\frac{r_{out}}{GV_{in}} = \frac{r_{out}}{1 + GB}.} \qquad (7.26)$$

The better the performance of a voltage source, the lower is its output impedance. The feedback in this circuit stabilises the output voltage, and the output impedance is decreased by the factor $(1 + GB)$. This result demonstrates rule (b).

DC power supply

Figure 7.24 shows how to make a simple stabilised DC power supply using voltage feedback. The source V_S may be a battery or an unstabilised supply, and R_S can be large, so as to draw little

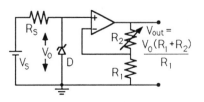

Fig. 7.24. A simple stabilised power supply.

current. The element D is a Zener diode, as discussed in Chapter 1. It conducts in the opposite direction to the arrow when the voltage across it exceeds some well defined threshold value V_0 (see figure 1.9(a)). Providing $V_S > V_0$, a current flows through R_S and maintains a voltage V_0 at the input of the operational amplifier. This is amplified and $V_{out} = V_0(R_1 + R_2)/R_1$. The output voltage may be controlled by varying R_1 or R_2. The output impedance is low ($< 1\ \Omega$ if $GB > 100$). This is what is required of a power supply.

In practice, common operational amplifiers are limited in output current to 20–100 mA. If more current is required, a high-power transistor must be added to boost the current. When operational amplifiers are used alone to drive a low-impedance load such as a loudspeaker (standard values are 4, 8 and 12 Ω), the output current is likely to be be limited by what the chip will supply. Most chips are protected internally against overload and will survive. However, a small AC signal will be amplified correctly, while a large one is limited by the chip. Large AC signals are then distorted.

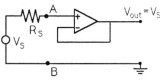
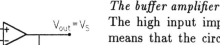

Fig. 7.25. A buffer amplifier.

The buffer amplifier

The high input impedance which results from voltage feedback means that the circuit draws very little current and acts as an almost ideal voltmeter. This may be a very desirable property. An extreme example of this is shown in figure 7.25, where *all* of the output voltage is fed back to the inverting input. In this case, $B = 1$ and the voltage gain is $A = 1$. This is called a **buffer amplifier** or **voltage follower** or **instrumental amplifier**. The input impedance of the circuit is $R_{in} \simeq Gr_{in} \simeq 10^5 \times 10^6\ \Omega = 10^{11}\ \Omega$. The output impedance is $R_{out} \simeq r_{out}/G \simeq 10^2/10^5\ \Omega = 10^{-3}\ \Omega$. It is possible, for example, to apply a standard cell to the input, giving $V_{in} = V_{st} = V_{out}$. Normally, no significant current can be drawn from a standard cell without polarising it. But the buffer amplifier is capable of supplying a large current at this reference voltage with negligible voltage loss in its output impedance. It is acting as a current amplifier. In Chapter 18, the emitter follower will be discussed; this is a very common circuit which acts as a buffer amplifier in just the same way (except that G is generally lower).

There is a snag associated with the large input impedance. It is impossible to avoid stray capacitances, both between the input terminals of the operational amplifier and between the non-inverting input and earth. The very small current drawn from the source charges these capacitances only very slowly, so the bandwidth is poor. Suppose, for example, the source is connected across AB of figure 7.25 via a coaxial cable, which has a capacitance of typically 150 pF per metre. The associated bandwidth is $1/CR'$, where R' is the resistance appearing across the capacitance; from the

Thevenin equivalent, R' is the parallel combination of R_{in} with R_S. If both R_S and R_{in} are high, the bandwidth is poor. For example, if $R_S = 10^9\ \Omega$ and $R_{in} = 10^{11}\ \Omega$ and $C = 50$ pF, the bandwidth is only 20 rad/s!

A partial remedy is to reduce B to 0.01 (thereby increasing the voltage gain to 100) or to reduce r_{in} by putting a lower resistor in parallel with it. However, these are palliatives. If the bandwidth is limited by CR' to unacceptable values, another type of circuit is required. Shunt feedback greatly reduces the problem, and will be discussed shortly.

7.13   Stabilised Current Supplies

So far, feedback signals have been derived from the output *voltage*. It is possible to arrange instead that the feedback signal is derived from the output *current*. If so, the output *current* rather than the output voltage is stabilised. This is the way to make a stabilised current supply, which might be used, for example, in supplying a magnet.

A simple way of doing this is shown in figure 7.26. In outline the circuit works as follows. Because $V_+ \simeq V_-$, the voltage across R_F is $\simeq V_{in}$ and current V_{in}/R_F flows through it. Very little of this current is supplied from V_- because of the high input impedance of the operational amplifier; almost all of it flows through R_L. The current through this load is stabilised at the value V_{in}/R_F, regardless of the magnitude of R_L.

An algebraic treatment gives this result, but also examines how good the approximations are; these approximations dictate the output impedance, i.e. how good a current source the circuit is.

The voltage fed back to the inverting input is

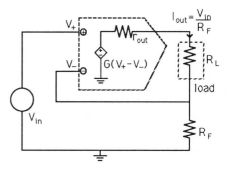

Fig. 7.26. Stabilisation of output current by voltage feedback.

$$V_- = R_F I_{out}$$
$$\simeq \frac{R_F G(V_+ - V_-)}{R_L + R_F + r_{out}}.$$

So

$$V_-(R_L + R_F + r_{out} + R_F G) = R_F G V_{in}.$$

Then

$$V_{out} = G\left(V_{in} - \frac{R_F G V_{in}}{R_L + R_F + r_{out} + R_F G}\right)$$
$$= G V_{in}\frac{R_L + R_F + r_{out}}{R_L + R_F + r_{out} + R_F G}$$
$$I_{out} = \frac{G V_{in}}{R_L + R_F + r_{out} + R_F G} \qquad (7.27)$$
$$= \frac{V_{in}}{R_F[1 + (R_L + R_F + r_{out})/R_F G]} \simeq \frac{V_{in}}{R_F}.$$

The output current is insensitive to the load R_L and the output impedance r_{out} of the amplifier.

The calculation of input impedance goes precisely as before for voltage feedback, giving the result $r_{in}(1 + GB)$ where $B = R_F/(R_F + R_L + r_{out})$. The reason, as before, is that the feedback reduces the voltage across r_{in}.

We would expect the output impedance to have increased because the output current has been stabilised, i.e. the output circuit acts as a better current generator than in the absence of feedback. To demonstrate this, the standard procedure is followed by constructing the Thevenin equivalent circuit across R_L:

$$R_{out} = \frac{\text{open circuit voltage}}{\text{short circuit current}}.$$

With R_L replaced by an open circuit, the voltage is GV_{in}, since there is no feedback. With R_L short circuited, the current is, from (7.27), $GV_{in}/(R_F + r_{out} + R_F G)$. So the output impedance of the circuit feeding R_L is

$$R_{out} = R_F + r_{out} + R_F G. \tag{7.28}$$

In the absence of a feedback connection to V_-, the output impedance viewed from the generator is $r_{out} + R_L + R_F$. With feedback, it is $r_{out} + R_L + (G+1)R_F$. Setting $B = R_F/(R_L + r_{out} + R_F)$ for series feedback, i.e. $R_F = B(R_L + r_{out} + R_F)$, the output impedance has been increased by a factor

$$\frac{r_{out} + R_L + R_F + GR_F}{r_{out} + R_L + R_F} = 1 + GB.$$

The load in figure 7.26 is not connected to earth. This may be inconvenient. Can you see a simple modification which earths the lower end of the load?

7.14   Input Impedance with Shunt Feedback

The circuit for shunt feedback is drawn in greater detail in figure 7.27(a), showing both r_{in} and r_{out}. This whole circuit is now to be replaced with a Thevenin equivalent, shown in figure 7.27(b). The calculation of output impedance goes the same way as for voltage feedback and gives the same result; it is reduced by a factor $1 + GB$. This is simply a consequence of the output voltage being stabilised.

A new result is that the input impedance is considerably reduced by shunt feedback. It is immediately obvious that this must be so. In the absence of feedback, the input impedance is $R_1 + r_{in}$. With

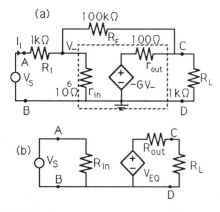

Fig. 7.27 (a) Amplifier with shunt feedback and (b) its equivalent circuit.

feedback, the inverting input of the amplifier is a virtual earth, so the input impedance is approximately R_1. This is an important result worth remembering when linking circuits together. Clearly, the feedback must be shorting r_{in} and it is desirable to follow how this happens. This can be done algebraically from equations (7.9) to (7.12), but it is more enlightening to follow a diagrammatic approach.

In figure 7.28(a), the components of figure 7.27(a) are redrawn in a more symmetrical configuration. Consider first the input impedance across the terminals AB. It is tempting to say that this will be R_1 in series with a parallel combination of r_{in} and R_F (if r_{out} can be neglected). This is not correct, because the output generator GV_- produces a voltage which depends on V_- and Thevenin's theorem cannot be applied so simply. However, with some reorganisation of components, we can allow for this. In (b), the output generator and r_{out} are replaced by their Norton equivalent and in (c) the parallel resistors R_L and r_{out} are combined into $R_0' = r_{out}R_L/(r_{out} + R_L)$. Then (d) returns to the Thevenin equivalent with $G' = GR_0'/r_{out}$.

The point of these manoeuvres is that the voltage at X is known to be $-G'V_-$, so the voltage across $(R_F + R_0')$ is $(G'+1)V_-$. Along this arm, earth potential is reached after resistance $(R_F+R_0')/(G'+1)$. So from V_- to earth, the resistance is r_{in} in parallel with $(R_F + R_0')/(G'+1)$, as shown in figure 7.29. This latter resistance is very small, of order $1\ \Omega$ if $R_F = 10^5\ \Omega$ and $G = 10^5$; it bypasses r_{in} and is responsible for lowering the input impedance across the operational amplifier to a very low figure. Note, however, that G varies with frequency, and so does the resistance of the path bypassing r_{in}.

We need to distinguish carefully between the input impedance to the circuit across AB and the input impedance of the amplifier, across XB in figure 7.29(a). The former gives the fraction of the input signal delivered to the circuit by a source of resistance R_S, as in figure 7.29(b).

Suppose there is stray capacitance C across the input of the operational amplifier itself. The limitation from this on bandwidth is $\omega_0 = 1/CR'$, where R' is given by the parallel combination of R_1, r_{in} and $(R_F + R_0')/(G' + 1)$. Because the last of these is very small, there is hardly ever a real limitation compared with the natural bandwidth of the operational amplifier. In this respect, shunt feedback is superior to voltage (series) feedback, which increases the input impedance of the operational amplifier.

We now show that feedback reduces the input impedance by a factor $B(G'+1)$. In the absence of feedback, the impedance across the input terminals of the operational amplifier is r_{in} in parallel with R_1, and if the former is high the result is just R_1. With

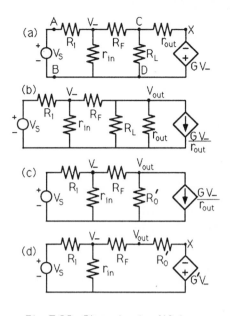

Fig. 7.28. Steps in simplifying the circuit.

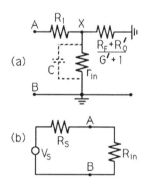

Fig. 7.29 (a) Input impedance with shunt feedback, (b) loading of the source.

feedback, it becomes (to a good approximation)

$$R_F/(G'+1) = \frac{R_1 R_F}{R_1(G'+1)} = \frac{R_1}{B(G'+1)}.$$

This implies a reduction by a factor $B(G'+1)$.

7.15* Cancelling Input Bias Currents

Earlier, a rule was given for choosing resistor R_3 of figure 7.15(a) so as to cancel input bias currents in the shunt feedback configuration. This rule will now be derived.

It is straightforward to add to figure 7.27 the current generators which are responsible for input bias currents I_B^+ and I_B^-, and to trace their effects. This is done in figure 7.30(a), where V_S and R_1 are also replaced with a Norton equivalent. Let us now find the equivalent circuit across r_{in}. Combining I_B^- with the generator V_S/R_1 and returning to the Thevenin equivalent form gives figure 7.30(b). There

$$V_- = v(1 + R_3/r_{in}) + I_B^+ R_3.$$

If the approximation is made that no current flows into r_{in}, currents through R_1 and R_F are the same. Then

$$\frac{V_S + I_B^- R_1 - V_-}{R_1} = \frac{V_- + G'v}{R_F + R_0'}$$

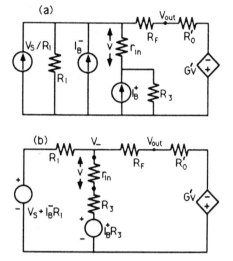

Fig. 7.30. Effects of I_B^+ and I_B^-.

$$V_S + I_B^- R_1 = \frac{R_1 G' v}{R_F + R_0'} + \left[v\left(1 + \frac{R_3}{r_{in}}\right) + I_B^+ R_3\right]\left(1 + \frac{R_1}{R_F + R_0'}\right).$$

For the effects of I_B^+ and I_B^- to drop out of v (and hence out of V_{out}) requires

$$I_B^- R_1 = I_B^+ R_3\left(1 + \frac{R_1}{R_F + R_0'}\right).$$

In practice, operational amplifiers are almost always designed so that $I_B^- \simeq I_B^+$. Then the requirement is that

$$R_3 = \frac{R_1(R_F + R_0')}{R_1 + R_F + R_0'}$$

i.e. R_3 should be equal to the parallel combination of R_1 with $(R_F + R_0')$. Generally R_0' can be neglected.

7.16* Current Amplifiers

The final type of feedback arises when a feedback current is used which is proportional to output current. An idealised form of this is shown in figure 7.31. Inside the amplifier, indicated by broken lines, the current I_{in} is amplified by a factor $-\beta$, the minus sign indicating an inversion of direction between I_{in} and I_2. We envisage some scheme (the current mirror, Chapter 18) which arranges that the current in the feedback arm is $I_F = BI_2$, where B is a constant. The feedback current flows away from the input terminal of the amplifier, so the feedback is negative.

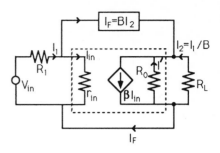

Fig. 7.31. A current amplifier.

It should come as no surprise that the current gain I_2/I_1 is stabilised at the value $1/B$, independent of the load R_L applied across the output. The algebra follows the same lines as previous examples. At the input,

$$I_{in} = I_1 - I_F = I_1 - BI_2. \tag{7.29}$$

At the output,

$$\beta I_{in} = I' + I_2 + I_F = I' + I_2 + BI_2. \tag{7.30}$$

The same voltages appear across R_0 and R_L so

$$R_0 I' = I_2 R_L \qquad \text{or} \qquad I' = I_2 R_L/R_0. \tag{7.31}$$

In order to find I_2/I_1, we eliminate I' and I_{in} by substituting (7.29) and (7.31) into (7.30):

$$\beta(I_1 - BI_2) = I_2(1 + B + R_L/R_0)$$
$$\frac{I_2}{I_1} = \frac{\beta}{1 + B(1 + \beta) + R_L/R_0}. \tag{7.32}$$

For large current amplification β,

$$I_2 \simeq I_1/B \tag{7.32a}$$

and the current gain is independent of the properties of the amplifier and of the load R_L.

The calculation of input and output impedances follows the lines of previous examples and presents no surprises. The output impedance of the amplifier becomes $R_0[1 + B(1 + \beta)]$, because output current has been stabilised. The input impedance is reduced by current feedback and becomes $r_{in}/(1 + B\beta)$. These results are derived in exercise 15.

7.17 Exercises

1. An amplifier has amplification $A = G/(1 + GB)$. Show that the change δA due to a change δG in gain is $\delta A = \delta G/(1 + GB)^2$ providing $\delta G/G \ll 1$. Find $\delta A/A$ for $B = 10^{-1}$, 10^{-2}

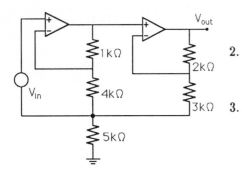

Fig. 7.32.

and 10^{-3} if $G = 10^5$ and $\delta G = 0.1G$. (Ans: 10^{-5}, 10^{-4}, 10^{-3}.)

2. Find $A = V_{out}/V_{in}$ for the circuit of figure 7.32, assuming the amplifiers have high input impedance and low output impedance. (Ans: 65/12.)

3. Demonstrate the relations given in figures 7.33–7.41. In every case, you may assume that the amplifier has a very high input resistance and a very high gain G, so that the voltage difference between the two inputs is very small. Note that the circuit of figure 7.38 can be used to generate a very large effective feedback resistance; this is useful if R_1 is large, and values of R_F required in figure 7.7 are too large to be readily available.

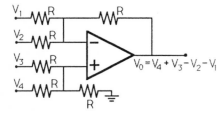

Fig. 7.33.

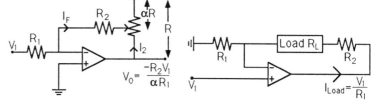

Fig. 7.34. Assume $I_2 \gg I_F$.

Fig. 7.35.

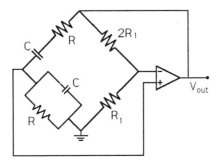

Fig. 7.36. Wien Bridge Oscillator, $f = 1/(2\pi CR)$.

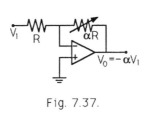

Fig. 7.37.

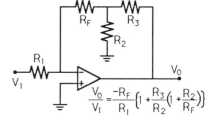

$$\frac{V_0}{V_1} = \frac{-R_F}{R_1}\left(1 + \frac{R_3}{R_2}\left(1 + \frac{R_2}{R_F}\right)\right)$$

Fig. 7.38.

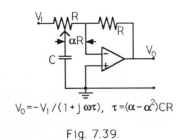

$V_0 = -V_1/(1+j\omega\tau), \quad \tau = (\alpha - \alpha^2)CR$

Fig. 7.39.

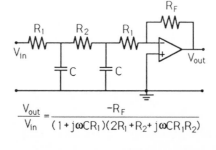

$$\frac{V_{out}}{V_{in}} = \frac{-R_F}{(1+j\omega CR_1)(2R_1 + R_2 + j\omega CR_1R_2)}$$

Fig .7.40

4. (King's College 1974) In the circuit of figure 7.42, the operational amplifier has a voltage gain A_1 of about -10^4, and A_2 is a difference amplifier whose output voltage is $A_2(V_1 - V_2)$ for inputs V_1 and V_2, where $A_2 \simeq +10^4$. SQ is a squaring circuit, whose output voltage is SV^2 for input V; $S \simeq +0.1$. Derive the relation which connects the output voltage x with the input voltage e. Assuming that $S = a$, $R_F/R_2 = b$ and $e = cR_1/R_F$, state the type of mathematical equation which this circuit solves and discuss any limitations which occur to you. (Ans: $ax^2 + bx + c = 0$; limitations: (i) requires $R_2/R_F \ll A_2$ and $A_1 \gg (R_F/R_1 + R_F/R_2)$, (ii) it is not clear how to get the circuit to flip between the two solutions, and (iii) if $b^2 - 4ac < 0$, the roots of the equation are complex and the input $-(ax^2 + bx + c)$ to A_2 is not zero, so A_2 saturates positively or negatively.)

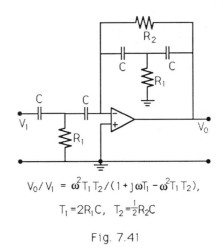

$$V_0/V_1 = \omega^2 T_1 T_2 / (1 + j\omega T_1 - \omega^2 T_1 T_2),$$

$$T_1 = 2R_1 C, \quad T_2 = \tfrac{1}{2} R_2 C$$

Fig. 7.41

5*. (QMC 1984) The amplifier in figure 7.43 has a large voltage gain G and very high input impedance. The input voltage at the left of the circuit is an AC signal of angular frequency ω. Show that the impedance Z of the network enclosed by the broken lines is $Z = (2R + 1/j\omega C)/(1 + j\omega CR)$, and hence show that the voltages V_1 and V_2 at points X and Y are related by $V_2 = V_1/(2 + 1/j\omega CR)$. Show that $V_1(2 + R/Z) = V_{in} + V_{out}$. Hence, neglecting terms of order $1/G$, find V_{out}/V_{in}. Sketch the frequency dependence of V_{out}/V_{in}. (Ans: $V_{out}/V_{in} = 1/(4 + j\omega CR + 2/j\omega CR)$, figure 7.44.)

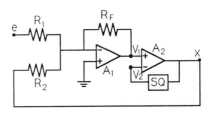

Fig. 7.42

6. (Adapted from IC 1988) Show that the current I_L through the load resistor of figure 7.45 is proportional to the input current I_{in}, but independent of the value of R_L. The output current can be set using voltage V_{in} connected to point X via an input resistor R_{in}. Show that the input impedance of the circuit is $Z_{in} = V_{in}/I_{in} = R_{in} - R_F R_L/R_1$.

7. If in figure 7.46 the amplifier has a gain $G = 10$, an input impedance of only 1 kΩ and negligible output impedance, find the voltage gain of the circuit and its input impedance. How do these results change if $G = 100$? (Ans: -1.85, $R_{in} = 4.05$ kΩ, $A = -6.94$, $R_{in} = 3.55$ kΩ.)

8. (QMC 1985) What is meant by the bandwidth of an amplifier? An operational amplifier has high input impedance, a voltage gain of 10^5, and a bandwidth from angular frequency 0 to 10^3 rad/s. How can this amplifier be used to provide (a) amplification with a narrow bandwidth peaked at $\omega = 250$ rad/s, and (b) amplification with bandwidth increased to cover the range $\omega = 0$ to 10^6 rad/s and an input impedance of 2.2 kΩ? In the latter case, what is then the voltage gain at low frequencies? (Ans: (a) tuned circuit as

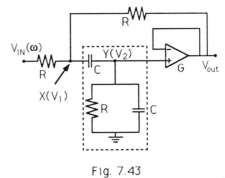

Fig. 7.43

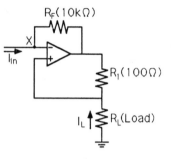

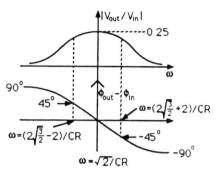

Fig. 7.44

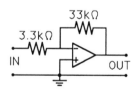

$R_F(10k\Omega)$

I_{in}

X

$R_1(100\Omega)$

I_L R_L(Load)

Fig. 7.45

$33k\Omega$

$3.3k\Omega$

IN   OUT

Fig. 7.46.

load with $LC = 1.6 \times 10^{-5}$, (b) $R_1 = 2.2$ kΩ, $R_F = 220$ kΩ, (c) 10^2.)

9.   If, in figure 7.13, $R_F = 330$ kΩ, $R_1 = 3.3$ kΩ, $r_{in} = 1$ MΩ and $G = 10^5$, find the effect on V_{out} of an input offset voltage of 1 mV. If $I_B^+ = 20$ nA and $R_3 = 33$ kΩ in figure 7.15(a), what is the effect on V_{out}? (Ans: 100 mV, 0.66 V.)

10*.   Using equations (7.9)–(7.12), show algebraically that the input resistance of the shunt feedback amplifier of figure 7.7 is R_1 in series with the parallel combination of r_{in} and $R_F/(G+1)$. Neglect r_{out}.

11*.   A sensor detects 1 mV signals from a nerve cell. Its output impedance is 10^7 Ω. It feeds into an operational amplifier having gain 10^4, input impedance 10^5 Ω, output impedance 120 Ω, and unity gain bandwidth 5×10^6 rad/s. If voltage feedback is applied to the amplifier with $B = 0.01$, what is the magnitude of the output voltage? What is the bandwidth if the input capacitance from the positive terminal of the operational amplifier to earth is 100 pF? How can a bandwidth of 10^4 rad/s be achieved, and what is then the voltage gain? Can 2 V output signals be achieved using two of these amplifiers in sequence with appropriate feedback fractions? (Ans: 50 mV, 2×10^3 rad/s, using $B = 0.002$ and putting 10^5 Ω across the input of the operational amplifier to reduce r_{in} to 5×10^4 Ω; yes.)

12.   (RHBNC 1989) Show that the circuit in figure 7.47 has an input impedance $Z_{in} = V_{in}/I_{in}$ given by $Z_{in} = j\omega C R_1 R_3 R_4/R_2$. Use the properties of operational amplifiers to relate the voltages at points B and D to the input voltage. If C is replaced by an inductance of 10 mH and $R_1 = R_3 = R_4 = 1$ kΩ and $R_2 = 100$ kΩ, calculate the effective input capacitance. (Ans: 1 μF.)

13.   (QMW 1990) In the circuit of figure 7.48, V_{in} is an AC signal $V_0 \cos \omega t$ and the operational amplifiers have a gain of 10^5, input impedance 10^6 Ω and output impedance 100 Ω. Find V_X separately for the positive and negative parts of the cycle and show that $V_{out} = |V_{in}|$. What mean power is delivered to the load resistor? (Ans: $V_0^2/2R_L$.)

14*.   The objective of this exercise is to demonstrate that effects due to r_{out} in the feedback amplifier of figure 7.7 are extremely small. Simplify figure 7.28(d) further by (i) redrawing V_S and R_1 in Norton equivalent form, (ii) combining R_1 and r_{in} in parallel to R_1', and (iii) redrawing R_1' and the current source in Thevenin equivalent form. Show that the current through the feedback resistor R_F is $I_F = (G'+1)V_-/(R_F + R_0')$. Show that $V_- = V'(R_F +$

$R_0')/(R_1' + R_F + R_0' + G'R_1')$ where $V' = V_S R_1'/R_1$. Hence show that the voltage gain is

$$\frac{V_{out}}{V_S} = -\frac{(R_1'/R_1)(R_F - R_0'/G')}{R_1' + (R_F + R_0' + R_1')/G'} \rightarrow \frac{R_F}{R_1} \quad \text{as} \quad G' \rightarrow \infty.$$

Using values of components from figure 7.7 and $G = 10^5$, what is the fractional effect due to (a) R_0'/G' in the numerator, and (b) R_0'/G' in the denominator? (Ans: one part in 10^8 reduction in gain, one part in 10^6 reduction.)

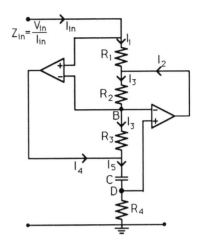

Fig. 7.47. The gyrator.

15*. Using equation (7.32), show that the output impedance across R_L is R_L in parallel with $R_0(1+B+B\beta)$. From V_{in}/I_1, show that the input impedance to the circuit of figure 7.31 is $R_1 + r_{in}(B + 1 + R_L/R_0)/(B\beta + B + 1 + R_L/R_0)$.

16*. Figure 7.49 may be used as a crude current amplifier. Show that

$$\frac{I_2}{I_1} = (\beta - r_{in}/R_F)$$

$$\times \left[1 + \frac{R_L}{R_0} + \frac{R_L}{R_F} + \frac{r_{in}}{R_F}\left(1 + \frac{R_L}{R_0}\right) + \frac{\beta R_L}{R_F}\right]^{-1}$$

where r_{in} and R_0 are the input and output impedances of the operational amplifier and $-\beta$ its current gain. From this result, show that the output impedance across R_L is R_0 in parallel with $(R_F + r_{in})/(1+\beta)$. This latter result shows that the current gain is much less stable than that of figure 7.31 against variations in R_L. Why is this? (Ans: $B = R_L/R_F$ is not independent of R_L and drops to zero as $R_L \rightarrow 0$.)

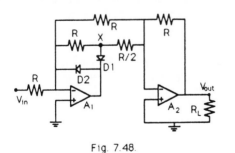

Fig. 7.48.

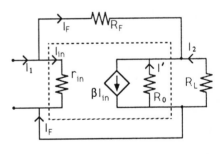

Fig. 7.49.

8

Integration and Differentiation

8.1 Integration

Consider the circuit shown in figure 8.1. We shall find that, to a good approximation, it integrates the input voltage:

$$V_{\text{out}} \propto \int V_{\text{in}} \, dt.$$

This is useful in many applications. It can, for example, be used in analogue computers to do integrals and solve differential equations. Secondly, if V_{in} is constant, it generates a linearly rising output which may be used to provide the timebase of an oscilloscope. A third example arises in the detection of ionising particles. Solid state detectors produce pulses of the form shown in figure 8.2 from the ionisation produced by a particle. The energy deposited is proportional to the area under the curve.

To see roughly how the circuit works, suppose the operational amplifier has an input impedance r_{in} large compared with R_1 and a low output impedance. Then the current I_{in} into the amplifier itself is negligible and all of the current through R_1 flows into the feedback capacitor. Suppose, for simplicity, that this capacitor has been discharged initially by temporarily closing and opening the switch S before the input pulse V_{in} arrives. The point P at the inverting input of the amplifier acts as a virtual earth and

$$I \simeq V_{\text{in}}/R_1$$

$$\boxed{V_{\text{out}} \simeq -\frac{Q}{C} = -\frac{1}{C} \int_0^t I \, dt' = -\frac{1}{CR_1} \int_0^t V_{\text{in}} \, dt'.} \qquad (8.1)$$

The circuit is called an **integrator** and we refer to the **integration** of the input pulse. In a practical circuit the switch S is electronic and additional logic arranges to discharge the capacitor before use.

It is instructive to try this circuit experimentally. Suitable component values are $R_1 = 10 \text{ k}\Omega$ and $C = 1 \text{ } \mu\text{F}$. When V_{in} is zero, you

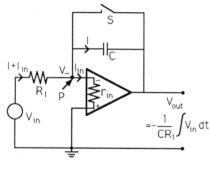

Fig. 8.1. Integrator.

$$=-\frac{1}{CR_1} \int V_{\text{in}} \, dt$$

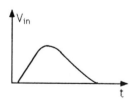

Fig. 8.2. Pulse from a solid state detector.

will find that the output voltage drifts slowly positive or negative because of the input offset voltage or input bias currents. You can balance this drift with the arrangement described in the previous chapter in figure 7.15.

Suppose you now apply square waves to the circuit. If their amplitude is perfectly symmetrical about zero, you should observe a triangular output $V_{out} = -(1/CR) \int V_{in} \, dt$. Over the positive half of the input, V_{out} goes negative linearly with time (figure 8.3); during the negative half of the input, V_{out} should recover to zero with the opposite slope. In reality, the input is never accurately symmetrical about zero, so V_{out} drifts rapidly positive or negative and saturates. A diode placed across C will prevent this, but you may have to try both polarities to find the one which holds the DC level constant.

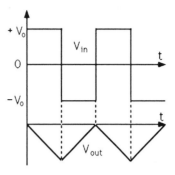

Fig. 8.3. V_{out} for an input square wave symmetrical about zero.

How is the diode functioning?

How good are the approximations in equation (8.1)? In fact, they turn out to be good if the duration of the input pulse is significantly less than the risetime of the circuit. The input impedance r_{in} of the amplifier will be included, but not its output impedance; the latter makes the algebra messy and adds no new features. We proceed in standard fashion, expressing the voltage V_- at point P in as many ways as possible:

$$V_- = V_{in} - (I + I_{in})R_1 \qquad (8.2)$$

$$= I_{in} r_{in} \qquad (8.3)$$

$$= V_{out} + (1/C) \int I \, dt' \qquad (8.4)$$

$$= -V_{out}/G. \qquad (8.5)$$

In the last equation, the output impedance of the amplifier has been neglected.

In order to solve for V_{out}/V_{in}, it is necessary to eliminate I_{in} and I. From (8.3) and (8.5),

$$I_{in} = -V_{out}/Gr_{in} \qquad (8.6)$$

and from (8.2), (8.5) and (8.6),

$$IR_1 = V_{in} + \frac{V_{out}}{G}\left(1 + \frac{R_1}{r_{in}}\right).$$

Finally, from (8.4) and (8.5),

$$V_{out}\left(1 + \frac{1}{G}\right) = -\frac{1}{CR_1} \int \left[V_{in} + \frac{V_{out}}{G}\left(1 + \frac{R_1}{r_{in}}\right)\right] \, dt'. \qquad (8.7)$$

Corrections to the simple equation (8.1) are of order $1/G$. Faithful integration of a pulse therefore requires use of an *amplifier* with gain G extending to high frequency, so that the correction terms can be neglected.

If V_{in} is a step rising from zero to a constant value at $t = 0$ and if the capacitor is initially uncharged, equation (8.7) may be differentiated:

$$(G+1)\frac{dV_{out}}{dt} + \frac{1}{CR_1}\left(1 + \frac{R_1}{r_{in}}\right)V_{out} = -\frac{GV_{in}}{CR_1} \qquad (8.7a)$$

and solved with the result

$$V_{out} = \frac{-GV_{in}}{1 + R_1/r_{in}}\left(1 - e^{-\gamma t}\right) \simeq \frac{-tV_{in}}{CR_1}(1 - \tfrac{1}{2}\gamma t)$$

where

$$\gamma = \frac{1 + R_1/r_{in}}{CR_1(G+1)}. \qquad (8.8)$$

The shape of V_{out} against t is shown in figure 8.4, assuming no saturation. Deviations from a straight line occur for times of order $1/\gamma$. With $C = 1\ \mu\text{F}$, $R_1 = 10^3\ \Omega$ and $G = 10^5$, this time is 100 s. In reality, the output is likely to saturate long before this when V_{out} reaches the limit the supply voltage provides. For the simple CR integrator of Chapter 3, γ was larger by a factor G. The amplifier has improved the linearity by the large factor G through the use of feedback. This is important in providing an accurately linear timebase for radar or an oscilloscope. The quantity γ is the reciprocal of the risetime of the output, and is equal to the bandwidth of the *circuit* (not the amplifier), as will be demonstrated explicitly below. Good linearity is therefore synonymous with small *circuit* bandwidth.

Are you clear about the distinction between circuit bandwidth and amplifier bandwidth? Remember the difference between the input impedance of the circuit and from the point P to earth.

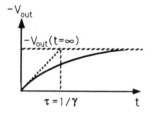

Fig. 8.4. $-V_{out}$ for constant V_{in}.

8.2   The Miller Effect

The integrator is a circuit where V_{out} depends on the past history of V_{in}, via the charge Q which has built up on the feedback capacitor. In a normal amplifier, this is undesirable; there the requirement is that V_{out} should be proportional to V_{in} itself, rather than its integral. The capacitance C_2 between input and output of the amplifier in figure 8.5 can seriously degrade the performance of the amplifier at high frequencies, where C_2 short circuits R_2. This is known as the **Miller effect**.

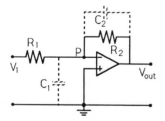

Fig. 8.5. Capacitance in a real amplifier.

To investigate it, suppose the input is an AC signal $V_1 e^{j\omega t}$. The output will be $V_0 e^{j\omega t}$, with V_0 complex. Ignore C_1 for the moment: it will become clear how to put it in later. If R_2 were absent, the relation between V_{out} and V_1 could be obtained from equation (8.7a):

$$V_0 = -\frac{GV_1}{1 + R_1/r_{\text{in}} + j\omega C_2(G+1)R_1}$$
$$= -\frac{GV_1/R_1}{1/R_1 + 1/r_{\text{in}} + j\omega C_2(G+1)}. \qquad (8.9)$$

The bandwidth ω_0 of the circuit is given by the value of ω for which

$$1/R_1 + 1/r_{\text{in}} = \omega_0 C_2(G+1)$$
$$\omega_0 = \frac{(1/R_1) + (1/r_{\text{in}})}{C_2(G+1)} = \gamma. \qquad (8.10)$$

This verifies that the bandwidth of the integrator is γ.

The numerator of equation (8.10) can be interpreted in terms of a Thevenin equivalent. Viewed from the point P of figure 8.5, the resistance to earth (with R_2 absent) is R_1 in parallel with r_{in}. Call this R_1'. Then

$$\boxed{\omega_0 = 1/C_2 R_1'(G+1).}$$

The important point about this formula is that the circuit is behaving as a filter where the capacitance C_2 is scaled by the factor $(G+1)$. This is the Miller effect. The factor $(G+1)$ has the same origin as the factor $(G'+1)$ appearing in figure 7.29(a). Transistors often act as voltage amplifiers with $G > 100$. Their high-frequency behaviour is then limited by capacitance from input to output. It is immediately apparent that an essential feature of high-performance transistors is low capacitance C_2 between input and output.

In figure 8.5, it is now clear how to handle the capacitance C_1; the input impedance of the operational amplifier becomes r_{in} in parallel with C_1. Finally, the effect of the feedback resistor R_2 may be included by replacing the admittance $j\omega C_2$ of C_2 with the admittance of C_2 in parallel with R_2. Then equation (8.9) becomes

$$V_0 = -\frac{GV_1/R_1}{1/R_1' + j\omega C_1 + (j\omega C_2 + 1/R_2)(G+1)}.$$

At low frequencies, if the last term in the denominator dominates,

$$\frac{V_0}{V_1} \simeq -\frac{R_2}{R_1}\frac{G}{G+1} \simeq -\frac{R_2}{R_1}$$

which is the elementary result. The bandwidth ω_0 is given by

$$\omega_0\left(C_2 + \frac{C_1}{G+1}\right) = \frac{1}{R_2} + \frac{1}{R_1'(G+1)}. \qquad (8.11)$$

Can you understand this result in terms of the parallel capacitances and resistances from the point P to earth?

For large G, $\omega_0 \simeq 1/C_2 R_2$. Thus, putting R_2 in parallel with C_2 (shunt feedback) has had the important and desirable effect of increasing the bandwidth by eliminating the factor $(G + 1)$. Ultimately, G falls at high frequencies and C_1 and R'_1 then come into play.

If you wish to demonstrate the Miller effect yourself on the oscilloscope, suitable component values are $R_2 = 1\ \mathrm{k\Omega}$, $C = 0.01\text{–}1\ \mu\mathrm{F}$.

8.3   Compensation

A common problem in pulse amplifiers is that stray capacitance distorts the leading edge of the pulse, either rounding it off or making it overshoot. Figure 8.6 shows possible responses to an input square pulse. The question arises how to cure this problem. In figure 8.6(d), a shunt feedback amplifier is shown with impedances Z_1, Z_2 and Z_in denoting parallel combinations of resistors and capacitors; e.g. Z_1 is C_1 in parallel with R_1. The gain of the circuit is given from the first form of equation (8.9) by

$$\frac{V_\mathrm{out}}{V_\mathrm{in}} = -\frac{G}{1 + (G+1)Z_1/Z_2 + Z_1/Z_\mathrm{in}}$$

$$= -\frac{Z_2/Z_1}{1 + (1/G)(1 + Z_2/Z_1 + Z_2/Z_\mathrm{in})}. \qquad (8.12)$$

If the gain is to be as independent of frequency as possible, it is necessary that Z_2/Z_1 and Z_2/Z_in should be independent of frequency. This requires that $C_1 R_1 = C_2 R_2 = C_\mathrm{in} R_\mathrm{in}$. Can you interpret this condition physically? In a practical situation, this condition can be achieved by adjusting small capacitors across two of the three locations until the best frequency response is obtained. If a square pulse is fed into the amplifier, these **trimming** capacitors are adjusted until the output shows the sharpest behaviour (figure 8.6(a)), without overshooting; see exercise 5.

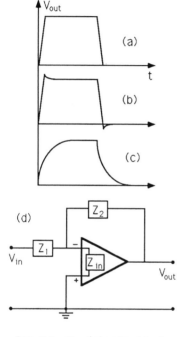

Fig. 8.6. V_out (a) with ideal adjustment, (b) non-ideal, overshooting, (c) non-ideal, reduced bandwidth, (d) compensation in the feedback amplifier.

8.4   Differentiation

The circuit of figure 8.7 has the capacitor C and resistor R of figure 8.1 interchanged. It acts so as to differentiate the input waveform, as will now be demonstrated. Making the usual assumption that the amplifier has large input impedance and small

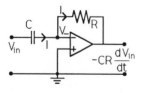

Fig. 8.7. Differentiating circuit.

output impedance,

$$V_- = -V_{out}/G \simeq 0$$

$$V_{out} = V_- - IR \simeq -IR$$

$$V_{in} = V_- + (1/C) \int I \, dt' \simeq (1/C) \int I \, dt'$$

$$= -(1/CR) \int V_{out} \, dt'.$$

Differentiating this last equation

$$V_{out} \simeq -CR dV_{in}/dt. \qquad (8.13)$$

There is, however, a problem with this arrangement. If there is noise at the input to the circuit or noise across any capacitance between the inverting input and earth, the circuit differentiates it, producing a noisy output. Use of this circuit is not recommended. If you want to solve a differential equation by analogue techniques, it is best to express the equation in terms of integrating circuits (exercise 7).

Can you see physically why this circuit is noise sensitive?

8.5   The Charge-sensitive Amplifier

If the source driving an amplifier has a large resistance, the circuit has a large time constant for charging any capacitance. We have already seen that this creates a problem with bandwidth for series voltage feedback. The origin of the problem is that the input current is very small. A weak source may provide only a few electrons, so it is appropriate to think in terms of current and charge, rather than voltage.

Figure 8.8(a) shows a circuit which circumvents this problem. The source is expressed in Norton form, and (b) gives its equivalent circuit. If R_S is very large, all of the current $V_S/R_S = I_S$ flows into C_1 and C_2 (making the usual assumption that r_{in} of the operational amplifier is very large). Then

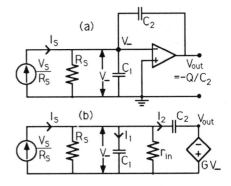

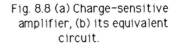

Fig. 8.8 (a) Charge-sensitive amplifier, (b) its equivalent circuit.

$$V_- = (1/C_1) \int I_1 \, dt$$

$$V_-(1 + G) = (1/C_2) \int I_2 \, dt$$

$$\int I_S \, dt = \int (I_1 + I_2) \, dt = V_-[C_1 + C_2(1 + G)].$$

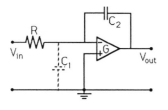

Fig. 8.9.

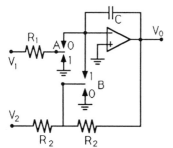

$$V_0 = -\int_0^t \frac{V_1\,dt}{CR_1} - V_2$$

Fig. 8.10.

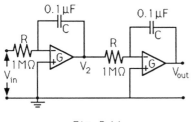

Fig. 8.11.

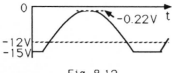

Fig. 8.12.

The output voltage is

$$V_{\text{out}} = -GV_- = -\frac{G}{C_1 + C_2(1+G)}\int I_S\,dt = -\frac{GQ}{C_1 + C_2(1+G)} \tag{8.14}$$

where Q is the charge generated by the source. For large G, $V_{\text{out}} \simeq -Q/C_2$.

This result is easy to understand. The impedance from the negative terminal to earth through C_2 is $1/j\omega C_2(G+1)$. The factor $(G+1)$ makes this much smaller than the impedance of C_1, so nearly all of the current I_S flows to C_2 and creates a voltage $-Q/C_2$. The result is independent of C_1 so long as it is small compared with $C_2(1+G)$. If the source is connected to the amplifier via a coaxial cable, V_{out} is insensitive to the capacitance C_1 of the cable.

Such an amplifier is called a **charge-sensitive amplifier**. It is capable of amplifying very fast pulses, e.g. from a nuclear radiation detector, providing the transistors in the amplifier are fast enough. There is no bandwidth problem from capacitance. The only problem is that bias currents are liable to charge up C_2. It is necessary to cancel these out carefully and put a resistor across C_2, so as to discharge it with a time constant longer than the pulses of interest. This restores the DC level of the output after each pulse and prevents 'pile-up'.

8.6 Exercises

1. In figure 8.9, $R = 1$ kΩ, $C_2 = 100$ pF and $G = 200$. Find (a) the time constant of the output when V_{in} is a step function, and (b) the bandwidth when V_{in} is an AC signal. If $C_1 = 10$ pF and a resistor $R_2 = 10$ kΩ is placed across C_2, what is the amplification of the circuit at low frequency and the new bandwidth? (Ans: (a) 2×10^{-5} s, (b) 5×10^4 rad/s; $A = -10$, bandwidth $= 1.05 \times 10^6$ rad/s.)

2. Demonstrate the relations given in figure 8.10, making the usual assumptions about the properties of operational amplifiers.

3**. (QMC 1967) The amplifiers in figure 8.11 are DC wide-band inverting amplifiers. What output is produced by a 100 mV square pulse of duration 1 s? Hints:

(i) Consider each half of the network separately. Ignoring the input impedance of the amplifier, show that

$$V_2\left(1 + \frac{1}{G}\right) = -\frac{1}{CR}\int\left(V_{\text{in}} + \frac{V_2}{G}\right)\,dt.$$

(ii) Solve the problem for $t = 0$ to 1 s neglecting terms of order $1/G$.

(iii) Solve the equation derived in (i) for V_2 exactly, with the trial substitution $V_2 = Ae^{\beta t} + B$, and show that the time constant is $CR(1 + G)$.

(iv) Discuss qualitatively the output of the second stage for $t > 1$ s. Will either amplifier saturate? If so, will it recover eventually and what will then happen?

(Ans: (ii) $V_2 = -t$ V, $V_{\text{out}} = 5t^2$ V, (iii) $B = -GV_{\text{in}}$, $\beta = -1/CR(G + 1)$, (iv) $V_2 = -1\exp[-(t - 1)/0.1(G + 1)] \simeq -1$; V_{out} continues to grow linearly to saturation ($G_{\text{sat}} \simeq -V_{\text{sat}}/V_2$); eventually V_2 has decayed sufficiently that V_{out} stops saturating; V_{out} drops away from V_{sat}, rapidly at first, corresponding to a low value of G, and more slowly later as G returns to its normal value.)

4. (QMC 1985) A two-input analogue adder has a feedback resistance of 1 kΩ and operates from a ±15 V supply. Find the input resistance values for an output of AC component 5 V RMS and DC level −3 V with 300 mV RMS AC at one input and 100 mV DC at the other. Sketch the output waveform if the input voltages are changed to 500 mV RMS AC and +400 mV DC. Find the input resistance value for an integrator with a 33 nF capacitor to give a triangular waveform output of 3 V peak-to-peak for a symmetrical 100 mV peak-to-peak square wave input with 100 μs pulse width. (Ans: AC input via a 60 Ω resistor, DC input via 33 Ω; waveform of figure 8.12, 50 Ω.)

5. (QMC 1987) Figure 8.13 shows the equivalent circuit of an oscilloscope probe. R is the resistance of the oscilloscope and C the total capacitance of the oscilloscope input and probe cable; V_1 is the probe input voltage and V_0 the voltage appearing on the oscilloscope. If $R_p = nR$ and $C_p = C/n$, show that (a) $V_0/V_1 = 1/(n + 1)$ and is independent of frequency, (b) the impedance between the probe input terminals is equivalent to a resistance of $(n + 1)R$ in parallel with a capacitance of $C/(n + 1)$. **Show that

$$C_p \frac{dV_1}{dt} + \frac{V_1}{R_p} = V_{\text{out}} \left(\frac{1}{R_p} + \frac{1}{R} \right) + \frac{dV_{\text{out}}}{dt}(C + C_p).$$

If V_1 is a square wave, sketch V_{out} if (c) $C_p > C/n$, and (d) $C_p < C/n$. (Ans: figure 8.14.)

6**. (IC 1988) The circuit shown in figure 8.15 is fed with a square-wave input of amplitude 5 V centred about ground and of frequency 100 kHz. Describe the operation of the circuit and

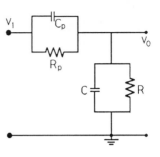

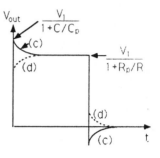

Fig. 8.13.

Fig. 8.14.

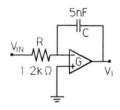

Fig. 8.15.

Fig. 8.16.

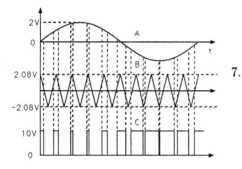

Fig. 8.17. Pulse width modulation.

relate its output voltage to its input voltage. Make a sketch showing both the input and output waveforms in the correct relative phase and calculate the amplitude of the output waveform.

A sine wave of frequency 10 kHz and amplitude 2 V centred on ground is fed to input A of the comparator of figure 8.16 and the output from figure 8.15 to input B. You may assume the sine wave passes zero when the output of the first circuit is at its peak positive value. Sketch the signals appearing at points A, B and C for one cycle of the sine wave, assuming that the comparator swings from 0 to +10 V and that hysteresis may be neglected. (Ans: figure 8.17.)

7. (Westfield 1974) Show how operational amplifiers can be used to perform the following mathematical operations: (a) multiplication by a constant, (b) addition, (c) integration, and (d) differentiation. **Devise a circuit to solve the following differential equation assuming all initial values are zero:

$$\frac{d^3x}{dt^3} + A\frac{d^2x}{dt^2} + B\frac{dx}{dt} + Cx = f(t).$$

Hint: make the input to one operational amplifier $d^3x/dt^3 = F(t) - Cx - B\,dx/dt - A\,d^2x/dt^2$. (Ans: figure 8.18).

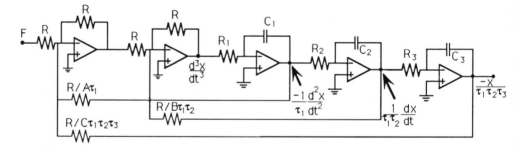

Fig. 8.18.

8**. (QMC 1972) Show that the high-frequency cut-off of an amplifier consisting of N identical resistance–capacitance coupled stages in cascade is $(2^{1/N} - 1)^{1/2}$ times the high-frequency cut-off of a single stage.

An amplifier is required to have a bandwidth of 3 MHz and a mid-band gain of 70 db. Four stages are available with a gain at their low-frequency cut-off of 19 db, input resistance of 1 kΩ, a shunt capacitance of 100 pF and an output impedance of 254 Ω. Is it possible to achieve the requirement? (Ans:

Yes, just. Stages 2, 3 and 4 have a gain of 10.13 and an effective resistance of 203 Ω in parallel with the capacitance; with 330 Ω in parallel with the input of the first stage, its gain is 3.175 and the resistance across the capacitance is 250 Ω.)

9**. If in figure 8.6(d) Z_1 is a parallel combination of C_1 with R_1 and Z_2 and Z_{in} are parallel combinations of C_2 with R_2 and C_{in} with R_{in}, discuss the effects of uncompensated C_1, C_F and C_{in} on the phase of V_{out} with respect to V_{in}. Show that the primary effect of C_1 is to advance the output phase and of C_2 and C_{in} are to retard it.

9

Diodes and Transistors as Switches

9.1 Introduction

This chapter gives an outline of the solid state physics involved in the operation of pn diodes and transistors. From a practical point of view, only those involved in developing new devices or those designing integrated circuits really require quantitative understanding of the details nowadays. Users can assemble integrated circuits into systems without an intimate knowledge of their internal workings. However, just as it pays a driver to have an idea how his or her car works, so it is useful for the average physicist or engineer to have a qualitative picture of how a transistor functions, along with an idea of typical performance figures. The qualitative picture is described in this chapter while the mathematics is given in appendix C. The physics going on is actually rather involved and is the subject of many lengthy books. If you wish to grapple with the subtleties, it is necessary to refer to one of these specialised discussions, for example, D H Navon, 1986, *Semiconductor Microdevices and Materials* CBS Publishing Japan Ltd.

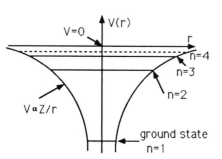

Fig. 9.1. Energy levels in an atom.

9.2 Conductors

In a single atom, the Coulomb field of the nucleus produces a potential $V \propto Z/r$, where Z is the nuclear charge and r the distance from the nucleus. Electrons go into well-defined (quantised) energy levels, illustrated by the horizontal lines in figure 9.1. These electrons actually shield the nuclear charge to some degree and modify the Coulomb field from a $1/r$ dependence, but that is a detail.

In a solid, the situation is broadly similar, except that electrons close to one atom feel a potential due to neighbouring atoms as well. Low-lying, tightly bound levels are affected very little, but the higher levels broaden into **bands**, illustrated in figure 9.2.

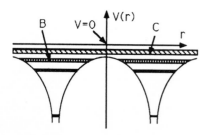

Fig. 9.2. Energy levels in a solid.

In a **conductor**, the least tightly bound electrons, called **valence** electrons, only partially fill a high-lying band called the **conduction band**, labelled C in figure 9.2. This band is not confined to a single atom. Electrons in it are free to travel through the solid under the action of an accelerating electric field and hence carry current.

At absolute zero, the electrons occupy the lowest available energy levels up to an energy E_F, called the **Fermi energy**. All states below E_F are occupied, and all above E_F are empty. Above absolute zero, some electrons are excited by thermal energy into higher levels. The probability F_e that an energy level is occupied is shown in figure 9.3. This curve is the **Fermi–Dirac distribution**, derived from statistical mechanics; it is given by

$$F_e(E) = \frac{1}{1 + \exp[(E - E_F)/kT]} \qquad (9.1)$$

where T is the absolute temperature and k is Boltzmann's constant. The value of E_F is such that $F_e(E_F) = 0.5$. At room temperature $kT \simeq 0.025$ eV. At low temperatures, the curve rises steeply through E_F; as the temperature rises, more electrons are excited, leaving vacancies in the levels below E_F.

Electrons in levels well below the Fermi energy (by an amount $\gg kT$) move through the conductor without making collisions; such a collision would require transfer of energy and momentum from an atom (or another electron) sufficient to excite the electron to an unoccupied level. Thermal energy is rarely enough. However, electrons near the Fermi energy can transfer energy freely with an atom in a collision, sometimes gaining energy and sometimes losing it. For electrons at the Fermi energy, the mean free path for collisions is $\lambda = 4 \times 10^{-6}$ cm in copper, i.e. hundreds of atomic diameters. The energy E_F is ~ 7 eV in copper, giving an electron a very high velocity v_e equivalent to what it would acquire thermally at a temperature of $\sim 85\,000$ K. A conductor is a hive of activity.

Let us now consider conduction under the action of an applied electric field $\boldsymbol{E}$. Along a wire, the energy levels of all electrons at coordinate x are altered by energy $eV(x)$, where e is the charge of the electron and

$$V(x) - V(x = 0) = -\int_0^x \boldsymbol{E}\, \mathrm{d}x.$$

The equilibrium Fermi energy slopes down the wire as shown in figure 9.4. An electron at $x = 0$ which happens to move to the right has extra energy above the Fermi level. In a subsequent collision, it is then more likely to lose energy than gain it, so on average it falls towards the Fermi energy and the atom with which it collides gains vibrational energy. This energy dissipation accounts for the heating of the wire due to resistivity.

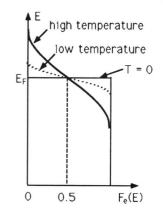

Fig. 9.3. The occupation probability $F_e(E)$ of energy levels in the conduction band v. energy E.

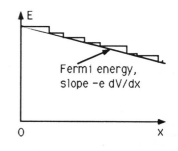

Fig. 9.4. Illustrating conduction.

By this procedure, the electron drifts gradually down the gradient making frequent collisions. The mean drift velocity $\bar{v}$ is the average velocity imparted by the electric field between collisions: it is given by

$$m\bar{v} = e\boldsymbol{E}\tau$$

where τ is the mean time between collisions and m is the mass of the electron; the mean free path is

$$\lambda = v_e\tau.$$

The value of $\bar{v}$ is very much less than v_e, so the drift is a small effect superimposed on the underlying random motion of the electrons. We write

$$\bar{v} = \mu\boldsymbol{E}$$

where μ is called the **mobility** of the electron. The current in a wire of area A containing n free electrons per unit volume is

$$I = eAn\bar{v} = eAn\mu\boldsymbol{E}. \tag{9.2}$$

For a uniform wire of length ℓ, $\boldsymbol{E} = V/\ell$, so the conductivity σ ($= 1/\text{resistivity } \rho$) is

$$\sigma = en\mu.$$

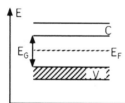

Fig. 9.5. The energy gap in an insulator or intrinsic semiconductor.

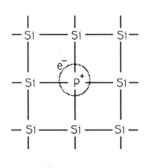

Fig. 9.6. n-type doping with phosphorus.

9.3   Semiconductors and Doping

In an insulator, the valence electrons fill a band, labelled V in figure 9.5, and the energy gap E_G to the conduction band is so large (several eV) that there is a negligible number of excited electrons reaching unoccupied levels in the conduction band. This is why the material is an insulator.

Semiconductors are materials where the energy gap is small enough that a significant number of electrons is excited thermally to the conduction band. For silicon, $E_G = 1.12$ eV and for germanium 0.67 eV. Pure semiconductors are therefore characterised by high but finite resistivity.

When an electron is excited from the valence band it leaves a vacancy there. The probability distribution for vacancies is

$$F_v(E) = 1 - F_e(E) = \frac{\exp[(E - E_F)/kT]}{1 + \exp[(E - E_F)/kT]}$$
$$= \frac{1}{1 + \exp[(E_F - E)/kT]}. \tag{9.3}$$

This is the same as equation (9.1) except for the reversal in sign of $(E_F - E)$. In figure 9.5, the Fermi level lies midway between the top of the valence band and the bottom of the conduction band.

This is because the few electrons excited into the conduction band leave a balancing distribution of vacancies in the valence band.

Doping

The resistivity of semiconductors in the pure state is too high for them to be useful except for special applications. However, the conductivity can be raised to an interesting level by **doping**. This makes a semiconductor where the conductivity is due primarily to the impurities. Doping levels are typically one part in 10^6–10^8.

Suppose a crystal of silicon is doped with a small amount of a pentavalent material: phosphorus, arsenic or antimony. The silicon forms a regular lattice in which each atom makes four covalent bonds to its neighbours. The impurity atoms are similar in size to silicon atoms, so they fit into the lattice by replacing them, as in figure 9.6. However, only four covalent bonds are made to neighbouring atoms. The fifth valence electron is very loosely bound by the unit charge of the impurity atom. In a hydrogen atom, the binding energy is 13.4 eV. But in silicon the dielectric constant is $\epsilon = 11.9$, so the binding energy is expected to be roughly $13.4/\epsilon^2 \simeq 0.1$ eV. The energy level is this amount E_N below the conduction band. The energy level is shown as a broken line in figure 9.7 to indicate that the electron is localised at the impurity atom. Experimentally, E_N varies from 0.039 to 0.049 eV.

At absolute zero, the electrons fill energy levels up to and including the impurity level. The Fermi energy is roughly halfway between this and the conduction band. At room temperature, electrons are very easily ionised from the impurity level. Remember that at room temperature $kT \simeq 0.025$ eV, compared with $\frac{1}{2}E_N \simeq 0.02$ eV. Ionisation produces a fixed positively charged phosphorus atom and an electron in the conduction band, free to move through the crystal. The Fermi distribution of figure 9.7 has been moved up close to the conduction band, implying much higher conductivity than for undoped material.

Semiconductors doped in this way are called **n-type** materials. The n denotes the fact that the impurities **donate** negative charges (electrons) to the conduction band. Be careful to distinguish this from the positive charge left on the lattice; the n refers to the mobile carrier.

p-type material

The converse happens if the silicon is doped with a trivalent material: boron, aluminium, gallium or indium. The impurity atom now steals an electron from a neighbouring atom, which in turn steals one from another atom and so on (figure 9.8(a)). The result is what is called a **hole**: a missing electron. As the deficiency moves from atom to atom, the hole moves just like a positive charge. Such

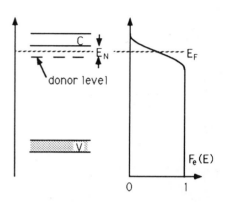

Fig. 9.7. Energy levels and $F_e(E)$ for an n-type material.

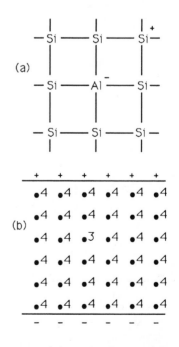

Fig. 9.8 (a) p-type doping, (b) a model of a lattice bounded by charged surfaces.

a material is referred to as **p type**, because it generates positively charged carriers.

Draw a lattice of atoms with four electrons on all atoms except one, which has three (figure 9.8(b)). Throw dice to decide which way the hole moves. What happens when the hole reaches a boundary which is positively or negatively charged? What happens if two holes arrive at neighbouring sites?

At absolute zero, electrons fill energy levels up to the top of the valence band (figure 9.9) and the impurity atoms, called **acceptors**, are unable to steal an electron. The impurity levels now sit empty ∼0.05 eV above the valence band. At non-zero temperatures, electrons are easily excited into them from the valence band leaving vacancies or holes. The Fermi energy is roughly midway between the top of the valence band and the acceptor level; the probability distribution for vacancies, $F_v(E)$ of equation (9.3), is as shown in figure 9.9. If an electric field is applied to the material, the hole moves just like a positive charge, but with a mobility μ_h which is a factor of three smaller in silicon than for electrons and a factor of two smaller in germanium than for electrons.

In a pure semiconductor, the density of electrons and holes is the same. At room temperature it is about 10^{10} cm^{-3}. This type of semiconductor is called an **intrinsic** semiconductor. On the other hand, in n-type material the free electrons dominate overwhelmingly over free holes. There the electrons are **majority carriers** and the holes **minority carriers**. In p-type material, the roles are reversed. These materials, where densities of holes and electrons differ, are called **extrinsic** semiconductors.

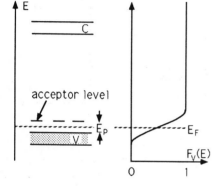

Fig. 9.9. Energy levels and $F_v(E)$ for a p-type material.

9.4   The pn Junction Diode

A diode may be created from the junction between p-type and n-type materials. The current in the diode varies with voltage according to figure 9.10, repeated from Chapter 1. It rises exponentially with voltage V according to

$$I = I_0(e^{eV/kT} - 1). \tag{9.4}$$

This formula will be derived from the solid state physics. It arises essentially from the Fermi–Dirac distribution. In a bipolar transistor, the base current follows the same dependence on voltage V_{BE} between base and emitter. Those of you who do not wish to follow the intricacies of the solid state physics may skip this section and accept equation (9.4) and figure 9.10 empirically.

Both n- and p-type materials are made by heating pure silicon and then exposing its surface to the impurity with which it is

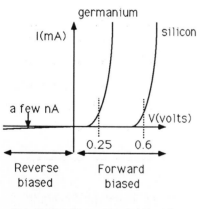

Fig. 9.10. Characteristics of pn diodes.

to be doped. The dopant may take the form of a hot gas or a liquid (indium melts at 156 °C). At elevated temperatures, atoms diffuse through a solid in similar fashion to a coloured dye diffusing through a liquid.

Suppose the silicon is initially exposed to donors, forming an n-type material. This is called the **substrate**. Then suppose the impurity is switched to acceptors. After a time these will outnumber donors near the surface of the silicon, converting it back to a p-type material (figure 9.11(a)). This outlines how a pn junction is made. The physics of the interface between the two materials plays the decisive role. To simplify the discussion (and the mathematics of appendix C) it will be assumed that there is a sharp transition at the interface between p-type material, having a concentration of P acceptor atoms per unit volume to n-type material, with concentration N donor atoms per unit volume. It is not necessary that these doping concentrations should be the same. Of course, in a real material the junction is not sharp.

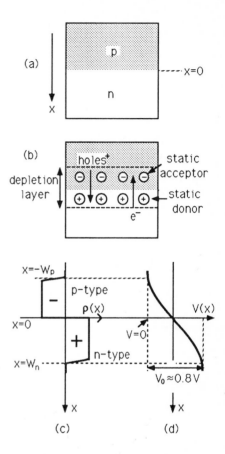

Diffusion of carriers

At room temperature, the impurity atoms themselves no longer diffuse but are immobile. However, electrons and holes diffuse readily across the junction. As shown in figure 9.11(b), electrons diffuse from the n-type region into the surface of the p-type material and recombine there with holes. Conversely, holes diffuse into the surface of the n-type material and recombine there with electrons. In the immediate vicinity of the junction there is a shortage of carriers of either type, because of recombination. This is called the **depletion layer**. In view of the shortage of carriers, it is a region of high resistivity and will support a large potential gradient.

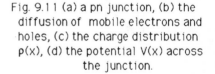

Fig. 9.11 (a) a pn junction, (b) the diffusion of mobile electrons and holes, (c) the charge distribution $\rho(x)$, (d) the potential $V(x)$ across the junction.

Charge distribution

The diffusion of carriers leaves a surface layer of negative static charges in the p-type region (figure 9.11(c)); these are the charges left on immobile acceptor atoms after holes have diffused away across the junction. Likewise, there is a surface layer of positive charge in the n-type region due to immobile donor atoms which have lost electrons. Diffusion proceeds until these charges build up a potential difference V_0 across the junction sufficient to balance the diffusion. An electron which has diffused into the p region feels an electrostatic force pulling it back again. An equilibrium is established between diffusion one way and drift of carriers in the reverse direction due to the electric field created by the static charge distribution.

Potential distribution

Using elementary electrostatics, it is shown in appendix C that these charge distributions dictate the potential across the junction. Suppose the zero of potential is defined to be in the p-type material far from the junction, i.e. at the top of the diagram; this zero is arbitrary, since results depend only on the potential difference. Then $V(x)$ rises in the depletion layer approximately parabolically over the width W_p of the layer, as shown in figure 9.11(d); within the p region

$$V(x) \simeq eP(x + W_p)^2 / 2\epsilon\epsilon_0. \tag{9.5}$$

It is zero at $x = -W_p$. The quantity $e = 1.6 \times 10^{-19}$ C is used to denote the *magnitude* of the charge of the electron. Remember that P is the concentration of acceptor atoms and ϵ the dielectric constant. Likewise, on the n-type side of the depletion layer, $V(x)$ approaches V_0 parabolically over the width W_n of the layer:

$$V_0 - V(x) \simeq eN(W_n - x)^2 / 2\epsilon\epsilon_0. \tag{9.6}$$

Here V_0 is the potential across the junction.

The potential acts as a barrier for charges of either sign to climb. Let us consider the charge distribution $n(x)$ of free electrons as a function of x and the charge distribution $p(x)$ of holes. Appendix C shows that

$$p(x) = P \exp(-eV(x)/kT) \tag{9.7}$$

$$n(x) = N \exp[e(V(x) - V_0)/kT]. \tag{9.8}$$

These densities vary rapidly with x and are shown on a logarithmic scale in figure 9.12. Both change by a factor $\exp(-eV_0/kT)$ over the width of the junction. If $V_0 = 0.8$ V, this is a factor 10^{14} at room temperature. A useful result from equations (9.7) and (9.8) is that the product $p(x)n(x)$ is independent of x:

$$p(x)n(x) \simeq PN \exp(-eV_0/kT). \tag{9.9}$$

This shows that either $p(x)$ or $n(x)$ must be small, and in the middle of the depletion layer both are small.

The charge density $\rho(x)$ of figure 9.11(c) is given by

$$\rho(x) = N + p(x) - P - n(x). \tag{9.10}$$

In checking the signs, remember that donor atoms when ionised have positive charge and the ionised acceptor atoms have negative charge. In the p region, $N = 0$ and $n(x)$ is small. For the p region outside the depletion layer, $p(x) \simeq P$ so $\rho(x) \simeq 0$. Within the depletion layer, $p(x)$ falls rapidly, and $\rho(x) \simeq -P$. In the n region the situation is reversed and $p(x)$ is very small. Within the n region of the depletion layer, $n(x)$ is small, so $\rho(x) \simeq N$. Finally, in the n region outside the depletion layer, $n(x) \simeq N$ and $\rho(x) \simeq 0$.

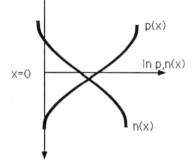

Fig. 9.12. p(x) and n(x).

The origin of the formulae for charge densities $p(x)$ and $n(x)$ is the distortion of the energy bands by $V(x)$ (figure 9.13). In equilibrium, the Fermi energy E_F must be flat and continuous across the junction. In the p region, E_F is close to the valence band V. In the n region it is close to the conduction band C. In between the bands follow the shape of $-V(x)$, so as to make the Fermi energy E_F flat. Clearly $V_0 \simeq E_G - \frac{1}{2}(E_P + E_N)$ if no external voltage is applied. The result is that $p(x)$ is very small in the n region, because the probability of excitation from the valence band across the energy gap is small; likewise for $n(x)$ in the p region. Both are small in the middle of the depletion layer because the Fermi energy there is midway between valence and conduction bands; this is the origin of the depletion.

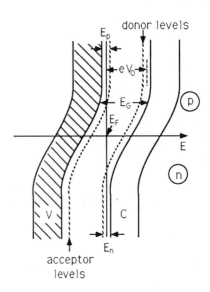

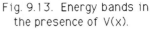

Fig. 9.13. Energy bands in the presence of V(x).

Currents

The objective of this discussion is to find the current in the diode as a function of applied voltage, so we now turn to the currents flowing across the junction (figure 9.14). The currents originate from two effects: diffusion due to the concentration gradient and drift in the reverse direction due to the electric field. At equilibrium these balance. If A is the area of the junction, the electric field creates a hole current, cf equation (9.2),

$$I_p(\text{drift}) = eA\mu_p p(x)E(x). \qquad (9.11)$$

Since $\boldsymbol{E} = -dV/dx$ is negative (figure 9.11(d)), this contribution to I_p is upwards.

Next consider diffusion. Holes originate from the p-type side where the concentration $p(x)$ is high. The number of holes diffusing away from a particular x value is proportional to $p(x)$ and the number diffusing away from $x+dx$ is proportional to $p(x)+(dp/dx)dx$. The net diffusion between these points is proportional to the difference $(dp/dx)dx$; if dp/dx were zero, there would be no net diffusion. The diffusion coefficient D_p for holes is defined so that

$$I_p(\text{diffusion}) = -eAD_p\frac{dp}{dx}. \qquad (9.12)$$

Here dp/dx is negative (figure 9.12), and diffusion is downwards.

Adding equations (9.11) and (9.12), the total hole current is

$$I_p = eA\left(\mu_p p\boldsymbol{E} - D_p\frac{dp}{dx}\right). \qquad (9.13)$$

By similar arguments, the current due to electrons is

$$I_n = -eA\left(\mu_n n\boldsymbol{E} + D_n\frac{dn}{dx}\right). \qquad (9.14)$$

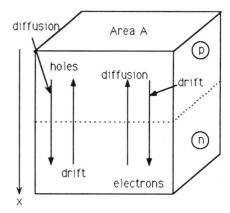

Fig. 9.14. Directions of diffusion and drift currents.

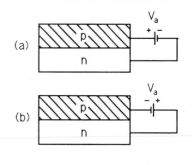

Fig. 9.15 (a) forward, (b) reverse bias.

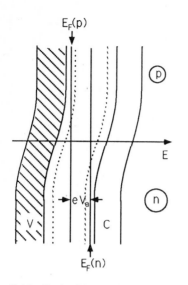

$E_F(p)$

(p)

E

(n)

eV_a

C

$E_F(n)$

Fig. 9.16. Reduction of the potential across the junction by forward bias.

Here the signs may be checked against figure 9.14, bearing in mind that current flow is opposite to electron flow because of the negative charge of the electron and also that dn/dx has the opposite sign to dp/dx.

In the absence of an applied voltage, I_p and I_n are both zero, otherwise charge distributions would change. Diffusion and drift currents balance. The potential V_0 across the junction does not drive current round the external circuit, since balancing contact potentials arise at the terminals between the metallic contacts and the p- and n-type materials.

Effect of an applied voltage

Having considered equilibrium, we are at last ready to consider the effect of applying a battery to the diode (figure 9.15). Suppose a positive potential V_a is applied to the p side. This is called **forward bias**. It reduces the potential of figure 9.11(d) across the junction from V_0 to $V_0 - V_a$ (figure 9.16). The density of electrons reaching the p side of the junction by diffusion rises from $N \exp(-eV_0/kT)$ to $N \exp[-e(V_0 - V_a)/kT]$, i.e. by a factor $\exp(eV_a/kT)$. A very few of these electrons actually make it to the terminal and flow round the external circuit. However, the vast majority neutralise in the p region with holes flowing in from the positive terminal of the battery; but the net result is still a current through the external circuit. Likewise, the density of holes reaching the n side from the p region rises by the same factor. Diffusion currents of both holes and electrons contribute and the total current across the junction is

$$I = I_0[\exp(eV_a/kT) - 1].$$

(9.15)

Apart from the 1 on the right-hand side, ensuring that $I = 0$ when $V_a = 0$, I rises exponentially with V_a.

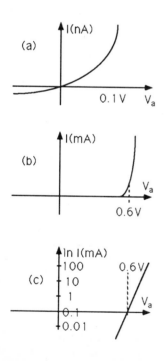

(a)

$I(nA)$

$0.1\,V$ V_a

(b)

$I(mA)$

V_a

$0.6\,V$

(c)

$\ln I(mA)$

100 $0.6\,V$
10
1
0.1
0.01

V_a

Fig. 9.17. Characteristics of a silicon pn diode (a) on a nA scale, (b) on a mA scale, (c) plotted logarithmically.

Diode characteristics

The resulting current is sketched in figure 9.17. It does actually pass continuously through zero at $V_a = 0$ if plotted on a nA scale, as in (a). However, on a mA scale, the current appears to rise extremely rapidly around $V_a = 0.6$ V for silicon. For germanium, I_0 is larger because the energy gap is smaller; the rise on the mA scale for forward bias occurs for germanium at $V_a \simeq 0.25$ V.

Under forward bias, the source of the current is diffusion of holes and electrons (majority carriers) across the junction. Essentially what happens is that the applied voltage reduces the potential barrier they have to climb, so they diffuse across the junction more freely. On the other hand, the drift current due to the electric field changes very little since there is negligible change in the static charge distribution, which dominates $\rho(x)$ of figure 9.11(c) and hence $V(x)$. The drift current is responsible for the 1 in equation (9.15).

Now consider reverse bias (figure 9.15(b)). The potential across the junction is increased (figure 9.18). Because this potential acts as a barrier, the diffusion currents drop. If the bias is sizable, the diffusion current drops essentially to zero and the reverse current is due to drift alone. On the p side of the junction there is a very small number of electrons, according to equation (9.9). They see an electric field of the correct sign to accelerate them across the junction. Likewise, holes from the n side of the junction are accelerated across it. The drift current is small because the concentration of these minority carriers is extremely tiny. In germanium, V_0 is less than for silicon, so the concentration of minority carriers (which depends exponentially on V_0) is greatly increased and so is the drift current, i.e. the current under reverse bias. The reverse current is typically nA in silicon and μA in germanium.

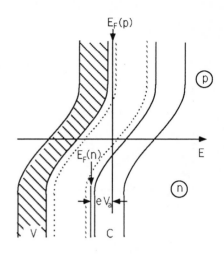

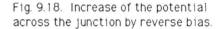

Fig. 9.18. Increase of the potential across the junction by reverse bias.

The PIN diode

This reverse current may be put to good use. Within the depletion layer, thermal processes spontaneously generate electron–hole pairs:

$$\text{atom} \rightarrow \text{ion} + \text{electron} \equiv \text{hole} + \text{electron}.$$

Both feel an electric field which sweeps them rapidly out of the depletion layer, so contributing to the reverse current. In the p–i–n, or PIN, diode, the depletion layer is deliberately made of undoped intrinsic semiconductor. Appendix C shows that the width of the depletion layer is proportional to $[(V_0 - V_a)/(N + P)]^{1/2}$, so a high reverse bias V_a and low (or no) doping makes a wide layer. If light with an energy greater than the energy gap E_G is shone on to this depletion layer (figure 9.19), photons convert to electron–hole pairs and the reverse current measures the light intensity. This makes a **photodiode**. Its output impedance is high because almost all charges are swept out of the depletion layer, regardless of applied voltage.

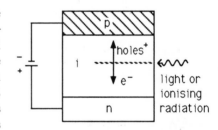

Fig. 9.19. Ionisation detector.

Solid state counters

Alternatively, if an ionising particle traverses the depletion layer, it generates electron–hole pairs by ionisation. The carriers swept out of the depletion layer by the electric field create a pulse. The size of the pulse measures the energy deposit. In Ge–Li and Si–Li semiconductor detectors, this is used to count particles and in particular to measure energies with high resolution (1–2%). The pulse height is very insensitive to the magnitude of the reverse bias, which simply acts to sweep pairs out of the production region. Nearly pure silicon is used to minimise recombination. For Ge–Li and Si–Li detectors, the depletion layer can be several mm or even cm wide (at a cost!) if the applied voltage is high enough (several kV).

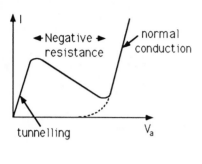

Fig. 9.20. Tunnel diode characteristic.

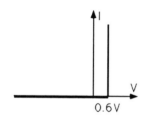

Fig. 9.21. Idealised diode characteristic (silicon).

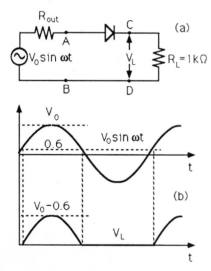

Fig. 9.22 (a) Half-wave rectifier circuit, (b) V_S and V_L v. t.

Zener diode

Two other processes affect the reverse current. If the mean free path of the carriers in the depletion layer is high, they may accelerate to sufficient energy to ionise atoms and generate an avalanche. Secondly, suppose the reverse bias is sufficient to bring the valence band of the p region level with the conduction band of the n region (figure 9.18). If the doping level is very high, the depletion layer is narrow enough (<100 Å) that electrons from the valence band on the p side can tunnel quantum mechanically across the junction to the conduction band on the n side, creating a current. Both tunnelling and the avalanche process are used in the Zener diode to produce a sudden increase in reverse current above a well-defined threshold voltage (see figure 1.9).

Tunnel diode

The tunnelling process is also used in the tunnel diode. This is doped so heavily that acceptor levels in the p region are level with the conduction band on the n side at *zero* bias. The characteristic is shown in figure 9.20. For small forward bias, the current is due to tunnelling. As the forward bias rises, the levels move out of alignment and the current drops. At higher voltages it rises again due to normal conduction. In between, the diode has a negative resistance which can be used for special purposes, for example to make an oscillator.

9.5   The Diode as a Switch

We now review briefly some of the applications of the diode. Many uses depend solely on the fact that I is large for positive V and very small for negative V, i.e. it has a small forward resistance and large backward resistance. In this approximation the diode acts simply as a switch, allowing current to flow one way, but not the other. The idealised characteristic is then as shown in figure 9.21. It requires 0.6 or 0.65 V to switch on a silicon diode and about 0.25 V to switch on a germanium diode.

Half-wave rectifier

An example of the use of a diode is in the half-wave rectifier shown in figure 9.22. Suppose the magnitude V_0 of the applied voltage $\gg 0.6$ V. When the diode is forward biased, the voltage across it is negligible, and almost all the voltage from the supply is applied to the load. On the other half-cycle, the diode is reverse biased, very little current flows, and all the supply voltage appears across the diode, and very little across the load.

A variant on this circuit is shown in figure 9.23(a). There, the diode in the first arm conducts when the supply voltage V_S is positive and greater than $V_1 + 0.6$ V; the diode in the second arm conducts when V_S is negative and less than $-(V_2 + 0.6)$ V. This clips the tops off V_S, shown dotted in figure 9.23(b).

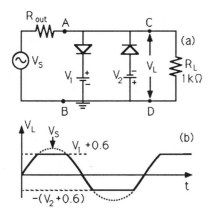

Fig. 9.23. Clipping with diodes.

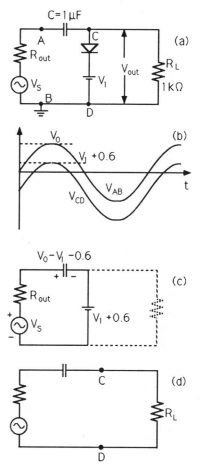

Fig. 9.24 (a) the diode clamp, (b) waveforms at AB and CD, (c) and (d) equivalent circuits when the diode does and does not conduct.

Diode clamp

Figure 9.24(a) shows the **diode clamp**. If R_L is large, the action of the circuit is to clamp the *top* of the output waveform V_{CD} at $V_1 + 0.6$ V (figure 9.24(b)). The way this comes about may be understood by drawing separate equivalent circuits for the part of the cycle when the diode conducts (figure 9.24(c)), and the part where it does not (figure 9.24(d)). In the former, the capacitor charges rapidly (with time constant CR_{out}) towards voltage $V_0 - V_1 - 0.6$ V. In the other half of the cycle, the diode does not conduct, and the capacitor discharges slowly through the load R_L. The result is that the capacitor charges up over a few cycles to $V_0 - V_1 - 0.6$ V and thereafter the diode does not conduct at all, except to replenish charge lost through R_L.

If the diode and V_1 are reversed, what happens to the output waveform?

It is instructive to try this circuit experimentally, observing the waveforms at A and C on two traces of an oscilloscope.

A **full-wave rectifier** is shown in figure 9.25. The direction of current flow may be followed by drawing separate paths through

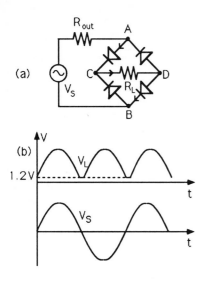

Fig. 9.25 (a) Full-wave rectifier,
(b) waveforms.

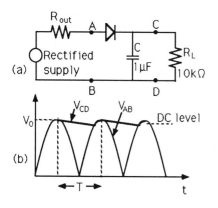

Fig. 9.26 (a) CR filter,
(b) its waveforms.

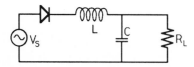

Fig. 9.27. Further filtering
with an inductor.

the diodes for the two cases $V_S > 0$ and $V_S < 0$. The arrows in figure 9.25(a) indicate the current flow when $V_A > V_B$, and the waveforms are shown in (b). The diodes used in ordinary electronics carry currents of up to 10 or 100 mA. However, large diodes capable of carrying tens of thousand of amps can be made and are used in solid state rectifiers for DC generators.

Power supplies

In a DC **power supply** it is necessary to smooth the ripple from figure 9.25(a) using filters. The crudest way of achieving this is shown in figure 9.26. A diode charges a large capacitor with time constant CR_{out}. When V_{AB} begins to drop from its maximum, the diode switches off and the capacitor begins to discharge through the load. If $CR_L \gg CR_{out}$, the discharge current is approximately $I = V_0/R_L$ and in the time interval T between one input cycle and the next, the output voltage drops by $V_0 T/CR_L$. This is a measure of the ripple, i.e. the fluctuation in the output voltage (though electrical engineers sometimes use instead the RMS ripple).

The ripple of a DC power supply may be smoothed further by using an inductor, as in figure 9.27. The inductor presents zero impedance to direct current but impedance $j\omega L$ to AC components of angular frequency ω.

The output of these circuits depends on the magnitude V_0 of the supply voltage. If a **regulated** voltage supply is required, where the voltage is independent of V_0, this may be achieved as in figure 9.28 using a Zener diode, which breaks down at an accurately defined voltage V_1. The difference between V_0 and V_1 is dropped across a large resistor R_1. Zener diodes may also be used to protect the input of a circuit against applied signals which are too large. For voltages below the threshold, no current flows through them and the circuit behaves as if they were not there; above the threshold they conduct and bypass the rest of the circuit.

9.6   The npn Bipolar Transistor

The physics of the bipolar transistor is closely related to that of the diode; this is why so much attention was given to the physics of the diode. Figure 9.29 shows schematically the layout of the npn bipolar transistor. It consists of a p-type **base** and n-type regions called **emitter** and **collector**. The gap between the emitter and collector is made very thin (~ 1 μm).

Typical characteristic curves are shown in figure 9.30. Here V_{CE} is the voltage between collector and emitter and V_{BE} the voltage from base to emitter. The device is controlled by the latter voltage. The base–emitter junction acts like a diode, so when the transistor is 'on' and current flows, the voltage at B is typically 0.6 V

above that at E. The characteristic diode curve appears in figure
9.30(b). However, little current flows to or from the base. Most
($\geq$ 99%) flows between the emitter and collector, which sucks up
all the electrons injected into the base from the emitter. The base
acts as a control terminal which governs the emitter current. In
fact, it will emerge that the ratio β of collector current to base current is approximately constant over the normal operating range; β
varies from one make of transistor to another and even between
transistors of the same make over a range typically 50 to 300. The
bipolar transistor can therefore be thought of as a current amplifier, or even better as a diode with current amplification (and
voltage amplification in the right circuit). Both sets of characteristic curves are insensitive to V_{CE}, providing it is positive and above
about 0.25 V.

How does the transistor work? The emitter is very heavily
doped, so that most of the current flowing across the junction with
the base is due to electrons diffusing from emitter (n-type) to base
(p-type) rather than in the reverse direction. If recombination in
the base region is ignored for the moment, the density of electrons
$n_B(x)$ in the base is given by the full line of figure 9.31 for forward
bias between base and emitter. On this scale, the electron density
in both emitter and collector regions is very high and out of sight
at the top of the diagram. The collector–base junction is strongly
reverse biased (for example, $V_C \simeq V_B + 10$ V). Consequently $n_B(x)$
falls to a very low value, essentially zero, at the edge of the depletion layer BC. All electrons reaching this point feel a strong electric
field accelerating them through this depletion layer into the collector. (The collector operates analogously to a vacuum cleaner for
electrons.) The diffusion current of electrons in the base is then

$$I = -|e|AD_n \frac{dn_B}{dx}$$

i.e. to the right on figure 9.31. The voltage between base and
emitter controls in a delicate way the density $n_B(x)$ of electrons
reaching the left-hand edge of the base region by diffusion through
the depletion layer EB; hence it controls the slope of the line on
figure 9.31 and the current reaching the collector.

The electric field across the base is negligible if the base is uniformly doped, because of its good conductivity. In this case the
drift current in the base region is negligible. However, it is common
to dope the base so as to create non-uniform doping, and hence
create an electric field accelerating electrons towards the collector; this reduces the transit time and improves the high-frequency
performance. We shall ignore this complication.

How must the base be doped to achieve this?

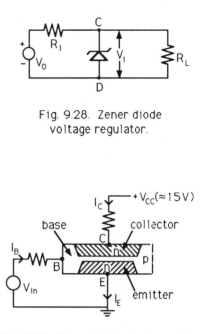

Fig. 9.28. Zener diode
voltage regulator.

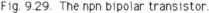

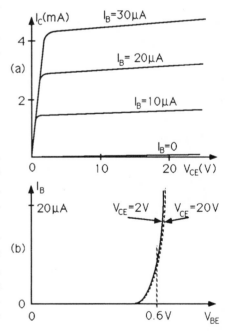

Fig. 9.29. The npn bipolar transistor.

(a)

(b)

Fig. 9.30. Characteristics of the
bipolar transistor.

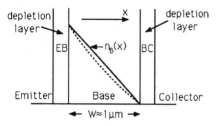

Fig. 9.31. The electron density in the base region.

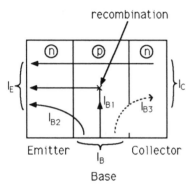

Fig. 9.32. Base currents in the npn transistor.

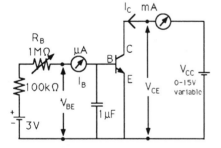

Fig. 9.33. Measuring transistor characteristics.

Currents

The voltage applied to the base terminal is normally much lower than that at the collector. So the concentration gradient in the base region is overwhelmingly from emitter to collector rather than from emitter to base terminal. Correspondingly, the electron diffusion current to the base terminal is very small. The drift current to the base terminal is also very small because the base is a reasonably good conductor and does not support any significant electric field. So why is there any current to the base at all? There are two reasons.

Firstly, electrons in the base region are minority carriers: $p_B(x)$ is much higher than $n_B(x)$. In consequence, some electrons collide with holes and annihilate. The actual electron density is therefore given by the slightly curved broken line of figure 9.31. This annihilation contributes a current between base and emitter; holes flow into the base from its terminal to replenish holes which have annihilated with electrons. This base current is in direct proportion to the current flowing to the collector. It is shown schematically on figure 9.32 as I_{B1}.

There is a second component I_{B2} to the base current due to diffusion of holes from the base to the emitter. Because the base is lightly doped, this hole current is much smaller than the electron current. The hole current and I_E are both proportional to $[\exp(eV_{BE}/kT) - 1]$, so again $I_{B2} \propto I_E$. The conclusion is that I_B is proportional to I_E but much smaller. The collector current is $I_C = I_E - I_B$, so the collector current is an amplified version of the base current. In practice this amplification is typically 50–300.

Characteristic curves

If you wish to measure the characteristics of a bipolar transistor yourself, this is readily done with the circuit of figure 9.33. The symbol for the npn transistor is indicated in the middle of the figure, where collector, base and emitter are labelled C, B and E. The arrow indicates the controlling diode action between base and emitter. In this circuit, the transistor is said to be in the **common emitter configuration**, because the emitter is common to the input V_{BE} and output V_{CE}. The currents may be measured with AVOs or multimeters and the voltages with an oscilloscope or multimeter. The behaviour is very sensitive to V_{BE} and the current into the base should not exceed 40 or 50 μA for most transistors, so the 100 kΩ resistor is included as a safety resistance. A change in R_B alters I_B and the meter reading I_C shows a correspondingly much larger change in I_C. This demonstrates immediately that the bipolar transistor acts as a current amplifier in this configuration. A practical point is that stray feedback from output to input can make the circuit oscillate at high currents; this is eliminated by the 1 μF capacitor which acts as a low-impedance path for high-frequency signals.

Alternatively the curves may be displayed on a **curve tracer**. This is a device where the applied voltages are programmed to sweep rapidly through the required range and display the results on an oscilloscope. Any transistor can be plugged into it and the required range of variables is selected by varying control knobs.

In figure 9.30(a), collector current I_C is proportional to base current I_B to a good approximation, for the reasons just discussed, providing V_{CE} (the voltage between collector and emitter) is above about 0.25 V. Below this voltage, the collector–base junction is becomingly significantly forward biased; remember that $V_{BE} \simeq 0.6$ V. Under such conditions, a significant current I_{B3} begins to flow from collector to base (see figure 9.32); this contributes positively to I_B and negatively to I_C, so β falls. Incidentally, in germanium transistors (now obsolete), even with V_{CB} positive the reverse bias current I_{B3} makes a significant contribution to I_B. Further details of these currents are discussed in appendix C.

Figure 9.30(b) is just the usual diode curve relating I_B with V_{BE}. It shows a tiny dependence on V_{CE}, so small that it is often obscured by temperature effects. However, the curves of I_C against V_{CE} (figure 9.30(a)) show a definite upward slope against V_{CE}. This is easy to understand. At large values of V_{CE} the depletion layer BC expands and the slope of $n_B(x)$ increases slightly, as shown in figure 9.34. Accordingly the diffusion current I_C across the base increases.

In summary, the current delivered by the emitter is controlled very delicately by the voltage V_{BE} between base and emitter. However, because I_C/I_B is approximately constant, it is more convenient to think of this type of transistor as a current amplifier. A small change in I_B goes with a very small change in V_{BE} and a large change in I_C. The power in the collector circuit $V_{CE}I_C$ is very much larger than $V_{BE}I_B$; typically they are in the ratio 1000 to 3000. The essential virtue of the transistor is this power amplification: a small power in the base circuit controls a large power in the collector circuit.

As an aside, you might wonder whether the collector and emitter are interchangeable. In fact, they are to a limited extent. If you mistakenly invert their connections, the collector–base junction now acts as a diode and the transistor will function, but in an inferior way; this is because the collector is lightly doped compared with both base and emitter.

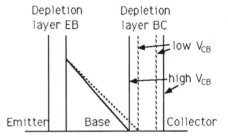

Fig. 9.34. Expansion of the depletion layer between base and collector for large V_{CE}.

9.7 Simple Transistor Circuits

Emitter follower
More quantitative details of the operation of the transistor will be

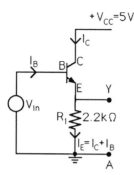

Fig. 9.35. A simple
current amplifier.

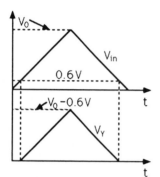

Fig. 9.36. Response to a
triangular input voltage.

deferred to a later chapter. For the moment the essential point is that quite small changes of voltage between base and emitter result in large changes in current through the collector. Figure 9.35 shows a circuit called the emitter follower; it acts as a current amplifier, but gives no voltage amplification.

If V_{in} is zero or negative, the diode is off and no current flows through the transistor, so $V_Y = 0$. If V_{in} is raised above 0.6 V the transistor switches on. Because the base–emitter diode conducts, $V_{BE} \simeq 0.6$ V and $V_Y \simeq V_{in} - 0.6$ V. This voltage appears across R_1 so

$$I_E = I_C + I_B = V_Y/R_1 = (V_{in} - 0.6)/R_1.$$

Figure 9.36 shows the resulting waveforms for a triangular input; V_Y follows V_{in} with a difference of 0.6 V. This is called **emitter follower** action. In this circuit there is no voltage amplification, but there is *current* amplification because $I_C = \beta I_B$, and hence there is also power amplification.

If you wanted the emitter follower to transmit both positive and negative halves of an AC input signal $2\sin\omega t$ V, what change to the circuit would achieve this?

OR gate

A circuit very similar to figure 9.35 is used in digital logic. It is shown in figure 9.37. Two transistors share a common load resistor R_1. Inputs A and B are applied to the bases of the two transistors. If an input signal > 0.6 V is present on either input, the point Y will follow it with a drop of 0.6 V between base and emitter. If inputs are present at both A and B, the point Y follows the larger input. This makes a rudimentary OR gate: there is an output present if there is a positive input above 0.6 V at either A *or* B. Waveforms are shown in figure 9.37(b), where it is assumed arbitrarily that input B is larger than at A.

In practice, it is necessary to refine the circuit to eliminate the 0.6 V drop between input and output, otherwise a series of circuits like this would gradually attenuate the signal voltage. This is achieved by working with cleverly arranged voltage levels. However, apart from this detail, the commercial ECL gate looks very similar to figure 9.37.

AND gate

A second digital circuit is shown in figure 9.38. An input signal greater than 0.7 V has to be present at both A *and* B for current to flow through both transistors and create an output at Y. If the inputs are 5 V, the output is 4.4 V. This circuit is called an AND gate or coincidence circuit. If only one input is present (or none),

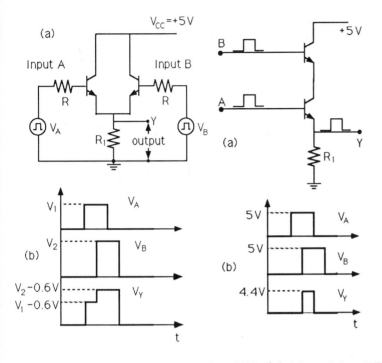

Fig. 9.37. (a) A rudimentary OR gate, (b) waveforms.

Fig. 9.38. (a) A transistor AND gate or coincidence circuit, (b) waveforms.

no current flows and the output is zero. Again, in the commercial version, refinements are necessary to bring the output up to 5 V.

9.8 Voltage Amplification

It is desirable to achieve voltage amplification as well as current amplification. The circuit of figure 9.39 achieves both. The output is now taken from the collector instead of the emitter. Waveforms are shown in figure 9.40. The resistor R_B is there just to protect the diode junction BE against excessive voltage; the base voltage V_B swings up to a maximum of 0.6 V. If V_{in} is positive, $I_B = (V_{in} - 0.6)/R_B$ and $I_C = \beta I_B$. Then $V_X = V_{CC} - I_C R_C$. As I_C rises V_X falls, and it may fall as low as 0.1 V if I_B is large enough. The transistor is then **saturated**.

This time, both current and voltage are amplified. Because of the voltage amplification, the switching action is more definite than that of the emitter follower and for this reason the circuit is often preferred in digital logic. With suitable choice of R_B and R_C, input signals switch the output decisively between the supply voltage $V_{CC} = 5$ V and 0.1 V. The output is suitable for application direct to subsequent circuits; the fact that the lower level is 0.1 V instead of strictly zero does not matter because it requires 0.6 V

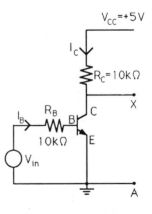

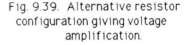

Fig. 9.39. Alternative resistor configuration giving voltage amplification.

to switch a later circuit. The difference between 0.6 V and 0.1 V is the **noise margin** for this type of circuit. The higher noise margin is one reason for preferring silicon-based devices over germanium-based ones.

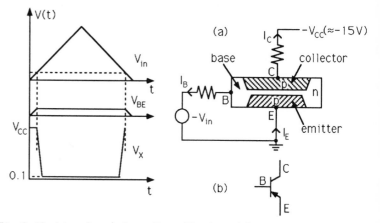

Fig. 9.40. Waveforms from the circuit of Fig. 9.39 for a triangular input.

Fig. 9.41 (a) the pnp transistor and (b) its circuit symbol.

There is one small penalty with this circuit. The output is inverted with respect to the input: V_X falls as V_B rises. Chapter 12 will explore how to cope with this logical inversion in digital circuits.

The pnp transistor has the reverse doping to npn. The base is n type and emitter and collector are p type (figure 9.41). Its operation is precisely the same as that of the npn transistor except that all voltages and currents are reversed in sign and except that the mobility of holes carrying the current is somewhat less than that of electrons in the npn transistor. Any circuit which is designed for npn transistors can equally well be implemented with pnp transistors simply by reversing all polarities.

9.9   Biasing

For digital logic the circuits of figures 9.35–9.39 are fine. They act as simple switches. However, suppose we want instead to amplify faithfully an AC input voltage $V = \sin \omega t$ V. In either of the circuits discussed so far the transistor comes on only when $V \geq 0.6$ V. The remedy is to superimpose the input signal on a DC level of, say, 2 V. Then the input swings between 3 and 1 V and the transistor stays on through the whole cycle. Figure 9.42 shows successive improvements in the way this may be achieved. The most primitive is with the circuit of figure 9.42(a). The battery V_0 drives current through R_2 followed by the parallel combination of R_1 and the

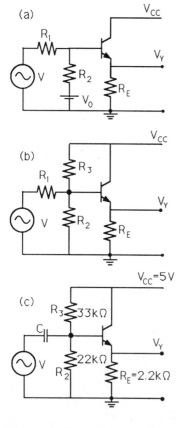

Fig. 9.42. Three ways of biasing the transistor for AC operation.

transistor, and by suitable choice of resistors the DC level at the base can be held at 2 V. The AC source drives current through R_1 and the parallel combination of R_2 and the transistor. The superposition theorem gives the general picture but is not accurate because the transistor is non-linear; later chapters will investigate how to do the algebra properly.

In practice it is inconvenient to use separate DC voltages for V_{CC} and V_0. A possible alternative is shown in figure 9.42(b). The power supply drives a current through R_3 followed by the parallel combination of R_1, R_2 and the transistor and sets up a DC level at the base. However, the circuit has a disadvantage. The AC source V drives R_1 and the parallel combination of R_2, R_3 and the transistor. There is a loss of voltage across R_1. This resistor cannot be set to zero without shorting the DC level at the base to earth.

In figure 9.42(c), R_1 is replaced by a capacitor. If this is chosen to be large, its impedance $1/j\omega C$ can be made negligible at the operating frequency so that the full AC signal is applied to the base of the transistor. It is conventional to choose R_2 and R_3 so that 90 % of the DC current supplied by V_{CC} flows through R_2 and 10% into the base of the transistor. Then R_1 and R_2 act approximately as a potential divider, and to achieve a 2 V level at the base requires $R_3/R_2 \simeq 3/2$. Suitable component values are shown in figure 9.42(c). This arithmetic will be done more precisely in Chapter 18.

You are recommended strongly to experiment with the circuits shown in figures 9.35, 9.37–9.39 and 9.42 to get a qualitative feeling for transistor operation before plunging into quantitative detail in later chapters.

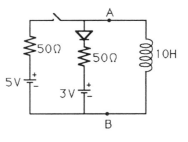

Fig. 9.43.

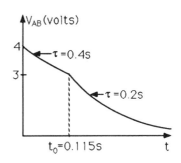

Fig. 9.44.

9.10   Exercises

1. (QMC 1969) In the circuit of figure 9.43, the switch is closed at time $t = 0$. By considering the Thevenin equivalent of the part of the circuit to the left of AB as a function of the voltage across AB, sketch the voltage across the inductor as a function of time. You may assume that the diode has zero forward and infinite reverse resistance, and that it switches at zero applied voltage. (Ans: For $V_{AB} > 3$ V, $V_{EQ} = 4$ V, $R_{EQ} = 25$ Ω; for $V_{AB} < 3$ V, $V_{EQ} = 5$ V, $R_{EQ} = 50$ Ω. See figure 9.44 for the waveform.)

2. (QMC 1973) In the circuit of figure 9.45, D_1 and D_2 are two identical diodes; V_A and V_B are inputs derived from square-wave generators of low output impedance and having quiescent outputs at earth. At time $t = 0$, square waves $V_A = +4$ V lasting for 1 ms and $V_B = +4$ V lasting for 2 ms

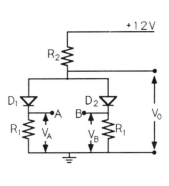

Fig. 9.45.

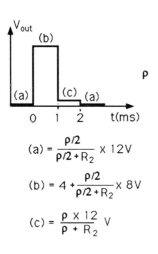

$$(a) = \frac{\rho/2}{\rho/2 + R_2} \times 12V$$

$$(b) = 4 + \frac{\rho/2}{\rho/2 + R_2} \times 8V$$

$$(c) = \frac{\rho \times 12}{\rho + R_2} \, V$$

Fig. 9.46.

are applied. Describe quantitatively the form of the output V_0, assuming that the diodes have a forward resistance ρ above zero applied voltage and infinite reverse resistance. (Ans: figure 9.46.)

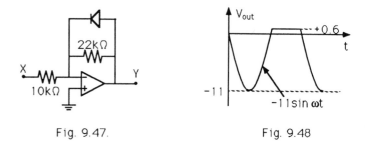

Fig. 9.47.   Fig. 9.48

3. (QMC 1985) In the circuit shown in figure 9.47, the operational amplifier has a large amplification G and high input impedance. An AC voltage $5\sin\omega t$ V is applied at point X. By considering the positive and negative parts of the cycle, find the output voltage at point Y. (Ans: figure 9.48.)

4**. An AC voltage $V_{in} = V_0 \sin\omega t$ is applied to the network shown in figure 9.49. Describe qualitatively the waveforms which appear at points A and B a long time after the input is switched on. (Ans: figure 9.50.)

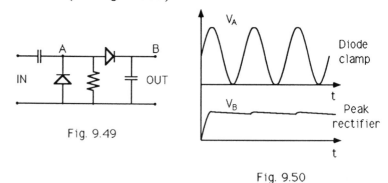

Fig. 9.49

Fig. 9.50

5. Find the peak current through a half-wave rectifier, figure 9.22(a), if the forward resistance of the diode is R_f. Take the applied voltage to be 250 V RMS, $R_{out} = 200\ \Omega$ and the load resistor 1 kΩ. What is the DC output current? (Ans: 0.295 A, 0.094 A.)

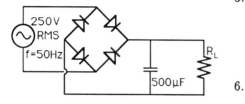

Fig. 9.51.

6. (QMC 1968) What is the minimum value of the load resistor R_L in figure 9.51 if the magnitude of the ripple is not to exceed 5% of the peak voltage. What will be the DC current

through the load when R_L has this minimum value? (Ans: 800 Ω, 0.44 A.)

7**. (QMC 1971) In the circuit of figure 9.52, an AC voltage $V_{in} = -V_0 \sin \omega t$ is applied, starting at time $t = 0$. Assuming that C_1 and C_2 are initially uncharged, describe qualitatively the waveforms which will appear at various points of the circuit. Deduce the form of the output voltage if (a) $C_1 \gg C_2$, (b) $C_1 \ll C_2$. (Ans: Initially C_1 charges through D_1 to voltage V_0, with A at zero. As V_{in} passes its most negative value, D_1 turns off, and A starts to rise, driving current through C_2. The charge on C_1 is shared with C_2, and the voltage at B rises to $2V_0 C_1/(C_1 + C_2)$. (a) $V_B \simeq 2V_0$ (voltage doubler), (b) B charges by an amount $2V_0 C_1/(C_1 + C_2)$ every cycle, producing the staircase potential shown in figure 9.53.)

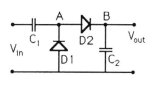

Fig. 9.52.

8. A positive voltage V_{in} is applied to the circuit of figure 9.39. Find (a) the current I_B through the base, (b) the voltage at X, assuming $I_C/I_B = 100$, and (c) the amplification for AC signals, $A = dV_X/dV_{in}$. What happens if V_{in} is negative? (Ans: $I_B = (V_{in} - 0.6)/R_B$, $V_X = 15 - 100R_C(V_{in} - 0.6)/R_B$, $A = -100R_C/R_B = -100$; $I_B \to 0$, $V_X \to 15$ V, $A \to 0$.)

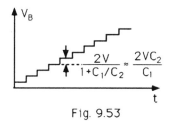

Fig. 9.53

9*. (RHBNC 1988) In the circuit of figure 9.54 (a totem-pole output circuit), estimate (i) the voltage at X, and (ii) the current through the point X when the switch A is open and when it is closed. State when the light emitting diode LED is on. (Ans: With A open, $T1$ is cut off and $T2$ on, LED on, $I_X \simeq (5 - 1.8)/1.1$ mA $= 2.9$ mA, $V_X \simeq 3.8$ V; with A closed, $T1$ is on, $T2$ off because $V_A \simeq 0.6$ V, $V_X \simeq 0$ V, LED off.)

10. Suppose that in the OR gate of figure 9.37, the supply voltage V_{CC} is 5 V, $V_A = 5$ V, $V_B = 0$, $R = 10$ kΩ and $R_1 = 100$ Ω. In addition there is a resistor $R_2 = 1$ kΩ between the power supply and the collectors of the transistors. Assuming $V_{BE} = 0.6$ V and $V_{CE} = 0.1$ V when the transistor is on, find the collector current I_C, the current I_A into the base of transistor A and the voltages at emitter and collector. How do these change when (a) $R_1 = 0$, (b) $R_2 = 0$? Why is the arithmetic unrealistic in the latter case? (Ans: 4.42 mA, 390 μA, 0.48 V, 0.58 V; (a) 4.9 mA, 440 μA, 0 V, 0.1 V; (b) 49.1 mA, -50 μA but the transistor does not saturate, so $V_{CE} \neq 0.1$ V.)

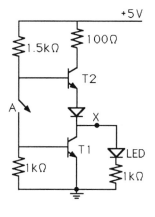

Fig. 9.54.

The Field-effect Transistor (FET)

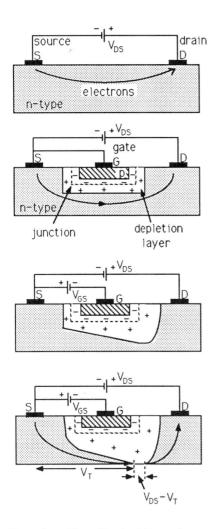

Fig. 10.1. The effect of the gate on electron flow in the n-channel, (a) no gate, (b) $V_{GS}=0$, (c) V_{GS} small and negative, (d) large negative V_{GS}.

10.1 Gate Action

We now come to the other important class of transistor; its mode of operation is quite different to that of the bipolar transistor. First a rough physical picture of its operation will be constructed and then actual characteristic curves will be discussed. First let us consider the **junction FET** or **JFET**.

For a start, consider a slice of lightly doped n-type material and suppose a voltage is applied between connections at the two ends (figure 10.1(a)). The material conducts and follows Ohm's law. The terminal at which electrons enter the material is called the **source** and the other terminal the **drain**.

Next, consider the effect of inserting a heavily doped p-type region (called the **gate**) between source and drain (figure 10.1(b)). Suppose initially that it is electrically connected to the source. Carriers diffuse across the pn junction and a depletion layer develops, where free charges from the two materials have neutralised. The static charges are positive in the n region and negative in the p region. The depletion layer extends further into the lightly doped n channel than into the heavily doped p region. The channel width in the n-type material is squeezed and the current from source to drain is reduced.

Thirdly, consider the effect of controlling the potential of the gate as in figure 10.1(c) by applying a voltage V_{GS} between it and the source. A negative value of V_{GS} repels mobile electrons into the n region and the depletion layer expands. This voltage controls the width of the channel and the current I_{DS} between drain and source. It is the basis for controlling a large current by means of a small gate voltage, i.e. an amplifier.

Pinch-off

Characteristic curves are shown in figure 10.2. The essential feature shown on the curves at the right is that the current I_D to the drain saturates for large V_{DS} at a value controlled in a sensitive way by V_{GS}. The curve at the left shows the saturation current against V_{GS}.

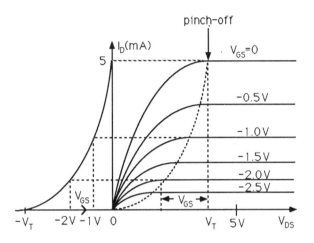

Fig. 10.2. Characteristic curves for the junction FET. The curve to the left shows the maximum (plateau) value of I_D v. V_{GS}.

It is shown in appendix C that the lateral extent of the depletion layer is proportional to $V^{1/2}$, where V is the total barrier height across the junction: natural + applied bias. For large enough negative gate voltage $(-V_T)$, the current is cut off. Next suppose the gate voltage is above this, allowing current to flow. How will the current vary with V_{DS}, the voltage between drain and source? Consider first $V_{GS} = 0$. For small V_{DS}, Ohm's law is obeyed and I_D rises linearly with V_{DS}. Along the length of the conduction channel, there is a voltage increase from source to drain, because of the ohmic resistance of the channel. So the voltage difference between gate and channel increases from a small value near the source to a larger value near the drain. Consequently, the depletion layer is wider at the end nearer the drain (figure 10.1(c)). As the depletion layer expands with increasing V_{DS}, I_D begins to turns over. Eventually, when V_{DS} reaches the critical value V_T, the depletion layer fills the channel except for a narrow region carrying the current. This is called **pinch-off**. For values of V_{DS} above this, no further current increase occurs; the length of the narrow channel increases to compensate the increase in V_{DS}. The voltage drop $(V_{DS} - V_T)$ occurs over a short distance (figure 10.1(d)).

Next consider what happens as V_{GS} is changed to, say, -0.5 V. This gate voltage reduces the channel width everywhere, so the current falls. Pinch-off is reached when the voltage in the channel is just sufficient to close it, i.e. when $V_{DS} = V_T + V_{GS}$; because V_{GS} is negative, this is a slightly lower value of V_{DS} than before.

For larger negative V_{GS}, the channel pinches off completely when $V_{GS} = -V_T$ and beyond this gate voltage negligible current flows. The value $-V_T$ is called the threshold voltage.

The left-hand curve in figure 10.2, showing the saturation current against V_{GS}, is approximately parabolic:

$$I_{DS}^{sat} \simeq K(V_{GS} + V_T)^2.$$

This curve is useful because the transistor is used mostly in the saturation region. The broken curve in figure 10.2 showing the value of V_{DS} for pinch-off has the same parabolic shape because, as explained in the previous paragraph, pinch-off occurs when $V_{DS} = V_T + V_{GS}$.

The full name of the device described here is the **n-channel depletion junction field-effect transistor**. The title 'field-effect transistor' (FET) for all the devices considered in this chapter arises from the fact that it is the electric field of the gate which controls the width of the depletion layer.

In the FET, the current in the n region is carried by **majority carriers** (electrons). This is in contrast to the bipolar transistor, where in the controlling base region of the npn transistor electrons are minority carriers.

In figure 10.1, the arrangement of source and drain is symmetric about the gate, and the source and drain are interchangeable. Some FETs are actually manufactured like this. However, in high-speed devices, the gate is positioned close to the source, so as to reduce the capacitance between gate and drain, and hence to reduce the Miller effect (Chapter 8).

The circuit symbol for the n-channel JFET is shown in figure 10.3(a). The arrow indicates a diode junction between p-type gate and n-type channel. A p-channel device is made using n-type gate and with reversed polarities for V_{DS} and V_{GS}. Its circuit symbol is shown in figure 10.3(b). Its properties are very similar to the n-channel device, except for the reversal of polarities.

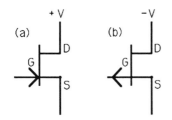

Fig. 10.3. Circuit symbols for (a) n-channel, (b) p-channel JFETs.

10.2   Simple FET Amplifiers

Source follower

With an appropriate network of voltage supplies and bias resistors, the JFET can be used to make amplifiers in a very similar way to the bipolar transistor. Figure 10.4 shows a **source follower**, the analogue of the emitter follower. In the absence of an input voltage, the gate G is at earth. The transistor conducts and the current through resistor R_S biases the source S to a potential above

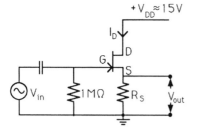

Fig. 10.4. The source follower.

earth. The voltage drop V_{GS} between gate and source is therefore negative; equilibrium is established when V_{GS} creates a current I_D such that $V_{GS} = -I_D R_S$. This situation is illustrated graphically in figure 10.5. The voltage across R_S is given as a function of I_D by the straight line, which is known as the **load line**. The **operating point**, marked by the cross, is given by the condition that $V_{GS} + V_S = 0$.

If an input signal V_{in} is now applied, $V_{GS} + V_S = V_{in}$. The operating point moves up or down the load line. The changes ΔV_{GS} and ΔV_S may be followed graphically. However, it is easier to do so algebraically using

$$\frac{dI_D}{dV_{GS}} = g_m \qquad (10.1)$$

which is called the mutual conductance. Then

$$V_{in} = \Delta V_{GS} + R_S \Delta I_D = \Delta V_{GS}(1 + R_S g_m)$$

$$\Delta V_S = R_S \Delta I_D = R_S g_m \Delta V_{GS} = \frac{R_S g_m}{1 + R_S g_m} V_{in}. \qquad (10.2)$$

Typical values of g_m are $(2\text{--}3) \times 10^{-3}$ mho (or siemen in SI units). If $R_S \gg 1/g_m$, i.e. $\gg 500\ \Omega$, V_S follows V_{in} closely.

Common source amplifier
The circuit just discussed provides no voltage amplification. If this is required, it is obtained as in figure 10.6(a) by putting a resistor R_D between supply and drain. The voltage change at the drain is given by

$$\Delta V_D = -R_D \Delta I_D = -\frac{R_D g_m}{1 + R_S g_m} V_{in}. \qquad (10.3)$$

The minus sign arises from the fact that as V_{in} goes positive I_D increases and V_D falls. It would improve the amplification if $R_S = 0$. This is not possible without upsetting the biasing of the transistor. However what can be done is to put a large capacitor across R_S, as shown by the broken symbol in figure 10.6. The term R_S is replaced by the impedance of R_S and C in parallel and equation (10.3) becomes

$$\Delta V_D = -\frac{R_D g_m V_{in}}{1 + R_S g_m/(1 + j\omega C_S R_S)}.$$

For AC signals with $\omega \gg 1/C_S R_S$ the denominator is now 1 for AC inputs. The large capacitor C_S is essentially holding the source S at a fixed DC voltage. The voltage amplification is then $-R_D g_m$.

In this simple analysis, the dependence of I_D on V_{DS} has been neglected, assuming the amplifier operates in the region where the

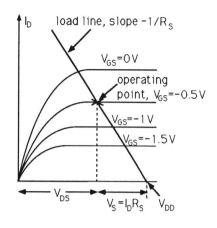

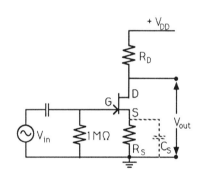

Fig. 10.5. The load line and determination of the operating point.

Fig. 10.6. A JFET voltage amplifier.

characteristic curves are flat. The effect of dI_D/dV_{DS} will be included in a fuller analysis in Chapter 17. For the moment it is more important to draw attention to the fact that the characteristic curves are not equally spaced, so $g_m = dI_D/dV_{GS}$ is not really constant. Thus the amplification will depend on the magnitude of V_{in} and the choice of operating point. It will be larger for positive inputs than negative, because of the non-uniform spacing of the characteristic curves. This does not matter in digital circuits, where the output voltage is high or low, and precise values of the amplification are not vital. To make a linear amplifier, negative feedback is required, as explained in Chapters 7 and 18.

Differences from the bipolar transistor

The JFET differs from the bipolar transistor in two important respects. Firstly, the input resistance is very high because the gate–source junction is reverse biased; the bipolar transistor however normally operates forward biased and has a fairly low input resistance (a few kΩ). The JFET has an input resistance of the order of 1–1000 MΩ, but other versions of the FET we shall meet soon have an even higher input resistance, which can be as much as 10^{14} Ω. Here lies a problem. It is easy for charge to build up on the gate because of a large time constant CR. This charge can generate a voltage large enough to damage the gate–source junction. The purpose of the 1 MΩ resistor in figures 10.4 and 10.6 is to prevent the gate voltage from floating because of the high input resistance.

Care is needed not to expose FETs to situations where charge build-up may cause damage, for example by rubbing them against clothing. Even electrostatic charge on the body can cause damage, so earthed wrist-straps are advisable when handling FETs. The FETs are generally delivered in wrapping with reasonable conductance, to prevent charge building up on the gate. When they are built into integrated circuits, Zener diodes are usually inserted across the inputs as a protection against large voltages of either polarity.

Secondly, the FET has much lower noise than the bipolar transistor. The latter has shot noise associated with the input base current. This is amplified by a factor $\beta \simeq 100$ in the collector circuit. In the FET, this source of noise is absent, though there is Johnson (voltage) noise across the 1 MΩ resistor; in order to reduce noise, this resistor may be reduced. The FET is therefore usually the choice for the first stage of a low-noise preamplifier.

10.3   MOSFETs

There are several variants on the physical construction of the FET, for example that shown in figure 10.7(a). The source and drain

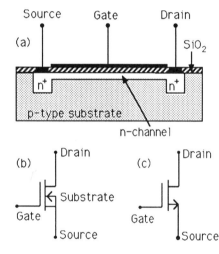

Fig. 10.7 (a) the depletion MOSFET, (b) and (c) its circuit symbols.

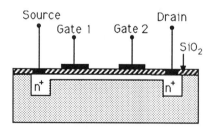

Fig. 10.8. MOSFET with two gates.

regions are heavily doped n-type regions (n^+ denotes heavy doping.) They are joined by a thin channel of lightly doped n region, but the gate is not connected directly to it. Instead, the gate is a metal layer separated by a thin ($\sim 10^3$ Å) insulating layer of SiO_2. If the gate is negatively biased, it creates an electric field which attracts holes from the p-type substrate into the n-channel region and which repels electrons. This forms a depletion layer in the channel, controlling its conduction properties in a way directly similar to the junction FET. Such a device is called a **depletion MOSFET**, standing for metal oxide semiconductor FET. The word 'depletion' signifies that the gate operates by depleting the n channel. Because the gate is insulated from the n channel, the input resistance is very high, typically 10^9 Ω and sometimes 10^{12}–10^{14} Ω. An alternative name is IGFET, denoting **insulated gate** FET.

Two common circuit symbols for this device are shown in figures 10.7(b) and (c). The separation of the gate symbol from the transistor clearly indicates the physical separation of the metal gate from the n-type channel. The arrow in figure 10.7(c) indicates the direction of current flow from drain to source. However, the official symbol shown in (b) does not indicate the current direction. Instead, the arrow indicates that the substrate is p type, making a pn junction between substrate (p) and channel (n). In practice, the p-type substrate must be physically connected to the source (or a more negative point), otherwise there will be a forward-biased diode from substrate to n region. This physical connection is indicated in figure 10.7(b), but the direction in which the arrow points is a possible source of confusion to the beginner.

It is straightforward to fabricate such depletion MOSFETs with more than one control gate at different points along the length of the n channel, as in figure 10.8. This may be useful in two ways. Firstly, the gates may carry signals which independently influence the logic of the current flow. Secondly, the gate near the drain may be held at a fixed potential with respect to the source; lines of electric field from the drain mostly end on this gate. This reduces the capacitance between the first gate and the drain, and hence reduces the Miller effect. Likewise, transistors with multiple source or drain or emitter or collector are employed frequently in integrated circuits.

Figure 10.9 shows a version of the MOSFET with *no* channel between drain and source. If, however, the gate is *positively* biased, it attracts electrons to the surface region and induces an n-type channel just beneath the surface. Its characteristics are shown in figure 10.9(b) and its circuit symbol in (c). It is called an **enhancement** MOSFET, since the positive voltage applied to the gate enhances current flow.

A continuous range of possibilities between enhancement and depletion exists. Figure 10.10 shows three sets of characteristics: (a) the depletion MOSFET where the gate is negatively biased for

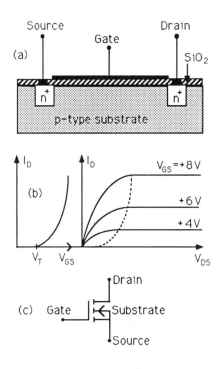

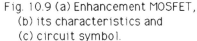

Fig. 10.9 (a) Enhancement MOSFET, (b) its characteristics and (c) circuit symbol.

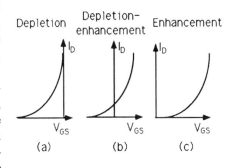

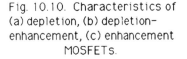

Fig. 10.10. Characteristics of (a) depletion, (b) depletion-enhancement, (c) enhancement MOSFETs.

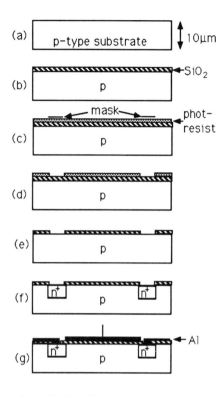

Fig. 10.11. Steps in fabricating an enhancement MOSFET.

operation, (b) the depletion–enhancement MOSFET, where the operating point is with the gate at or close to earth (in the former case making biasing unnecessary), and (c) the enhancement MOSFET.

10.4   Fabrication of Transistors and Integrated Circuits

The way in which an n-channel enhancement MOSFET may be made is indicated by the steps in figure 10.11. The process starts (a) from a lightly doped p-type substrate, typically 10 μm thick. For reasons we shall come to later, this is in turn created on an n-type substrate typically 100 μm thick.

At (b), the surface is oxidised to SiO$_2$ to provide an insulating barrier. At (c), the SiO$_2$ layer is covered with a light-sensitive material known as photo-resist. This is covered by a mask at points where the SiO$_2$ layer is to be etched away later. The photo-resist polymerises when illuminated, making it chemically resistant. The remaining area under the mask is dissolved at step (d) and at (e) the SiO$_2$ is etched away chemically, exposing the source and drain regions of figure 10.9(a). The n-type regions are now formed by diffusing in donors, (f). Similar steps are then used to deposit the aluminium gate and connections at the source and drain, (g). The bipolar transistors of figure 10.12 may be made by similar processes.

Integrated circuits

The most difficult part of the process is attaching leads to the outside world. Soldering is notoriously unreliable, particularly on tiny components. It is an obvious extrapolation of the technique to make whole **integrated circuits** in the silicon rather than individual transistors. Small resistors (<1 kΩ) may simply be channels with suitable doping. Larger resistors (up to 100 kΩ) are made from MOSFET transistors with fixed gate biases which govern the resistance. Small capacitors may be created like the gate of a MOSFET, using a surface deposit of metal as one plate and a channel region as the other. Small capacitors (<50 pF) may also be made from a pn junction with reverse bias; as shown in appendix C, the capacitance depends on the magnitude of the reverse bias, so the capacitance may be varied by adjusting the bias. However, large capacitors are expensive in surface area and are avoided in circuit design. For this reason, most circuits are designed to be DC coupled. Inductors are impossible and are substituted by the gyrator of exercise 7.12. The simplest transistor to fabricate in an integrated circuit is the enhancement MOSFET of figures 10.9 and 10.11 and this is therefore the one in commonest use. NMOS (n-channel) is used more commonly than PMOS (p-channel).

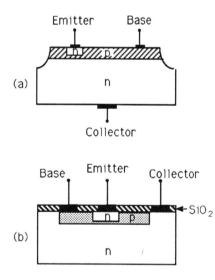

Fig. 10.12 (a) mesa and (b) planar type bipolar transistors.

A complication is that the p-type substrate must always be reverse biased, otherwise it will form a forward-biased pn junction with n regions. In a circuit where all MOSFETs are of enhancement type and all have their source at the same voltage, this condition is satisfied naturally. Otherwise, it is necessary to isolate individual transistors from one another. This is achieved by diffusing an n-type barrier through the p region as in figure 10.13, making an 'island' for the transistor. The whole circuit is then made on top of an n-type substrate typically 100 μm thick. Because enhancement MOSFETS require no isolation via islands, their packing density is a factor of twenty or so higher than for bipolar transistors.

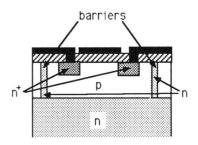

Fig. 10.13. Isolation of a transistor using n-type barriers.

In this way, whole integrated circuits are constructed. These become very cheap when mass produced, despite the capital cost of the production process. The limitations are (a) the yield of successful circuits in a batch process, and (b) problems of predicting precise circuit behaviour. Progressively larger circuits with smaller components have been developed in recent years. Large-scale integration (LSI) uses 1000–10 000 components per chip and very-large-scale integration (VLSI) uses more than 10 000. Computer memories are examples of VLSI circuits, where the emphasis is on massive repetition of a single simple unit. The physical dimensions of components are only a few tens of microns or less and the wavelength of light used in the masking procedure has become a real limitation. Present developments to even smaller dimensions replace optical maks by **electron beam lithography**. An example of an integrated circuit made by this process is shown in figure 10.14.

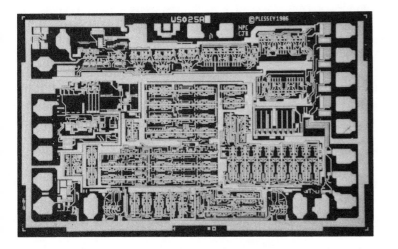

Fig. 10.14. An integrated circuit using electron beam lithography, reproduced with permission from the Rutherford-Appleton Laboratory.

10.5   SPICE and CAD

The design of any sizable circuit nowadays calls upon computer-aided design (CAD). Several programs are available to simulate circuit behaviour. Amongst these are SPICE, developed at Berkeley, and ECAP, developed by IBM. These are available in many versions depending on the size of the host computer (PC or mainframe). The available software is developing extremely rapidly.

SPICE predicts circuit behaviour starting from a circuit diagram and characteristics of active devices. It is organised on the basis of circuit nodes. Between any two nodes users can insert batteries, resistors, capacitors, inductors, diodes or current sources; between three nodes a transistor may be inserted. Diodes and transistors are treated on the basis of models outlined in appendix C, but with some further refinements to simulate their behaviour as precisely as possible. The computer uses equations reproducing the full non-linear characteristics (including temperature dependence). It also includes estimates of internal capacitances and their variation with voltage.

The program will calculate the DC operating points of transistors and voltages at all nodes, and hence currents can be calculated. It will calculate the response to an AC signal (i.e. the Bode plot) and response to pulses (risetime, falltime and propagation delay). Referring forward to Chapter 15, it will also evaluate Fourier series, i.e. the harmonic content of an AC signal.

The simplest versions of the program require the user to prepare a file of input parameters and a sequence of commands asking for specific output information. More elaborate versions allow the user to replace the input file with a circuit drawing and to produce graphical outputs and to work interactively. Some versions will call up the design of circuit blocks such as operational amplifiers or similar subsystems.

For integrated circuit design, it is necessary to simulate separately the doping distributions and junction locations in transistors. SUPREM, developed at Stanford, will model this process in one dimension (depth), starting from the diffusion cycle. Two-dimensional simulations are available but much slower. From these doping distributions, yet other programs will simulate fully the behaviour of one device, e.g. the precise shape of the depletion layer in a JFET or MOSFET. Such information can be used to provide model parameters for the transistors in SPICE.

When it comes to the final manufacturing stage, interactive programs are used to lay out the artwork of a circuit and calculate from it the required masking sequence.

The moment of truth comes when the circuit is ready for test. This can be a complicated procedure if the circuit fulfils many different logical functions or works over a wide range of parameters. Again the computer is used to run through the full range of test conditions. However, if circuit performance differs from expectation, it can be a tortuous process tracking down where the computer simulation used in the design is at fault. It is possible to make microprobes to check the behaviour of the circuit at accessible points. However, this is not to be undertaken lightly and the probe is anyway likely to modify circuit behaviour. For this reason, an important feature of design is that it should be tolerant of faults.

You will now understand why the absolute gain of an operational amplifier, for example, can vary by a factor of two. However, within a circuit it is possible to guarantee that a given pair of resistors or transistors close together in the silicon have the same characteristics to tight tolerances ($\frac{1}{2}\%$). This is a valuable feature of integrated circuit design and leads to good control over sensitivity to temperature or supply voltages.

You might ponder whether semiconductor man is likely to have evolved naturally somewhere within the Universe; perhaps we are now witnessing this.

10.6   CMOS

An example of a circuit in which resistors have been eliminated is shown in figure 10.15. The load resistor is replaced by a MOS-FET with its gate connected to the supply voltage. This transistor is permanently on, but by suitable choice of channel length and thickness it can be made to have any required resistance. A positive signal applied to the input switches on the lower transistor; if this is made with a lower resistance than the upper one, V_{out} falls to a low voltage.

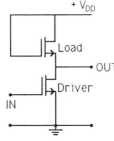

Fig. 10.15. A MOSFET inverter.

An important development is the use of complementary MOS-FETs in the same circuit. This puts to advantage the fact that n- and p-channel devices can easily be laid down on the same substrate. Figure 10.16 shows an example where the upper transistor is p channel and the lower one n channel. With the input zero, the upper transistor switches on and $V_{\text{out}} \simeq +V_{\text{DD}}$. With input $+V_{\text{DD}}$, the lower transistor switches on and the upper one off, so $V_{\text{out}} \simeq 0$.

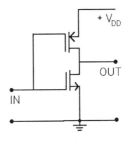

Fig. 10.16. A CMOS inverter.

This circuit has the virtue that one or other transistor is off in both cases, so the DC current through them is zero, except during switching. This minimises power consumption, typically 10 nW per component. All battery-operated calculators take advantage of CMOS circuits.

11

Gates

11.1 Number Systems

In a digital computer, numbers are expressed in **binary** form, of which examples are given in table 11.1. For instance, the decimal number 13 (written 13_{10}, where the subscript denotes the base) is represented in binary by

$$13_{10} = 1 \times 2^3 + 1 \times 2^2 + 0 \times 2^1 + 1 \times 2^0 \equiv 1101_2.$$

Each binary digit or **bit** can be held by an electronic element which is on or off.

Although computers work internally with binary representations, alternative more compact ways of *writing* numbers are octal (base 8) and hexadecimal (base 16) systems illustrated in table 11.1. These are simple contractions of binary. Computer printouts often list the contents in these forms. As examples,

$$13_{10} = 1 \times 8^1 + 5 \times 8^0 \equiv 15_8$$
$$99_{10} = 6 \times (16)^1 + 3 \times (16)^0 \equiv 63_{16} \equiv \$63.$$

The hexadecimal system uses A to F for decimal 10 to 15. In practice, hexadecimal is used much more commonly than octal.

BCD

We are so accustomed to decimal that many pieces of hardware use a mixed binary and decimal system called **binary coded decimal** (BCD). Scalers are often designed to count from 0 to 9 in binary, then carry to a new unit which counts the next decade 10 to 90, and so on. Each decimal digit 0–9 is individually coded by four binary digits from 0000 to 1001_2. Decimal 13 is represented in BCD by 0001 0011. Examples of BCD representations of numbers are given in table 11.1. This system is uneconomical in terms of the number of bits used, but it has the advantage of being fairly easy to read and to convert to decimal. The electronic logic required at the

Table 11.1   Representation of positive integers.

Decimal	Binary	Octal	Hexadecimal	BCD	Gray code
0	000 0000	0	0	0000	0000
1	1	1	1	0001	0001
2	10	2	2	0010	0011
3	11	3	3	0011	0010
4	100	4	4	0100	0110
5	101	5	5	0101	0111
6	110	6	6	0110	0101
7	111	7	7	0111	0100
8	1000	10	8	1000	1100
9	1001	11	9	1001	1101
10	1010	12	A 0001	0000	1111
11	1011	13	B 0001	0001	1110
12	1100	14	C 0001	0010	1010
13	1101	15	D 0001	0011	1011
14	1110	16	E 0001	0100	1001
15	1111	17	F 0001	0101	1000
16	1 0000	20	10 0001	0110	
19	1 0011	23	13 0001	1001	
20	1 0100	24	14 0010	0000	
28	1 1100	34	1C 0010	1000	
35	10 0011	43	23 0011	0101	
99	110 0011	143	63 1001	1001	

transition 9 to 10 is not complicated. BCD is also used internally in digital clocks, electronic calculators and digital voltmeters.

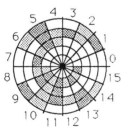

Fig. 11.1. Layout of a mechanical encoder for Gray code.

Gray code

The Gray code is a binary scheme in which only one bit changes when 1 is added to the number. The bit which changes is the least significant digit which will generate a new number. This system (also called **reflected code**) is often used in mechanical or optical encoders (see figure 11.1). Because only one bit changes at a time, small misalignments between the encoding elements of successive digits generate errors in the reading never more than one digit, whereas in binary many digits could be wrong (e.g. in going from binary 0111 to 1000).

11.2   Negative Numbers

A way in which a negative number might be represented in binary is to assign one bit, the leftmost one, as a sign digit. Then -1

would be represented by $1000\ldots001$. This is not used in practice because addition and subtraction are rather awkward with this representation. Instead, almost all computers use the **2's complement** representation. In this system, -1 is obtained by subtracting 1 from $0000\ldots000$ with the result $1111\ldots111$; then -2 is represented by $1111\ldots110$ and so on (see table 11.2).

Table 11.2   Representation of negative numbers in 2's complement and 1's complement form.

Decimal	2's complement	1's complement
0	0000 0000	0000 0000
-1	1111 1111	1111 1110
-2	1111 1110	1111 1101
-3	1111 1101	1111 1100
-4	1111 1100	1111 1011

Such numbers add according to precisely the same rules as positive numbers, except that the carry from the left-hand end is discarded. For example, in decimal $(-1)+(-2)=-3$; in binary, the arithmetic is as given in table 11.3 and the carry from the left-hand end is ignored.

Table 11.3   Addition of negative numbers in binary.

$$1111\ldots111$$
$$+\,1111\ldots110$$
$$=1111\ldots101$$

Suppose, for example, that the computer deals with numbers limited to eight bits. The largest positive number it could handle would be decimal 255, but this has precisely the same representation as decimal -1. It is conventional to regard the leftmost bit as a sign digit and limit the positive numbers to the decimal range 0 to 127 (binary 0111 1111) and negative numbers to the range -1 to -128 (binary 1111 1111 to 1000 0000). Then adding decimal (-1) to (-2) is the same as adding decimal 255 to 254 and disregarding the binary component 256; the result is $509-256=253$, which represents (-3). As a second example, in decimal $(-1)+(2)=+1$; in binary, this arithmetic is shown in table 11.4.

1's complement
There is a second system called **1's complement**, which is rarely used today. In this system, a negative number is formed from a

Table 11.4   Binary addition.

$$
\begin{array}{r}
1111\ldots 111 \\
+\ 0000\ldots 010 \\
\hline
=\ 0000\ldots 001.
\end{array}
$$

positive one by changing all 0's to 1's and vice versa. In this form

$$
\begin{aligned}
\text{decimal } +1 &= 0000\ldots 001 \\
\text{decimal } -1 &= 1111\ldots 110 \\
\text{decimal } +2 &= 0000\ldots 010 \\
\text{decimal } -2 &= 1111\ldots 101
\end{aligned}
$$

and so on. The snag with this system is that there are two representations for zero, namely $0000\ldots 000$ and $-0 = 1111\ldots 111$. The reason for mentioning it is that an easy way of forming the 2's complement of a number is to take the 1's complement (logical inversion of each bit) and add 1 to the right-hand end; for example,

$$
\begin{aligned}
\text{decimal } 3 &= 0000\ldots 0011 \\
\text{decimal } -3 &= 1111\ldots 1100 + 0000\ldots 0001 \\
&= 1111\ldots 1101.
\end{aligned}
$$

Most computers use this trick to do subtraction. To calculate $(A - B)$ they form the 1's complement of B, then add the result to A and carry in a 1 from the right-hand end.

11.3   Logical States

It is arbitrary whether a binary 1 is represented by the more positive or more negative voltage (or by a current which is on or off). The choice of the more positive (high) level for a 1 is called *positive* logic and the reverse choice of the more negative (low) level is called *negative* logic.

We shall use for illustration the TTL voltage levels $+5$ V and 0 V and positive logic. Then logical 1 is represented by $+5$ V and logical 0 by 0 V (figure 11.2(a)). If any binary digit is denoted by A, its complement, written $\overline{A}$ (or sometimes A') is its converse; if $A = 0$, then $\overline{A} = 1$ and is represented here by $+5$ V, and if $A = 1$, $\overline{A} = 0$ and is represented by 0 V. A circuit which converts A into $\overline{A}$ is said to execute the NOT operation: $\overline{A} = \text{NOT}(A)$. Electronically, it is described as **inversion** of A; this does not mean that the 5 V level representing 1 is inverted to -5 V, but that it is logically inverted to 0 V. The circuit symbol is shown in figure 11.2(b); the bubble on the output denotes the complement (i.e. logical inversion).

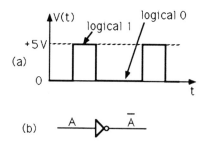

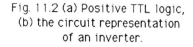

Fig. 11.2 (a) Positive TTL logic, (b) the circuit representation of an inverter.

11.4 Logical Combinations of *A* and *B*

AND

Suppose two binary digits *A* and *B* are to be multiplied together. Either can be 0 or 1, so there are four possible combinations giving the results shown in the table of figure 11.3(a). This is called a **truth table**. The operation of binary multiplication is called AND; it is written *A.B* (or often just *AB*). The result is 1 only if *A* and *B* are both 1. Electrically, it may be realised with the circuit of figure 11.3(b): current flows only if both switches *A* and *B* are closed. Electronically, the switches are replaced by transistors which may be switched in a few ns (10^{-9} s). Commercial units will be discussed shortly. Readers may follow this chapter and the next two without needing to know how transistors function, but for those who are interested figure 9.38 shows a rudimentary circuit performing the AND function and exercise 5 of this chapter discusses it.

NAND

A unit which performs the AND function is represented in circuit diagrams by the symbol shown in figure 11.3(a). Its complement, shown in figure 11.3(c), is called NAND (short for NOT AND). Its circuit symbol has a bubble on the gate output to indicate the complement. Electronically, the switching of NAND circuits is more decisive than AND (see Chapter 9), so commercial units are based upon NAND circuitry. If AND is required, it is formed either within the unit or externally by NAND followed by inversion to get back to AND.

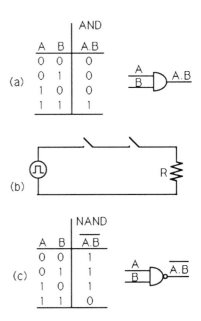

Fig. 11.3 (a) Truth table and circuit symbol for AND, (b) forming AND with switches, (c) truth table and circuit symbol for NAND.

OR and NOR

A second basic logical operation is called OR. Its truth table and circuit symbol are given in figure 11.4(a). It is denoted algebraically by $A + B$ and the result is 1 if either *A* or *B* (or both) is 1. The + symbol should not be confused with arithmetic addition; the way to do addition in terms of OR and AND will be discussed very shortly. Electrically, the OR operation may be realised with switches by the circuit of figure 11.4(b); current flows through the load R if *either* switch *A or B* is closed. Electronically, the switches are replaced by transistors. A circuit doing this is given in figure 9.37, but again it is not essential to understand the transistor operation in order to use commercial units. The complement of OR is called NOR and is written $\overline{A + B}$. Its truth table and circuit symbol are given in figure 11.4(a).

Circuits performing any of these logical operations are called **gates**. The origin of this name can be understood from figure 11.5(a). There, a long signal *A* 'opens a gate' to allow the

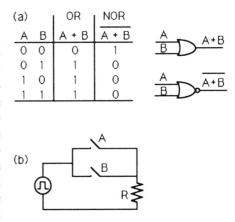

Fig. 11.4 (a) Truth tables and circuit symbols for OR and NOR, (b) the electrical principle of the OR gate.

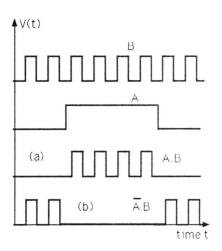

Fig. 11.5 (a) A.B and (b) $\overline{A}$.B.

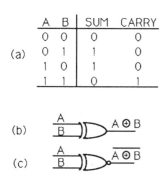

Fig. 11.6 (a) Truth table for a half adder, (b) and (c) circuit symbols for XOR and XNOR.

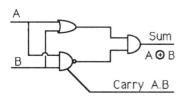

Fig. 11.7. The half adder.

pulse train from B through to the output by forming $A.B$. If $\overline{A}$ is substituted for A to make $\overline{A}.B$, as in figure 11.5(b), the pulse train is halted when $A = 1$, $\overline{A} = 0$. Names given to this operation are **veto**, **inhibit** and **busy**; these names should be self-explanatory.

Half adder

From the operations AND, OR and NOT, every operation required in digital logic can be constructed. Indeed, technically they can all be done with NAND gates only or with NOR gates only, though the contortions required to achieve this may be inconvenient. As an example of the use of gates, figure 11.6(a) gives the truth table for the arithmetic addition of two binary digits A and B. This is called a **half adder**. (A full adder has the further feature of adding in a carry from a lower digit, and will be discussed in the next chapter.) The last line of figure 11.6(a) expresses the fact that decimal 1 plus decimal 1 = decimal 2 ≡ binary 10. The carry is simply $A.B$ and may be formed with an AND gate. The sum, called **exclusive or** (XOR) and written $A \oplus B$, may be expressed as $A.\overline{B} + B.\overline{A}$; it is 1 if $A = 1$ and $B = 0$ or if $B = 1$ and $A = 0$, and is 0 otherwise. This is actually only one of several logical expressions for the exclusive OR. A different one is shown in figure 11.7; corresponding waveforms of A and B running through all combinations are shown in figure 11.8.

The complement of $A \oplus B$ is called XNOR or **equivalence**. It is equal to 1 if A and B are the same, otherwise it is 0: $\overline{A \oplus B} = A.B + \overline{A}.\overline{B}$.

Fan-in and fan-out

So far, the discussion has been limited to gates with just two inputs. Gates are available commercially with more inputs; layouts of several chips are shown in figure D.1 of appendix D. The number of inputs provided is called the fan-in of the circuit. The number of outputs which can be taken in parallel is called the fan-out. A four-input OR gate forms $A + B + C + D$ and gives an output 1 if any of A, B, C or D is 1. A four-input AND gate forms $A.B.C.D$ and gives an output 1 only if A, B, C and D are all 1. A four-input OR gate can be used with just two inputs to form $A + B$. A four-input AND gate will form $A.B$ only if inputs C and D are connected to logical 1 levels.

11.5   Asynchronous and Synchronous Logic; Hazards and Glitches

Care is needed over timing. Figure 11.9 shows a pair of signals A and B forming $A.B$. Suppose, however, that B is a little late;

the delayed signal will be denoted by B_{del}. The result is shown in figure 11.9(d). The first $A.B_{\text{del}}$ signal is correspondingly late, but the second one is narrower than it should be. If this happens once, the result may not be serious, but if it happens several times in sequence, the output might be whittled away to nothing.

At time t_5, the result is worse. The intention in lines (a) and (b) is that A rises at the same instant as B goes down. This is dangerous practice, and the result of B coming late is a 'glitch' in (d).

Suppose next $(A.B).\overline{B}$ is formed. This is a hypothetical situation, since logically B and $\overline{B}$ are opposite and the result should always be zero. However, suppose $\overline{B}$ travels a different route through the logic to B and arrives late for some reason. Waveform (f) shows narrow signals known as **static hazards**. They arise from the faulty timing.

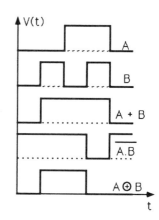
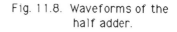

Fig. 11.8. Waveforms of the half adder.

Synchronous logic

The logic described so far is called **asynchronous**, meaning that there is no clock pulse to define times. An alternative is to use a narrow **clock** or **strobe** pulse C dedicated to timing, as in figure 11.10. The output $A.B.C$ is used to regenerate a standard output pulse timed correctly by the clock, despite the faulty timing of B. Such a system is called **synchronous**. It eliminates glitches and hazards, providing timing faults are not too large.

11.6   TTL, CMOS and ECL Gates

There are three standard commercial logic systems: TTL, CMOS and ECL. Their voltage standards are given in table 11.5. These systems have evolved gradually over the years, improving in power consumption and speed. Characteristics of modern versions are summarised in table 11.6. The propagation delay is the time between input and output. For further details, consult manufacturers' catalogues.

Table 11.5   Conventions used by common digital systems.

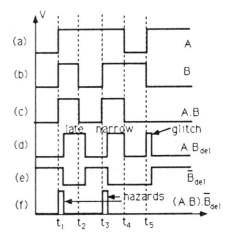
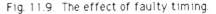

Fig. 11.9. The effect of faulty timing.

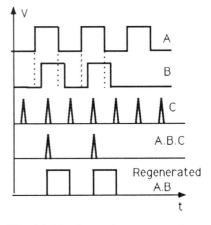

Fig. 11.10. A synchronous system.

System	Supply voltage (V)	Upper level voltage (V)	Lower level voltage (V)
TTL	5 ± 0.25	2.5 to 5	0 to 0.4
ECL	-5.2	-0.95 to -0.7	-1.9 to -1.6
CMOS	$V_0 = 3$ to 18	V_0	0 to 0.5

Table 11.6   Characteristics of standard digital systems.

System	Power dissipation (mW)	Propagation delay (ns)	Input current (mA)	Output current (mA)
TTL ALS	1.3	10	0.1	8
TTL AS	8	2	0.5	20
TTL F	5	3.5	0.6	20
ECL	25–40	1–2	0.1–0.5	40
CMOS HC	2.5 nW quiescent 0.5 at 1 MHz	9	0	8
CMOS AC	2.5 nW quiescent 0.5 at 1 MHz	3	0	24

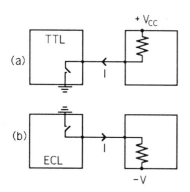

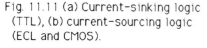

Fig. 11.11 (a) Current-sinking logic (TTL), (b) current-sourcing logic (ECL and CMOS).

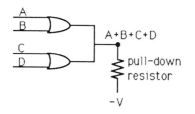

Fig. 11.12. Wired OR with ECL.

ECL

The ECL system is fastest, but its power consumption is relatively high, so it is used only where speed is a critical consideration. At the other extreme, CMOS uses only a tiny amount of power when quiescent. It is used in battery-driven calculators and memories. However, when it switches it draws current and at high speeds uses nearly as much power as TTL. It used to be the slowest of the three systems, but recent versions are catching up. TTL is intermediate in speed and power consumption, and is used widely. CMOS has the virtue that it can operate with any power supply from 3 to 18 V. However, it is becoming common practice to standardise on $V_0 = +5$ V to make CMOS and TTL compatible.

TTL

TTL logic is said to be **current-sinking**, because the output transistor draws or *sinks* current from the following circuit (figure 11.11(a)). This is in contrast to ECL and CMOS, which are **current-sourcing** (figure 11.11(b)); current flows *from* this type of gate.

A feature of the ECL system is that outputs of different gates may be simply connected together at a subsequent gate or at a load connected to $-V$, as in figure 11.12. The latter is called a **pull-down resistor** and the configuration is called a **wired OR**.

Standard TTL gates of table 11.6 may not be connected to a common load, because of conflicts between one output transistor trying to hold the output low while a second tries to drive it high. This may result in damage to an output transistor. However, there is a version called **open collector** where gates may drive a common load, forming a **wired AND**, as in figure 11.13. The resistor is called a **pull-up resistor**.

Buses

Suppose a large number of gates is to be connected to a single line or **highway**, e.g. in a memory. In this circumstance, the open collector gate suffers two disadvantages. Firstly it has a large time constant CR created by the capacitance of the highway and the load resistor. Secondly, trouble-shooting is difficult: a short to earth anywhere upsets the logic and can be very hard to find. To meet this problem, a third variety of TTL is available, called a **three-state gate**. Figure 11.14 shows the symbol for it and a typical arrangement where many gates are connected to a highway or **bus**. The three-state gate is designed so that in its third state it is simply disabled; it then presents a large impedance to the highway. It is activated by a signal addressing it; then both high and low output states have low output impedance and drive the highway with small time constant. However, it is necessary to ensure that only one gate at a time is enabled on to the highway, otherwise currents from one output may damage another.

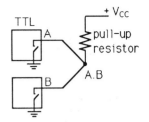

Fig. 11.13. Wired AND with TTL open collector units.

CMOS

CMOS draws very little input current; however, fan-out is limited to ~ 50 in practice because of deterioration of switching speed, due to the input capacitance of ~ 50 pF. The upper switching level is $\sim 70\%$ of the supply voltage and the lower switching level $\sim 30\%$. These provide a large margin against noise. CMOS gates cannot be joined in wired ORs or wired ANDs because of potential conflicts between output transistors.

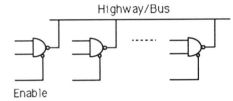

Fig. 11.14. Three-state gates connected to a highway or bus.

A few practical points about using TTL and CMOS gates, particularly about connecting them to other systems, are as follows. Firstly, inputs to TTL must not exceed 5.5 V; inputs to CMOS may only go beyond the range 0 to V_0 (the supply voltage) if the current is below 10 mA, the limit handled by the protection diodes at the inputs. Secondly, in driving TTL the source must supply adequate current; where one gate needs to drive a very large number of other gates, buffer drivers are used having large current capability. A third point concerns unused inputs to gates. If left floating, TTL inputs sit at 1.3 V and are susceptible to pick-up from neighbouring circuitry. It is essential to connect them either to a low or a high level. Completely unused TTL gates on a chip can be left floating. However, with CMOS on a large scale this is not good practice. The input of an unused CMOS gate floats to half the supply voltage and then draws current continuously; this is avoided if one input of the gate is held high or low.

A final general comment is that, with the development of large systems today, it is becoming increasingly important to design into the logic efficient procedures for diagnosing faults.

11.7   Exercises

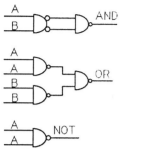

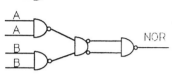

Fig. 11.15.

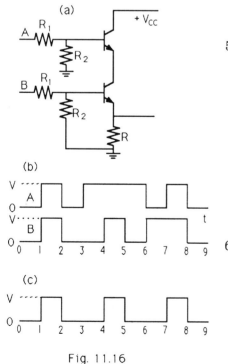

(a)

(b)

(c)

Fig. 11.16

1. Convert decimal 23 and 39 to binary, octal and hexadecimal and add them together in these systems. (Ans: $10111 + 100111 = 111110$; $27 + 47 = 76$; $17 + 27 = 3E$.)

2. (QMC 1972) Discuss the number systems and numerical codes which are used in digital computers. Illustrate your answer by writing the representation of 56.1_{10} in binary and BCD. (Ans: $111000.000110011\ldots, 01010110.0001$.)

3. (QMC 1986) State how the 2's complement of a binary number can be obtained directly from the binary digits. Using 2's complement, perform the following subtractions and interpret your results as signed decimal values: (a) 11011–10110, (b) 01101–11001. (Ans: complement and add 1 to the right-hand end; (a) 5, (b) −12.)

4. (QMC 1985) Explain, with the aid of truth tables, the meaning of the logical operations AND, OR, NOT, NOR, NAND. Give the logic circuit symbols for each and sketch the two-input transistor–resistor circuits that would implement these operations. Verify using truth tables that a NAND gate with the inputs connected together is a NOT gate and that a NAND gate with the inputs inverted is an OR gate. Hence give logic circuit schematic diagrams achieving AND, OR, NOT, NOR functions using two-input NAND gates only. (Ans: figure 11.15.)

5. (QMC 1987) The circuit of figure 11.16 can be used to perform a logical operation on digital pulses applied to inputs A and B. A signal of $+V$ volts is used for the logical '0' state and zero volts for the logical '1' state. The resistors R_2 have values several times larger than R_1. Give the name of the logical operation and its truth table. Describe how the circuit works. How would you represent the logical action of this circuit using switches? Draw the output that would result from input pulses shown in (b). If now the pulses are redefined so that $+V$ becomes the '1' state and zero volts the '0' state, what is the new function of the circuit? Draw a truth table for the inputs and output. (Ans: OR, figure 11.16(c), AND.)

6**. (QMC 1988) For the logic diagram shown in figure 11.17, derive the function f in terms of a minimal sum of x and y. What is this logic function? With an inverter used to derive $\bar{y}$ from an input y, carry out a waveform analysis of the circuit assuming each gate has a propagation delay of δ. Form the timing diagram assuming the inputs x and y cycle through the truth table for f, with x as the most significant bit. Show the links between the gate input pulse edges and the corresponding delayed pulse edges and comment on possible existence of glitches and absence of hazards, where these might have occurred, because of inputs changing in opposite directions. (Ans: $x.\bar{y} + \bar{x}.y$, XOR.)

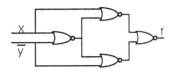

Fig. 11.17.

12

Combinational Logic

12.1 Boolean Algebra

This chapter and the one following will be concerned with the prin
ciples of how to design logical systems. Many of the examples a
in fact available commercially as integrated circuits (ICs), usual
in versions superior to the simple systems discussed here. In pra
tice, one would buy them where possible, very cheaply, rather th
making them. Examples are arithmetic units and counters. How
ever, it is necessary to understand the principles of their design,
as to be able to cope with modifications and with more difficu
non-standard problems.

So far, we have managed well enough with truth tables a
simple logical operations OR and AND. However, when it com
to the design of a complicated system, algebraic and graphic
methods are needed to help sort out the logic. Just such an algeb
was developed by George Boole, who in 1854 published a treati
entitled 'An Investigation of the Laws of Thought'. His algeb
is actually more general than is needed for binary systems, but
readily simplified to present needs.

Suppose A, B, C, etc, are symbols (or 'literals') denoting bina
digits, i.e. each of them is either 0 or 1. Rules of Boolean algeb
are given in figure 12.1. From truth tables, they are all obviou
In using them, the order in which Boolean expressions must
evaluated is as follows: parentheses first, then NOT, AND, a
finally OR.

Two statements are identical if they have the same truth tab!
The second of relations (ix) differs from the familiar result f
decimal numbers, but is verified in figure 12.2(a).

Try relations (vi) for yourself along the same lines.

12.2 De Morgan's Theorems

Relations (vii) give two famous theorems due to de Morgan; the

(i) $\left.\begin{array}{l} A + B = B + A \\ A.B = B.A \end{array}\right\}$ commutivity: the order is irrelevant

(ii) $\begin{array}{l} A + 0 = A \\ A.1 = A \end{array}$

(iii) $\begin{array}{l} A + 1 = 1 \\ A.0 = 0 \end{array}$

(iv) $\begin{array}{l} A + \overline{A} = 1 \\ A.\overline{A} = 0 \end{array}$

(v) $\overline{\overline{A}} = A$

(vi) $\left.\begin{array}{l} A + A.B = A \\ A.(A + B) = A \end{array}\right\}$ absorption – very useful

(vii) $\left.\begin{array}{l} \overline{A + B} = \overline{A}.\overline{B} \\ \overline{A.B} = \overline{A} + \overline{B} \end{array}\right\}$ de Morgan's theorems

(viii) $\left.\begin{array}{l} A + (B + C) = (A + B) + C \\ A.(B.C) = (A.B).C \end{array}\right\}$ order of operations irrelevant

(ix) $\left.\begin{array}{l} A.(B+C) = A.B + A.C \\ A + (B.C) = (A + B).(A + C) \end{array}\right\}$ distribution: how to evaluate expressions

Fig. 12.1. Rules of Boolean algebra.

(a)

A	0	0	0	0	1	1	1	1
B	0	0	1	1	0	0	1	1
C	0	1	0	1	0	1	0	1
B.C	0	0	0	1	0	0	0	1
A+(B.C)	0	0	0	1	1	1	1	1
A+B	0	0	1	1	1	1	1	1
A+C	0	1	0	1	1	1	1	1
(A+B).(A+C)	0	0	0	1	1	1	1	1

(b)

	A	0	0	1	1
	B	0	1	0	1
	A+B	0	1	1	1
(a)	$\overline{A+B}$	1	0	0	0
	$\overline{A}$	1	1	0	0
	$\overline{B}$	1	0	1	0
(a)	$\overline{A}.\overline{B}$	1	0	0	0
	A.B	0	0	0	1
(b)	$\overline{A.B}$	1	1	1	0
(b)	$\overline{A}+\overline{B}$	1	1	1	0

Fig. 12.2 (a) Verification that A+(B.C) = (A+B).(A+C), (b) proof of de Morgan's theorems.

are demonstrated in the truth tables of figure 12.2(b). They are much easier to remember in words. They imply that:

(a)   NOR ≡ AND between complementary signals, and
(b)   NAND ≡ OR between complementary signals.

Later a simple graphical way of expressing them will be given. These relations demonstrate a general property of all of relations (i) to (ix), known as **duality**. Suppose AND is changed to OR and vice versa in one relation of a pair, and suppose 0 is simultaneously changed to 1; the second relation of the pair is then obtained.

It is always possible to express any logical expression in terms of NOT, AND and OR, often in several different ways. Usually we proceed by constructing a truth table, then simplifying the result using algebraic or, more often, graphical methods. Some examples will indicate the process.

12.3   The Full Adder

Let us consider how to add two binary numbers together. For any one digit of the result, it is necessary to add two input binary digits A and B and a carry C coming from the previous digit. Results are given in figure 12.3 for all possible values of A, B and C. Algebraically, the ones in this table may be denoted by

$$\text{Sum} = \overline{A}.\overline{B}.C + \overline{A}.B.\overline{C} + A.\overline{B}.\overline{C} + A.B.C \qquad (12.1)$$

A	0	0	0	0	1	1	1	1
B	0	0	1	1	0	0	1	1
C	0	1	0	1	0	1	0	1
Sum	0	1	1	0	1	0	0	1
Carry	0	0	0	1	0	1	1	1

Fig. 12.3. The truth table of a full adder.

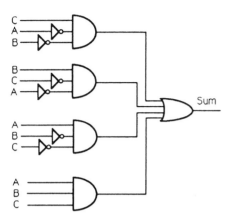

Fig. 12.4. A layout for forming Sum

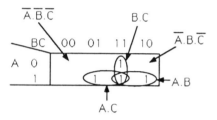

Fig. 12.5. Karnaugh map for the carry signal of a full adder.

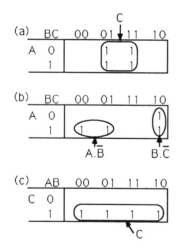

Fig. 12.6. Simplifications of Karnaugh maps.

$$\text{Carry} = \overline{A}.B.C + A.\overline{B}.C + A.B.\overline{C} + A.B.C. \qquad (12.2)$$

A circuit forming these logical expressions with OR and AND gates gives an output 1 wherever there is a 1 in figure 12.3; otherwise the outputs are zero. The logic therefore reproduces the truth table. These expressions are said to be in AND–OR form, since they are formed as ANDs followed by ORs. Figure 12.4 gives a circuit for the Sum signal.

An obvious question, however, is whether these expressions can be simplified. In the case of equation (12.2) this is so. It may be rewritten:

$$\text{Carry} = A.B.(C + \overline{C}) + B.C.(A + \overline{A}) + C.A.(B + \overline{B}). \qquad (12.2a)$$

The fact that some terms appear twice in (12.2a) and once in (12.2) is of no consequence: $A.B.C$ is the same as $B.C.A$ and $C.A.B$ (the order is irrelevant) and $A.B.C + A.B.C = A.B.C$ from Boolean relation (vi). Then equation (12.2a) gives

$$\text{Carry} = A.B + B.C + C.A. \qquad (12.3)$$

You may wish to verify this relation using a truth table.

12.4   The Karnaugh Map

An important graphical method of spotting such simplifications is the Karnaugh map, shown in figure 12.5. This diagram displays the eight possible values (reading top left to bottom right):

$$\overline{A}.\overline{B}.\overline{C}, \overline{A}.\overline{B}.C, \overline{A}.B.C, \overline{A}.B.\overline{C}, A.\overline{B}.\overline{C}, A.\overline{B}.C, A.B.C, A.B.\overline{C}.$$

Entries are arranged according to the Gray code so that only one bit changes between 0 and 1 from one entry to the next. In this example, entries are marked into the map corresponding to terms in the logical expression for Carry, equation (12.2). If two entries appear on neighbouring sites, they can be simplified by relations $A + \overline{A} = 1$ or $B + \overline{B} = 1$ or $C + \overline{C} = 1$. These simplifications are indicated in figure 12.5 by the loops drawn round neighbouring pairs, and we read off the result given by equation (12.3): for example, the bottom left loop gives $A.C.(B + \overline{B}) = A.C$.

As a second example, figure 12.6(a) shows the entries for $\overline{A}.\overline{B}.C$ (upper left) $+\overline{A}.B.C + A.\overline{B}.C + A.B.C$. In this case, two sets of neighbouring pairs may be combined using $A + \overline{A} = 1$ and $B + \overline{B} = 1$; C is one for all entries and is combined with all pairs of values of A and B. So the result is simply C.

Next consider the reverse procedure. Suppose the Karnaugh map is to be drawn for the expression $A.\overline{B} + B.\overline{C}$. How is this done? The answer is that $A.\overline{B} = A.\overline{B}.(C + \overline{C})$ and is represented by the two entries at the bottom left of figure 12.6(b); $A.\overline{B}$ is common to both. Likewise $B.\overline{C}$ is given by the entries at the right of the table, summing over the possible values $A = 0$ or 1.

It does not matter whether horizontal rows of the Karnaugh map display values of B and C or A and B or even A and C. Figure 12.6(c) shows the Karanugh map of figure 12.6(a) replotted with AB horizontal and C vertical. The map looks different from figure 12.6(a). However, after the simplifications have been made, combining neighbouring terms, the result is the same.

It is recommended that you get plenty of practice in manipulating such diagrams. They are almost invariably easier to use than Boolean algebra itself, though once an expression has been simplified using the Karnaugh map, it is good practice to check that the same result can be obtained algebraically, lest a mistake has crept in somewhere.

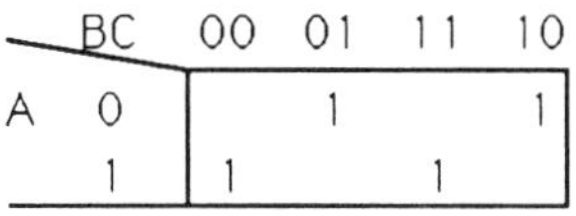

As a further example, figure 12.7 gives the Karnaugh map for the sum signal of the full adder, equation (12.1). Because no entries are adjacent, the expression cannot be simplified. Even so, the pattern of ones gives a valuable clue to a way of forming the required sum signal. The ones occupy alternate horizontal locations. This corresponds to the fact that the arithmetic sum of B and C is given by the exclusive OR, $B \oplus C$. The staggering of the ones between the two horizontal lines suggests that the required overall sum signal is given by $A \oplus (B \oplus C)$, and this is indeed so, as you can readily verify with a truth table. The Sum signal can therefore be constructed from two half adders, as shown in figure 12.8. The order in which the two exclusive ORs is carried out does not matter; $A \oplus (B \oplus C)$ or $(A \oplus C) \oplus B$ or $(A \oplus B) \oplus C$ give the same result.

The Karnaugh map is readily extended to expressions involving four characters A, B, C and D. As an illustration, suppose two two-digit binary numbers are to be multiplied. The required results are given in table 12.1.

BC	00	01	11	10
A 0		1		1
1	1		1	

Fig. 12.7. Karnaugh map for the sum signal of a full adder.

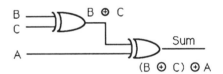

Fig. 12.8. Forming the sum from two half adders.

Table 12.1   Multiplication of two binary two-digit numbers.

$00 \times 00 = 0000 \;\; 00 \times 01 = 0000 \;\; 00 \times 10 = 0000 \;\; 00 \times 11 = 0000$
$01 \times 00 = 0000 \;\; 01 \times 01 = 0001 \;\; 01 \times 10 = 0010 \;\; 01 \times 11 = 0011$
$10 \times 00 = 0000 \;\; 10 \times 01 = 0010 \;\; 10 \times 10 = 0100 \;\; 10 \times 11 = 0110$
$11 \times 00 = 0000 \;\; 11 \times 01 = 0011 \;\; 11 \times 10 = 0110 \;\; 11 \times 11 = 1001$

Let AB and CD represent the two binary numbers and $WXYZ$ the binary digits of the result. Then the ones of table 12.1 give

$$W = A.B.C.D$$
$$X = A.\overline{B}.C.\overline{D} + A.\overline{B}.C.D + A.B.C.\overline{D} \qquad (12.4)$$
$$Y = \overline{A}.B.C.\overline{D} + \overline{A}.B.C.D + A.\overline{B}.\overline{C}.D + A.\overline{B}.C.D$$
$$\qquad + A.B.\overline{C}.D + A.B.C.\overline{D} \qquad (12.5)$$

$$Z = \overline{A}.B.\overline{C}.D + \overline{A}.B.C.D + A.B.\overline{C}.D + A.B.C.D. \quad (1\!?$$

The Karnaugh maps for X, Y and Z are shown in figure 12.9.

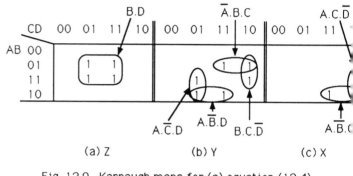

(a) Z   (b) Y   (c) X

Fig. 12.9. Karnaugh maps for (a) equation (12.4),
(b) (12.5) and (c) (12.6).

The simplifications are

$$Z = B.D$$
$$Y = A.\overline{C}.D + A.\overline{B}.D + \overline{A}.B.C + B.C.\overline{D}$$
$$= A.D.(\overline{C} + \overline{B}) + B.C.(\overline{A} + \overline{D})$$
$$X = A.\overline{B}.C + A.C.\overline{D} = A.C.(\overline{B} + \overline{D})$$

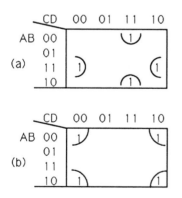

Fig. 12.10. Closing the Karnaugh
map top and bottom
and at the sides.

as can be verified algebraically from equations (12.4)–(12.6). I a straightforward matter to arrange NOT, AND and OR gates as to create these signals.

In order to consider a further point, figure 12.10(a) shows expression

$$f = A.B.\overline{C}.\overline{D} + \overline{A}.\overline{B}.C.D + A.\overline{B}.C.D + A.B.C.\overline{D}. \quad (12$$

Algebraically, the first and fourth terms reduce to $A.B.\overline{D}$ and second and third to $\overline{B}.C.D$. To get these results from the Karnaı map, we must imagine it wrapped around a horizontal axis so th top and bottom edges join; then the entries $\overline{A}.\overline{B}.C.D$ at the and $A.\overline{B}.C.D$ at the bottom are adjacent and may be combin to $\overline{B}.C.D$ using $A + \overline{A} = 1$, as shown by the semicircles on map. Likewise, we must imagine the map wrapped around a v tical axis, so that left- and right-hand edges join; this leads to simplification $A.B.\overline{C}.\overline{D} + A.B.C.\overline{D} = A.B.\overline{D}$. In figure 12.10(wrapping around both horizontal and vertical axes simplifies map to $\overline{B}.\overline{D}$; all combinations of A and C appear and may summed.

If all entries of the Karnaugh map are 1, what is the result?

Σ notation

An expression like equation (12.7) is called **a sum of products**. When written like this in terms of the basic elements of the truth table, it is said to be in **standard form**. Each term in the expression is called a **minterm or standard product**. There is a shorthand notation for an expression such as equation (12.7). Suppose the binary number $ABCD$ is converted to decimal; the entries of the Karnaugh map may be numbered in decimal as shown in figure 12.11. Then the entries in figure 12.10(a), for example, may be designated by $\Sigma(3,11,12,14)$, where the Σ stands for a sum of products.

CD AB	00	01	11	10
00	0	1	3	2
01	4	5	7	6
11	12	13	15	14
10	8	9	11	10

Fig. 12.11. Decimal notation for entries of the Karnaugh map.

12.5 Don't Care or Can't Happen Conditions

Suppose a counter works in BCD code. It counts from 0 to 9 in binary, then resets to 0. Suppose also that the numbers are to be displayed as shown in figure 12.12(a) on segments of light emitting diodes (LEDs). A Karnaugh map may be drawn for every segment of the display in terms of the four binary digits of the BCD number (exercise 5).

As an illustration, the logic required to light up the top left-hand vertical segment will be derived. This segment occurs in the numbers 0, 4, 5, 6, 8 and 9. The BCD number is represented by the four binary digits A, B, C and D and the required logic is $\Sigma(0,4,5,6,8,9)$. However, a further simplification is possible. The numbers 10 to 15 cannot arise (unless the counter fails). In figure 12.12(b), these 'can't happen' conditions are marked by X's. We can choose to make them 1's if it simplifies the Karnaugh map or we can choose to ignore them if it does not. Setting them to 1, the blocks ringed in the figure give the result $A+\overline{C}.\overline{D}+B.\overline{C}+B.\overline{D}$.

A chip to display *all* the segments is available commercially (the 4511 decoder); it has the additional feature of giving a blank output if the input is invalid, i.e. if it is presented with any of decimal 10 to 15.

A second example arises in the logic of traffic lights, which should follow the sequence red → red + yellow → green → yellow → red. This example actually involves holding the current values of red (R), yellow (Y) and green (G) in a memory and will be discussed more fully in the next chapter. However, for the moment we need only consider the combinational logic giving new values of red, yellow and green when a clock pulse (C) triggers a change. These conditions are shown in figure 12.13. For example, if the current state is red, entries are at the top right corner of each rectangle ($R.\overline{Y}.\overline{G}$) and the new state is red + yellow: $R_{\text{new}} = 1$, $Y_{\text{new}} = 1$,

(a)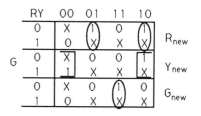

CD AB	00	01	11	10
00	0	1	3	2
01	4	5		6
11	X	X	X	X
10	8	9	X	X

Fig 12.12 (a) display of decimal numbers, (b) the Karnaugh map for the top left-hand vertical segment.

RY	00	01	11	10	
0	X	1	0	1	R_{new}
1	0	X	X	X	
0	X	0	0	0	Y_{new}
1	1	X	X	X	
0	X	0	1	0	G_{new}
1	0	X	X	X	

G

Fig. 12.13. Logic of traffic lights, allowing for "can't happen" conditions, labelled X.

$G_{\text{new}} = 0$. In the table, X indicates a condition which should not arise, e.g. red + yellow + green. In working out the logic, X may be counted as 1 or 0, whichever gives the more convenient result. This allows simplifications in the logic, which becomes

$$R_{\text{new}} = (\overline{R}.Y + R.\overline{Y}).C$$
$$Y_{\text{new}} = \overline{Y}.C$$
$$G_{\text{new}} = R.Y.C.$$

Of course, we might well decide that faulty combinations are not really 'don't care' but require special action, rather than being ignored. That would lead to different logic.

12.6   Products of Karnaugh Maps

There is one further trick which is useful in manipulating Karnaugh maps. Suppose we have a complicated product of expressions, such as

$$F = (\overline{A} + B + C).(\overline{B} + A.B + C.D).(\overline{A}.B + \overline{C}.D).(\overline{B} + A.C + D).$$

Multiplying this out is tedious and prone to errors. There is an alternative shown in figure 12.14.

When any two brackets are multiplied, the resulting Karnaugh map contains a 1 only where both brackets contain a 1. For example, the maps for the first two brackets are shown in figures 12.14(a) and (b). Each contains the term $\overline{A}.\overline{B}.\overline{C}.\overline{D}$ in the top left-hand corner. When we AND the two brackets, this term gives $\overline{A}.\overline{B}.\overline{C}.\overline{D}.\overline{A}.\overline{B}.\overline{C}.\overline{D} = \overline{A}.\overline{B}.\overline{C}.\overline{D}$. On the other hand, the bottom left-hand corner contains a 0 in the first bracket and a 1 in the second. The former is represented by $(A.\overline{B}.\overline{C}.\overline{D})$, so its product with the term $A.\overline{B}.\overline{C}.\overline{D}$ of (b) gives a zero. Thus, the Karnaugh map of the expression F may be formed by drawing the maps of all four individual brackets and then using the result that 1's are present in F only at points of the map where 1's are present in all four brackets. Another way of saying this is that there is a 0 in the result whenever there is a 0 in any of the individual brackets. The result is illustrated in figure 12.14(e). Only three locations have a 1 common to all brackets, so the result is

$$F = \overline{C}.D.(\overline{A}.\overline{B} + A.B) + \overline{A}.B.C.D.$$

This may be verified by multiplying out F using Boolean algebra.

At this point, you are advised to get plenty of practice with Karnaugh maps by tackling exercises 1 to 5.

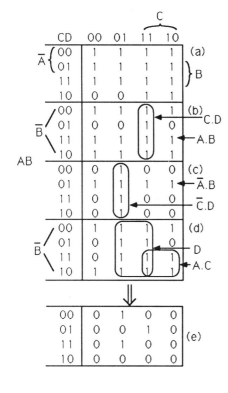

Fig. 12.14. Karnaugh maps for (a) $(\overline{A}+B+C)$, (b) $(\overline{B}+A.B+C.D)$, (c) $(\overline{A}.B+\overline{C}.D)$, (d) $(\overline{B}+A.C+D)$ and (e) their product.

12.7   Products of Sums

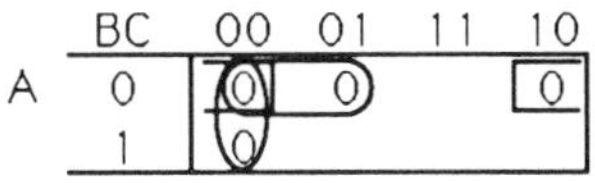

It may sometimes be convenient to work out logical expressions in terms of the zeros of the Karnaugh map, rather than the ones. This can happen, for example, if there are rather few zeros, but a second practical reason will appear below. To illustrate the point, suppose figure 12.5 is redrawn plotting the zeros rather than the 1's (figure 12.15). The simplified expression for these locations is $\overline{A}.\overline{C}+\overline{A}.\overline{B}+\overline{B}.\overline{C}$. Since this expression stands for the *zeros* of the Karnaugh map, the expression for the carry signal can be rewritten as its complement:

Fig. 12.15. Zeros for the carry signal of the full adder.

$$\overline{\text{Carry}} = \overline{A}.\overline{C}+\overline{A}.\overline{B}+\overline{B}.\overline{C}. \qquad (12.8)$$

This looks more complicated than before, until we realise that the expression may be simplified using de Morgan's theorem. Remember that this theorem states that every signal can be turned into its complement by interchanging AND with OR:

$$\text{Carry} = (A+C).(A+B).(B+C). \qquad (12.9)$$

This form is known as a **product of sums** or alternatively as an OR–AND form. As a little exercise, this expression is multiplied out to verify that it gives the same result as equation (12.3):

$$\begin{aligned}
\text{Carry} &= (A.A + A.C + A.B + B.C).(B+C)\\
&= (A + A.C + A.B + B.C).(B+C)\\
&= (A + B.C).(B+C) \quad \text{using rule (vi) of Boolean algebra}\\
&= A.B + B.C + A.C
\end{aligned}$$

which agrees with equation (12.3).

Equation (12.9) is a simplified expression. In standard form it reads

$$\overline{\text{Carry}} = \overline{A}.\overline{B}.\overline{C}+\overline{A}.B.\overline{C}+A.\overline{B}.\overline{C}+\overline{A}.\overline{B}.C$$
$$\text{Carry} = (A+B+C).(A+\overline{B}+C).(\overline{A}+B+C).(A+B+\overline{C}).$$

The individual terms in this expression, such as $(A+B+C)$, are known as **maxterms** or **standard sums**. An alternative notation for the result is:

$$\text{Carry} = \Pi(0,1,2,4).$$

This stands directly for the locations of the zeros in the Karnaugh map, and the symbol Π indicates that the product is to be formed of the sums to which they are related by de Morgan's theorem.

12.8   Use of NOR and NAND Gates

Some families of logic gates are only readily available (or cheaper) in the form of NAND and NOR gates. This raises the question of how to go about using inverted gates. This is actually easier than it may appear. figure 12.16(a) shows the logic for forming the Carry signal of a full adder in AND–OR form. Suppose every signal is inverted twice between the gates; this is indicated by the bubbles in figure 12.16(b). Now de Morgan's theorem states that inputs and outputs may be complemented if OR is changed to AND; this is shown diagrammatically in figure 12.16(c) and is often the simplest way of using this theorem. Figure 12.16(b) can be redrawn as in (d). This circuit achieves the same result as figure 12.16(a), but using NAND gates only. It is called a NAND–NAND form.

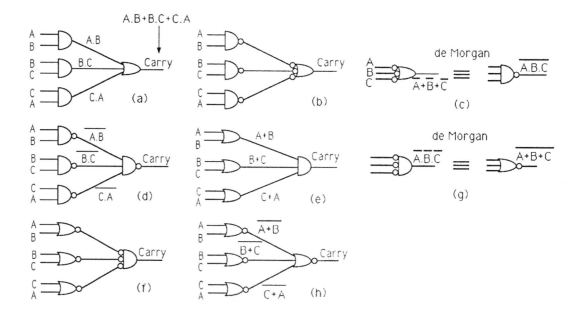

Fig. 12.16 (a), (b), (d), (e), (f) and (h), Various realisations of the carry of the full adder; (c) and (g) diagrammatic representations of de Morgan's theorems.

Let us now examine the relevance of the product of sums representation. Equation (12.9) leads to the circuit of figure 12.16(e). figures 12.16(f) to (h) then show how to convert this into a form realised purely in terms of NOR gates. This is called a NOR–NOR form. As an algebraic check, figure 12.16(h) forms $\overline{\overline{A+B}+\overline{B+C}+\overline{C+A}} = (A+B).(B+C).(C+A)$ by de Morgan's theorem.

The conclusion is that sums of products (derived from the 1's of the Karnaugh map) are convenient for obtaining a circuit in terms of NAND gates, while products of sums (derived from the 0's) are the starting point for obtaining logic in terms of NOR gates. However, a word of warning is appropriate here. Positive logic is much easier to follow and test. It may be worth a small extra expense or an extra gate or two to be able to express logic in terms of AND and OR gates when making a single circuit or a few off. The economies achieved with NAND and NOR gates may be essential on a large scale, and it is necessary to be conversant with the manipulations illustrated in figure 12.16 and to be able to decipher the logic diagrams. A potential source of confusion is that NAND gates which are performing the OR function are sometimes indicated in circuit diagrams by the symbol on the left of figure 12.16(c); likewise, NOR gates performing the AND function may be indicated by the symbol on the left of figure 12.16(g).

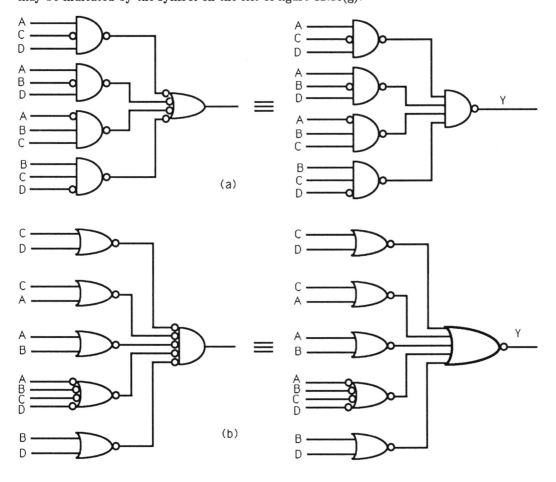

Fig. 12.17. Realisation of Y of Fig. 12.8(b) in terms of (a) NAND, (b) NOR gates.

As a second example, Y of figure 12.9(b) is given by $A.\overline{C}.D + A.\overline{B}.D + \overline{A}.B.C + B.C.\overline{D}$ and hence by the NAND gate configuration of figure 12.17(a). Alternatively, it is given by

$$\overline{Y} = \overline{C}.\overline{D} + \overline{C}.\overline{A} + \overline{A}.\overline{B} + A.B.C.D + \overline{B}.\overline{D}$$

or

$$Y = (C + D).(C + A).(A + B).(\overline{A} + \overline{B} + \overline{C} + \overline{D}).(D + B)$$

and the NOR gates of figure 12.17(b).

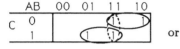

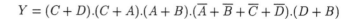

Fig. 12.18. A static 1 hazard condition.

12.9   Hazards

Suppose the logical function f shown in figure 12.18 is to be constructed. The two full loops show that the desired result can be formed from $B.C + A.\overline{C}$. The broken loop, representing $A.B$, may appear superfluous, since it is logically contained within the other two terms. Nonetheless it is good practice to add the term $A.B$, despite using another gate. The reason is that it eliminates the possibility of a static hazard.

Figure 12.19(a) shows waveforms describing possible combinations of A, B and C. Incidentally, a way of generating these waveforms is from the binary 4, 2 and 1 digits of a binary counter driven by a pulse generator. The signal C appears in $B.C$ and $\overline{C}$ appears in $A.\overline{C}$. Providing timing is perfect, everything works correctly. Suppose, however, the internal timing is faulty and $\overline{C}$ is early compared with C, as shown by the eighth line of the diagram. The result, shown by the ninth line, is that at time t_x, $A.\overline{C}$ goes down early, before $B.C$ comes up. The resulting waveform $A.\overline{C}_e + B.C$ in line 10 momentarily goes low, when it ought to stay high. This is called a static 1 hazard: it ought to be 1 but it is not. It is cured by adding $A.B$ to the logic, as shown by line 11.

Static 1 hazards arise when using 1's in the Karnaugh map, i.e. sums of products. The converse happens when using zeros in the Karnaugh map, i.e. products of sums. The zeros of figure 12.18 give the logical result

$$\overline{f} = \overline{C}.\overline{A} + C.\overline{B}$$
$$f = (C + A).(\overline{C} + B)$$

but there is the option of adding $\overline{A}.\overline{B}$ to $\overline{f}$ and forming

$$f = (C + A).(\overline{C} + B).(A + B).$$

Figure 12.19(b) shows the waveforms assuming faulty internal timing between C and $\overline{C}$; a static 0 hazard arises at time t_y, i.e. the

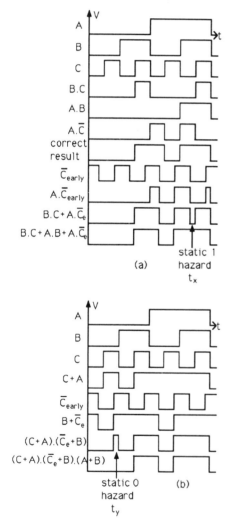

static 1 hazard
(a) t_x

static 0 hazard
t_y (b)

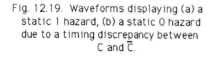

Fig. 12.19. Waveforms displaying (a) a static 1 hazard, (b) a static 0 hazard due to a timing discrepancy between C and $\overline{C}$.

output is 1 when it should be zero. This is cured by adding $(A+B)$ to the logic, line 8.

These hazards arise because of faulty relative timing of C and $\overline{C}$ in neighbouring minterms $A.B.\overline{C}$ and $A.B.C$ in the logic. The cure is to make sure that neighbouring terms in the Karnaugh map are always combined explicitly in the logic, even though this may appear to introduce some redundancy.

A second example is shown in figure 12.20. This repeats the Karnaugh map of the carry of a full adder, from figure 12.5. It might be tempting to make use of the fact that $A \oplus B$ is available from a half adder; the expression

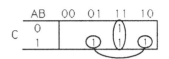

$$\text{Carry} = (A \oplus B).C + A.B \qquad (12.10)$$

may be expressed using the combinations encircled in figure 12.20. This, however, leaves the possibility of two static hazards, one between minterms 5 and 7 due to faulty timing of $\overline{A}$ with respect to A and the second between minterms 6 and 7 due to faulty timing of $\overline{B}$ with respect to B.

Although hazards are eliminated by the rules described here, glitches due to faulty *external* timing between A, B and C remain.

This is a good point to get some practice in manipulating gates (exercises 6 to 12).

12.10   Decoders and Encoders

So far logic has been assembled out of simple gates. This can be extremely tedious and also expensive if the logic involves many gates. Devices such as decoders and encoders are available commercially to make this more manageable and cheaper.

Two examples of **decoders** are shown in figure 12.21. When enabled by a signal on line E, the device in (a) gives an output on one of the four lines $D0$ to $D3$ according to the value of the binary input AB. For example, if A and B are both 1, an output 1 appears on line $D3$. Likewise, (b) decodes a three-digit binary number to give an output on one of the eight output lines. The two units work by simply forming all possible minterms, one for each output line; this is shown for the 74139 in figure 12.21(c).

Such devices are a very convenient way of doing sum of products logic since all minterms of the Karnaugh map are available as outputs. Suppose, for example, a full adder is to be constructed. For this, the Sum signal can be written (figure 12.7) Sum $= \Sigma(1,2,4,7)$ and the Carry (figure 12.5) as $\Sigma(3,5,6,7)$. These can be formed by feeding the 74138 decoder with inputs A, B and C and taking the required combination of the output lines to two OR gates, one for Sum $(1 + 2 + 4 + 7)$ and one for Carry $(3 + 5 + 6 + 7)$. When line E is enabled, the outputs of the two OR gates generate Sum

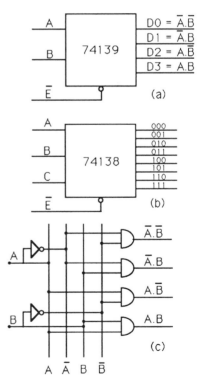

and Carry. This is much simpler than wiring up the large number of OR and AND gates required in figure 12.4, for example. Output line $D0$ of the decoder is unused, but could, for example, identify zero input.

Another use for the decoder is to address registers or devices in a computer. Suppose, for example, a computer has two line printers, a magnetic tape and a disc. These units might be addressed as 0 to 3. To select one of them, say unit 2, the computer puts the binary number 10 on its **address bus**, a set of lines distributed to all four devices. Each unit is equipped with its own decoder and uses one of the output lines $D0$–$D3$ of figure 12.21(a) to switch on. In a memory, decoders identify individual registers; this will be discussed in the next chapter.

Other types of decoder are available. For example, the 74145 drives a BCD display. A BCD number in the range 0–9 on the four input lines results in an output signal on one of ten output lines. If the binary input code is in the range 10–15, all outputs are zero. The output voltage and current are sufficient do drive visual displays. A second example is the 7447, which generates an output for driving a seven-segment display from a BCD input; it has four input lines and seven outputs. The similar 4511 ignores illegal inputs. Its layout is shown in figure D.4 of appendix D.

Encoders are devices working the other way round. For example, a decimal to binary converter has ten input lines plus an enable, and four output lines corresponding to the four binary digits. The 74148 priority encoder does this, with the additional refinement that if two or more decimal inputs are present simultaneously it chooses the largest to encode. A keyboard encoder has 84 or so input lines, corresponding to the number of symbols (upper and lower case combined) on the keyboard, and seven output lines for binary output code.

A technical detail is that encoders and decoders are usually activated by the enable signal going *low* rather than *high*. In circuit diagrams, this is denoted by a bubble on the enable line, to indicate a complementary signal. Such an input control line is said to be **active low**. The circuit is permanently enabled if the line E is earthed. The layout of the 74138 is shown in figure D.3 of appendix D.

12.11 Multiplexing

Suppose we have two slow streams of data, A and B. They can be gated alternately on to a single faster line using a control pulse C by forming $A.C + B.\overline{C}$: A goes on to the line when $C = 1$ and B goes on when $C = 0$. Then they can be unscrambled at the other end of the line by using the C pulse to route A and B to separate outputs:

$$A = C.(A.C + B.\overline{C})$$
$$B = \overline{C}.(A.C + B.\overline{C}).$$

More generally, many lines can be multiplexed on to a single line by making C an address from which the input is selected and to which the output is sent. As an example, a number of telephone conversations may be routed on to a single fast line by sampling each conversation at a rate above audio frequencies and then unscrambling them at the other end of the line. This technique is called **multiplexing** and the device which does it is called a **multiplexer** (MUX) or **data selector**. A 4 to 1 line MUX is shown in figure 12.22(a) and the way it works in (b). Its layout is shown in figure D.3 of appendix D. It is essentially an encoder with a single output line. The decoding system is called a **demultiplexer** (DEMUX); this is achieved simply by connecting the fast line to the enable input of a decoder chip and the appropriate address to the select lines (A and B of figure 12.21(a)).

Multiplexing is used extensively in data processing systems where many registers (or devices) may feed on to a common bus line; encoders and decoders arrange that each register may be addressed individually. Details will be given in the next chapter.

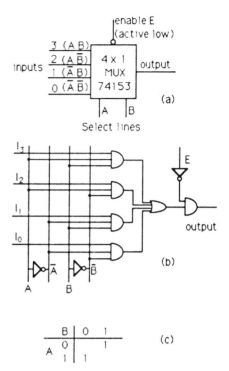

Fig. 12.22 (a) symbolic form and (b) gate layout of a 4 to 1 multiplexer; (c) using a multiplexer to generate $\overline{A}.B + A.\overline{B}$.

Use for logic

Like the decoder, a multiplexer can be used to generate logical expressions by putting control signals on lines 0, 1, 2 and 3. By putting 1's on any of these lines, a direct realisation of the Karnaugh map is generated at the output. For example, with a permanent input voltage on line 3 and the other three lines earthed, the output is $A.B$, so the multiplexer functions as an AND gate. Alternatively, with high input signals on lines 1 and 2, as in figure 12.22(c), it gives an output $A.\overline{B} + \overline{A}.B$, i.e. exclusive OR.

Multiplexers can also be used to form expressions of three or more variables. Suppose, for example, the logical expression (12.1) for the sum of a full adder is to be formed. This may be rewritten

$$S = (\overline{A}.\overline{B}).C + (\overline{A}.B).\overline{C} + (A.\overline{B}).\overline{C} + (A.B).C. \qquad (12.11)$$

The terms $\overline{A}.\overline{B}$, $\overline{A}.B$, $A.\overline{B}$ and $A.B$ are available using lines 0, 1, 2 and 3 of the multiplexer. So S may be formed by applying C to lines 0 and 3 and $\overline{C}$ to lines 1 and 2. More complicated expressions of four variables may be formed in a similar manner. For example, the lengthy expression (12.5)

$$Y = (\overline{A}.B).C + (A.\overline{B}).D + (A.B).(\overline{C}.D + C.\overline{D})$$

may be constructed by forming $\overline{C}.D + C.\overline{D}$ with ones on lines 1 and 2 of one multiplexer, then applying this to line 3 of a second multiplexer (whose select lines are controlled by A and B), together with D on line 2 and C on line 1.

Fig. 12.23. Measuring a trajectory in a magnetic field.

12.12   Table Look Up

Suppose a particle travels through a non-uniform magnetic field, and its trajectory is measured with three detectors, as shown in figure 12.23. Computing its momentum from the three coordinates involves a lengthy, iterative numerical calculation if the field is non-uniform. If you are doing an experiment, you may want to select particles within a certain band of momenta. This decision may have to be taken in much less time than is taken by even the fastest computer to do the calculation of the trajectory. A solution is to precompute the momenta for every possible trajectory and store in a memory the result that: 1, the track is of interest, or 0, it is not. The coordinates in the three detectors are used to address the memory. When an event is detected, the coordinates are used to look up in the memory whether or not the event is of interest (1 or 0). This can be done in a microsecond or less. The result is then used for accepting or rejecting the event. The technique is uneconomical in terms of memory, but memory is cheap nowadays.

This type of situation arises in a wide variety of applications, not only in scientific experiments (where it was developed), but also in business and control situations. It can be used to look up the logical outcome of a very complicated situation which might otherwise take thousands of logic gates to solve. If you want to experiment with different solutions to a control situation, you might have to rewire the logic every time you change your mind, whereas using a table in a computer the logic can be changed easily by means of a program.

Memories

In an experimental situation, the memory could be that of a computer which can be programmed freely. Once the logic has been sorted out, the memory can be programmed permanently, and it ought to be proof against the power going off. Applications arise in using microprocessors in cars, washing machines, electronic games and so on. At an early stage of testing a new product, the memory is likely to be EPROM (erasable programmable **read-only memory**) or EEPROM (electrically erasable PROM). The program can be loaded (by applying special voltages typically $+21$ to $+50$ V for 10–50 ms) into the memory, and can then be tested extensively in the field. If changes are required, EPROM can be erased by exposure to ultraviolet light for about 30 minutes, and EEPROM can be erased very simply by a voltage of -21 V. These types of memory are convenient but relatively expensive, and are therefore inappropriate for mass production. Once the product has been developed to perfection, it is worth the expense (£5–50 K) of getting a manufacturer to design a special chip with the program encoded; then copies can be manufactured by the thousand at a much lower unit cost. In all cases, the memory persists when power is off.

Such read-only memories (ROM) were originally invented to store the very complicated sequence of operations used in dedicated computer programs, for example, those providing the systems software of the manufacturer and beyond the control of the user.

FPLA

An alternative piece of hardware is the field programmable logic array (FPLA), for example the 82S100. This device has 16 input lines and 8 outputs. Internally, all 16 inputs are presented to 48 logic gates, which can be programmed by the user to create up to 48 minterms. These minterms may then be programmed in any required combinations of 8 output OR gates. Such a device is useful when the number of logical terms is large compared with what we would be prepared to wire up through individual logic gates but small compared with the contents of a ROM.

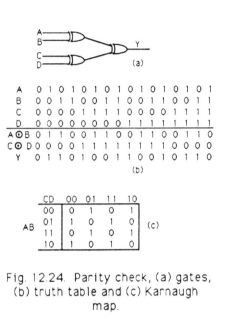

Fig. 12.24. Parity check, (a) gates, (b) truth table and (c) Karnaugh map.

12.13 Parity Checks

It is normally desirable to make checks against transmission errors in a communications system and in arithmetic operations; i.e. the system tests itself. This is a large subject, but figure 12.24 illustrates a simple and common form called the parity check. Suppose there are four lines holding binary digits A, B, C and D. If these are fed through a 'tree' of exclusive ORs, the truth table reveals that the output is 1 if there is an odd number of ones in the input. This result may be used by forming and transmitting the parity bit Y with $ABCD$, and checking it at the receiver. For even parity code, the parity bit is such as to make the parity of the whole word $YABCD$ even, as above. For odd parity, the parity bit is inverted, so that the parity of the whole word $\overline{Y}ABCD$ is odd.

In general, it is wise to transmit some redundant information; for example, if a 24-bit number is to be transmitted, it can be broken down into 4×6-bit numbers, as in figure 12.25, and the parity checks can be formed over rows and columns. If a single error occurs in transmission, as in (b), the parity bits point uniquely to the row and column where the error occurred. If an error occurs in transmitting the parity bit itself, this shows up because it is present in only the row or column.

More complicated and more efficient codes (for example, Hamming codes) are capable of detecting and correcting multiple errors.

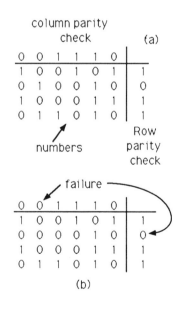

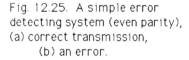

Fig. 12.25. A simple error detecting system (even parity), (a) correct transmission, (b) an error.

Sophisticated error correction is used, for example, in reading compact discs, hence protecting them against corruption by reading or recording errors. It is also used in digital communications, for example, in satellite links.

12.14   Equivalence

CD	00	01	11	10
00	1	0	1	0
01	0	1	0	1
11	1	0	1	0
10	0	1	0	1

AB (row labels)

Fig. 12.26. The Karnaugh map of XNOR.

The converse of figure 12.24(c) is XNOR or EQUIVALENCE, shown in figure 12.26, and introduced in the previous chapter. It has ones where an even number of inputs is the same. It may therefore be used to form a parity check which is 1 if there is an even number of ones in the input. It is also useful in comparing two numbers $AB\dots$ and $CD\dots$. If digits A and C are equal, $\overline{A \oplus C} = A.C + \overline{A}.\overline{C} = 1$. If and only if the two numbers are exactly equal, $\overline{A \oplus C} . \overline{B \oplus D} \dots$ is one. This is used in testing the equality of two numbers in a computer.

12.15   Addition and Multiplication of Several Bits

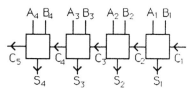

Fig. 12.27. A sequential array of 4 full adders.

The arithmetic operations considered so far have been the addition of just two binary digits with a carry. Suppose two four-digit numbers are to be added together. The simplest way of achieving this is by means of four full adders, one for each digit (figure 12.27). The snag is that the second adder has to wait for the carry C_2 to arrive from the first adder, and so on down the chain. This is called 'ripple through carry', since the carries ripple down the chain like a wave. The time taken for the whole addition is four times that required by one adder. Computing speed is important, so the question arises whether this can be speeded up,

Look-ahead carry

With some proliferation in the number of gates, this is indeed possible by providing separate logic to compute all the carries C_2 to C_5 and presenting them simultaneously to the full adders. To see how this is done, consider one adder. From figure 12.20 and equation (12.10),

$$C_{i+1} = G_i + X_i.C_i$$

where

$$G_i = A_i.B_i \tag{12.10a}$$

$$X_i = A_i \oplus B_i. \tag{12.10b}$$

A scheme achieving this is shown in figure 12.28. If this expression is written out explicitly for $i = 2$ to 5,

$$C_2 = G_1 + X_1.C_1$$
$$C_3 = G_2 + X_2.(G_1 + X_1.C_1)$$
$$C_4 = G_3 + X_3.(G_2 + X_2.G_1 + X_2.X_1.C_1)$$
$$C_5 = G_4 + X_4.(G_3 + X_3.G_2 + X_3.X_2.G_1 + X_3.X_2.X_1.C_1).$$

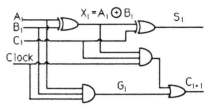

Fig. 12.28. The logic of one full adder. The clock pulse eliminates hazards.

All the X_i and the G_i are given directly in terms of A_i, B_i and C_1 by equations (12.10). It is possible to arrange a scheme shown in figure 12.29. At the first level, all X_i and G_i are formed simultaneously; at the second level, all of the C_i are formed and at the third level all the S_i. Thus *all* the S_i are available after three gate delays; in comparison, in the earlier scheme of figure 12.27, S_1 is available after two gate delays, S_2 after four, S_3 after six and S_4 after eight, and C_5 is available after only nine gate delays. There is therefore a substantial gain in speed to be achieved, at the price of the extra logic required to form the C_i simultaneously rather than iteratively. This process is called **look-ahead carry**. It is characteristic of the sort of elaboration required to achieve maximum speed in digital computers. Such a unit (74283) with look-ahead carry is available commercially for adding two four-bit numbers and an input carry. Its layout is shown in figure D.3 of appendix D. Several of these may be cascaded to make a 16-, 32- or 64-bit adder. A unit 74182 is available to provide look-ahead carry over four such units.

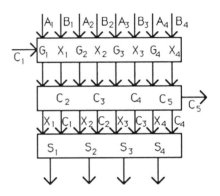

Fig. 12.29. Organisation of look-ahead carry.

Next consider multiplication of two four-digit binary numbers. An example is given in figure 12.30. The process involves shifting X left three times, multiplying by each digit of Y in turn and adding the results. Various tricks can be used to speed up this process. For example, C and D can be added together in one adder simultaneously with A and B in a second. If one of the two rows is zero, the addition can be skipped completely. Also, rather than shifting X one place at a time, the logic can be organised so that B is presented to the $(A + B)$ addition unit one place left, likewise for D and C, and finally $(C + D)$ can be presented two places left at the unit which forms $(A + B) + (C + D)$. Many such complicated designs exist for speeding up multiplication (and division) and the reader is referred to specialised texts for details.

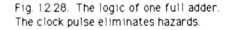

```
     1  1  0  1     X
     1  0  0  1     Y
     1  1  0  1     A
  0  0  0  0        B
0  0  0  0           C
1  1  0  1           D
1  1  1  0  1  0  1
```

Fig. 12.30. Binary multiplication.

The ALU

If you fancy building your own computer, a versatile arithmetic-logic unit, 74181, is available at a cost of a little over £2. It operates on two four-bit binary words A and B, under the control of three select lines, M, S_0 and S_1. If the first of these, M, is equal to 0, the unit performs one of four logical functions shown in figure 12.31(a); the function is chosen by the two select lines S_0

S_0	S_1	Output	
(a)			
0	0	$\overline{A}$	(NOT)
0	1	$A \oplus B$	(XOR)
1	0	$A + B$	(OR)
1	1	$A.B$	(AND)

S_0	S_1	Output			
(b)					
0	0	$\overline{B}_3$	$\overline{B}_2$	$\overline{B}_1$	$\overline{B}_0$
0	1	B_3	B_2	B_1	B_0
1	0	1	1	1	1
1	1	0	0	0	0

C	S_0	S_1	Output
(c)			
0	0	0	$A + \overline{B}$
0	0	1	$A + B + C$
0	1	0	$A - 1$
0	1	1	A
1	0	0	$A - B$
1	0	1	$A + B + C$
1	1	0	A
1	1	1	$A + 1$

Fig. 12.31. Functions fo the 74181 ALU:
(a) logical operations (M = 0),
(b) operations on B (M = 1),
(c) arithmetic operations (M = 1).

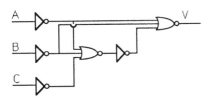

Fig. 12.32.

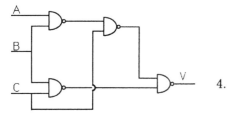

Fig. 12.33.

and S_1. If $M = 1$, it instead performs arithmetic. First it operates on B according to figure 12.31(b). Then it feeds the result to a four-bit adder, which also receives A and a carry C. According to the value of C, the results are as given in figure 12.31(c). Line 5 does subtraction by forming the 1's complement of B and adding in a carry of 1. Units may be cascaded to handle more than four bits. To use the unit, further logic must set up M and the select lines S_0 and S_1, according to a program.

12.16 Exercises

1. (RHBNC 1988) Take the Boolean expression $f = A.\overline{B}.\overline{C} + A.\overline{B}.C$ and simplify it using a Karnaugh map. Also simplify it algebraically and hence show that both methods are equivalent. (Ans: $f = A.\overline{B}$.)

2*. (Kings 1974) (a) Compile a truth table representing the logical operation of a scale-of-ten decimal counter made up of four binary digits A, B, C and D fed by a pulse P and give expressions for new values of A, B, C and D in terms of P and the old values; make use of don't care conditions. Determine the simplest logical circuit which will produce an output only when the seventh pulse in each set of ten is applied to the counter, assuming it starts from 0. (b) Establish a truth table for the logic circuit in figure 12.32 and hence determine the simplest logical equivalent of this arrangement. (c) Derive a Boolean algebraic expression representing the operation of the system in figure 12.33 and hence determine its simplest logical equivalent. (Ans: (a) $D = \overline{D}.P$, $C = P.(\overline{A}.\overline{C}.D + C.\overline{D})$, $B = P.(B.\overline{C} + B.\overline{D} + \overline{B}.C.D)$, $A = P.(A.\overline{D} + B.C.D)$; $P.B.C.\overline{D}$, (b) $A.B.C$, (c) C.)

3*. (QEC 1974) (a) Simplify the following logical expressions: (i) $A.B + A.\overline{C} + \overline{A}.\overline{B}.\overline{C}$, (ii) $(\overline{A} + \overline{B} + C).(A.B + D)$, (iii) $A + A.B + \overline{A}.C.\overline{D} + \overline{A}.\overline{B}.C.D + B.C.D$; (b) Use a Karnaugh map to minimise the following expressions: (i) $\overline{A}.\overline{B}.\overline{C} + \overline{A}.B.C + A.B + A.\overline{B}.\overline{C} + A.C$; (ii) $C.\overline{D} + A.C.D + \overline{B}.\overline{C}.\overline{D} + A.\overline{B}.D + \overline{A}.B.\overline{C}.\overline{D}$. (Ans: (a) $A.B + \overline{B}.\overline{C}$, $A.B.C + (\overline{A} + \overline{B}).D$, $A + C$; (b) $A + B.C + \overline{B}.\overline{C}$ and $A.\overline{B} + \overline{A}.\overline{D} + A.C$.)

4. (QMC 1972) Use the Karnaugh map to obtain the simplest minimal forms of the following functions: $f_1 = A.\overline{B}.\overline{C}.D + A.\overline{B}.C + A.\overline{B}.\overline{C}.\overline{D} + \overline{A}.\overline{B}.C.D + A.\overline{B}.C.\overline{D}$, $f_2 = \overline{A}.\overline{C}.\overline{D} + \overline{A}.B.\overline{C} + \overline{B}.\overline{C}.\overline{D} + \overline{A}.B.C + A.B.C$, $f_3 = (A + B + \overline{D}).(B + C + \overline{D}).(\overline{A} + B + \overline{C} + \overline{D}).(A + \overline{B} + \overline{D})$. What is the practical importance of this minimisation? (Ans: $f_1 = \overline{B}.(A + C.D)$, $f_2 = \overline{A}.B + \overline{B}.\overline{C}.\overline{D} + B.C + \overline{A}.\overline{C}.\overline{D}$, where the last term is desirable to eliminate hazards, $f_3 = A.B + \overline{D}$.)

5.   (RHBNC 1988) The seven-segment LED display showing a
     BCD count is of the form of figure 12.12(a). Develop a design
     for a combinational logic circuit to drive its three horizontal
     bars. (Ans: top $= A + D.(B + C) + \overline{B}.\overline{D}$, where the BCD
     number $= ABCD$, middle $= A + C.(\overline{B} + \overline{D}) + B.\overline{C}$, bottom$=$
     $C.(\overline{B} + \overline{D}) + \overline{B}.\overline{D} + B.\overline{C}.D).$)

6*. (RHBNC 1987) You are provided with a square-wave gener-
     ator G and a divide-by-eight counter which triggers on the
     trailing (negative going) edge of G; the counter has outputs
     A, B and C for 1, 2 and 4. Using these and their comple-
     ments and NAND logic gates only, design a circuit which will
     provide the output waveform in figure 12.34, assuming that
     the counter is initially reset to zero before the first G pulse
     arrives. (Ans: figure 12.35.)

7.   Show that $\overline{A}.B + C.D = \Sigma(3,4,5,6,7,11,15)$ and express it also
     as a product of maxterms. (Ans: $\Pi(0,1,2,8,9,10,12,13,14)$.)

8.   (RHBNC 1987) Draw a Karnaugh map for the sum of prod-
     ucts function $f = \Sigma(4,6,8,9)$ in the range 0 to 11. Minimise
     this and draw the final circuit using NAND and NOR gates.
     (Ans: $B.\overline{D} + A.\overline{C}$; figure 12.36.)

9**. (Bedford 1974) Give the truth tables for the logical expres-
     sions AND, OR, NOT and NOR. State de Morgan's theorems
     and show that the NOR operation can be used as a univer-
     sal decision function to replace the three primary operations
     specified above. Draw a logic circuit diagram to show how
     the function $f = (A + B.C).(B + \overline{C}).\overline{A} + A.\overline{B}$ may be imple-
     mented using NOR gates, assuming that the variables A, B
     and C are available as inputs. (Ans: figure 12.37.)

10*. (QMC 1985) (a) Simplify algebraically the function $f =$
     $\overline{w}.\overline{x} + \overline{w}.x.z + \overline{w}.x.\overline{z} + w.x.\overline{y} + w.\overline{y}.\overline{z}$. Draw the logic dia-
     grams of an implementation of f using NAND gates. You
     may assume that both the normal and complemented inputs
     are available. (b) Using a Karnaugh map, implement the
     function $f' = \Sigma(1,3,4,5,11,12)$ and $d = \Sigma(7,9,13,15)$ where d
     are don't care conditions, in both NAND and NOR forms.
     Draw logic diagrams of your solutions assuming all input lit-
     erals are available. (Ans: $\overline{w} + \overline{y}.\overline{z} + x.\overline{y}$, figure 12.38, (b)
     figure 12.39.)

11*. (QMC 1988) For the logic circuit shown in figure 12.40, de-
     rive the function F. Show, with the addition of AND and
     NOT gates only, how figure 12.40 can be converted first into
     a half adder and then used to make a full adder. State
     the Boolean expressions for the full-adder Sum and Carry-
     out. Write down and explain the truth table of a half sub-
     tractor. Derive Boolean expressions for the difference and

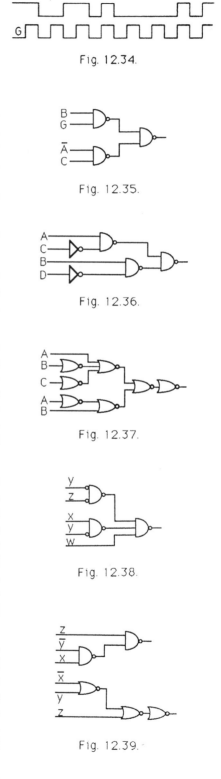

output

Fig. 12.34.

Fig. 12.35.

Fig. 12.36.

Fig. 12.37.

Fig. 12.38.

Fig. 12.39.

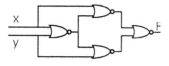

Fig. 12.40.

x	y	B	d
0	0	0	0
0	1	1	1
1	0	0	1
1	1	0	0

Fig. 12.41.

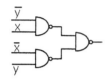

Fig. 12.42.

borrow outputs and show how the NOR gate half-adder circuit must be modified to become a half subtractor. (Ans: $F = x.y + \bar{x}.\bar{y} = \overline{x + y}$; figure 12.41, where $B = \bar{x}.y =$ output of top gate, $D = x + y$.)

12. (QMC 1986) From the canonical sum-of-products expression for the two-input exclusive-OR function, derive the minimal expression that describes this function in terms of NAND gates only. Draw the logic diagram for this implementation. (Ans: $((x.y')'.(x'.y)')'$, where $'$ denotes the complement; figure 12.42.)

13. (QMC 1985) You are to generate a seven-bit binary code for the decimal digits 0 to 9. The code words can be represented by $p_3 p_2 p_1 b_3 b_2 b_1 b_0$ where b_3–b_0 are the normal binary digits in 8421 BCD code and the digits p_1, p_2 and p_3 are even parity digits for the groups of three binary digits $b_3 b_2 b_1$, $b_3 b_1 b_0$ and $b_2 b_1 b_0$, respectively. Write down Boolean expressions for the parity digits in terms of the BCD digits and then list the ten code words. Convert the code words for decimal 2 and 9 to hexadecimal. (Ans: $b_3 \oplus b_2 \oplus b_1$, $b_3 \oplus b_1 \oplus b_0$, $b_2 \oplus b_1 \oplus b_0$; 72, 13, 54, 35, 26, 47, 38, 59.)

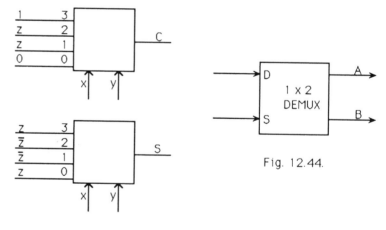

Fig. 12.43.

Fig. 12.44.

14. (QMC 1987) (a) Draw the internal gate-level implementation of a 2×4 decoder. The outputs are to be active low and the circuit must be enabled with two active low inputs. Write down the Boolean expressions for each of the outputs. (b) Draw the internal gate-level implementation of a 4×1 multiplexer. Write down the Boolean expression for the output in terms of the inputs. Implement a full adder using two 4×1

multiplexers, given that $S = \bar{x}.\bar{y}.z + \bar{x}.y.\bar{z} + x.\bar{y}.\bar{z} + x.y.z$ and $C_0 = x.y + x.z + y.z$. Hint: identify the output of each multiplexer with the minterms in these expressions using the Karnaugh map. (Ans: figure 12.43.)

15**. (QMC 1988) A 1×2 demultiplexer is shown in figure 12.44. Implement this circuit using AND and NOT gates only such that data fed into input D appears on output A when select input S is low and on output B when S is high. Redesign your circuit to implement it using NOR and NOT gates only. Using eight of your demultiplexers, design the data rotator circuit shown in figure 12.45. The outputs of the 1×2 DEMUX can be wired-OR and with the input data word equal to *dcba*, as shown in figure 12.45, the circuit's function table is given by

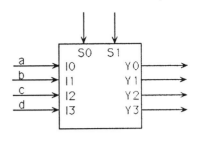

Fig. 12.45.

$S1$	$S0$	$Y3$	$Y2$	$Y1$	$Y0$
0	0	d	c	b	a
0	1	c	b	a	d
1	0	b	a	d	c
1	1	a	d	c	b

(Ans: $A = D.\bar{S}$, $B = D.S$; figure 12.46, where $x = S1$ and $y = S0$.)

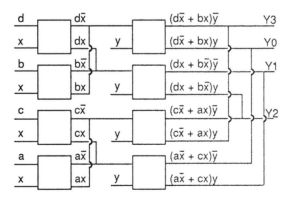

Fig. 12.46.

16*. (RHBNC 1988) A speedometer-style display (figure 12.47) is to be implemented by converting a three-bit binary code to the following code.

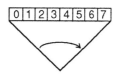

Fig. 12.47. Speedometer-type display.

input:		a	b	c	d	e	f	g
	0	0	0	0	0	0	0	0
	1	1	0	0	0	0	0	0
	2	1	1	0	0	0	0	0
output:	3	1	1	1	0	0	0	0
	4	1	1	1	1	0	0	0
	5	1	1	1	1	1	0	0
	6	1	1	1	1	1	1	0
	7	1	1	1	1	1	1	1

Here $a = 001$ and $g = 111$. Devise a code converter to achieve this, expressing your result (a) in AND–OR form, and (b) in NAND form. Consider whether the circuit would have been better implemented in one of the following forms: (i) using multiplexers, (ii) using demultiplexers, (iii) using PLAs, (iv) using PROMs. (Ans: output $7 = A.B.C$; $6 = 7 + A.B.\overline{C} = A.B$; $5 = 6 + A.\overline{B}.C$; $4 = A$; $3 = 4 + \overline{A}.B.C$; $2 = 4 + \overline{A}.B$; $1 = (\overline{A}.\overline{B}.\overline{C})'$; $0 = \overline{A}.\overline{B}.\overline{C}$. (i) little advantage using 7×3 input multiplexers; (ii) Easy with 3 input $\times$ 8 output demultiplexers followed by ORs; (iii) Easy with PLA; (iv) PROM rather elabaorate.)

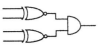

Fig. 12.48.

17*. (RHBNC 1988) Draw the circuit of a comparator for two one-bit numbers. It should output 1 if the two bits are identical and should output zero otherwise. Extend the design so that it can compare two two-bit numbers. (Ans: XNOR $= A.C + \overline{A}.\overline{C}$; figure 12.48.)

18*. Design a look-ahead carry for a 16-bit adder using four 4-stage adders.

19*. Show both algebraically and by rearranging OR and NAND gates that the exclusive OR $(A \oplus B)$ can be expressed in the following eight forms: $A'.B + A.B'$, $(A.B + A'.B')'$, $(A+B).(A'+B')$, $((A'B).(A.B')')'$, $((A+B)'+(A'+B')')'$, $(A.B)'.(A'.B')'$, $(A+B')'+(A'+B)'$ and $((A+B').(A'+B))'$, where $'$ denotes the complement.

20. (RHBNC, 1989). Starting with a one-bit comparator which has three outputs corresponding to $A = B$, $A > B$ and $A < B$, derive a circuit which will compare two two-bit numbers $A_1 A_0$ and $B_1 B_0$. (Ans: figure 12.49.)

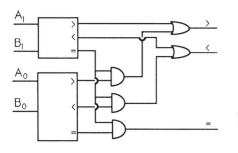

Fig. 12.49.

13

Sequential Logic

13.1 The *RS* Flip-flop

A flip-flop is a combination of gates having two alternative stable
states which can represent 0 and 1. A trigger pulse can flip it from
one state to the other. Until this trigger pulse arrives, the circuit
maintains its logical state indefinitely. The logic gates discussed
in the previous chapter were capable of making a decision; the
flip-flop memorises the decision.

The simplest variety is shown in figure 13.1. The convention will
be adopted here that a logical 0 is represented by a low level and
logical 1 by a high level (positive logic). Consider first the voltage
levels in the absence of any input pulses and suppose that gate 1
gives an output Q which is positive (high). This signal drives gate
2 so that its output $\overline{Q}$ is low (logical 0). This gate feeds a low signal
back to gate 1. In the absence of an input from R, the output Q
is high. The inversion of the signal at both gates provides overall
positive feedback: the signal fed back to the input of each gate
reinforces the existing output. Because of the symmetry of the
circuit, there is a second stable configuration with the opposite
polarity: gate 1 low and gate 2 high.

As long as there is no external input to either gate, either configuration is stable. However, if a large enough **reset** pulse R is
applied to gate 1, it drives Q low and $\overline{Q}$ high. The corresponding
waveforms are sketched in the figure. A positive **set** pulse S at
gate 2 restores the original situation, i.e. Q high, $\overline{Q}$ low.

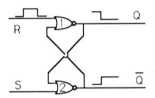

Fig. 13.1. The RS flip-flop and
waveforms when a reset is
applied.

State and excitation tables
This sequence of events can be traced through the two tables of
figure 13.2. The first shows the quiescent states with no input to S
or R, or at most a 1 on one of them. This is called the **state table**
of the flip-flop. The second is the **excitation table** showing the
changes between the output Q at some time and the output $Q(t+1)$
after switching. With a 1 on input R (reset) and a 0 on input S,

S	R	Q	$\overline{Q}$		S	R	Q(t + 1)
0	0	1	0		0	0	Q
0	0	0	1		0	1	0
1	0	1	0		1	0	1
0	1	0	1		1	1	?

(a)   (b)

Fig. 13.2 (a) state table,
(b) excitation table of
the RS flip-flop.

197

the flip-flop goes to $Q(t+1) = 0$ whatever its previous condition; conversely, with a 1 on input S (set) and a 0 on input R, it goes to $Q(t+1) = 1$. In the absence of any input (i.e. $S = R = 0$), it stays steady. The device is called a **set–reset (RS) flip-flop** or **latch**. If it is fed contradictory information, a 1 on both R and S input lines, both Q and $\overline{Q}$ temporarily go to 0; however, this is not a stable configuration, so the circuit settles eventually according to whichever input survives longest. This may not be well defined or reproducible. Set–reset circuits are available commercially with four or more latches per chip; one example is shown in figure D.3 of appendix D.

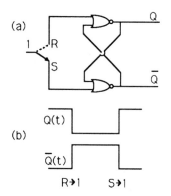

(a)

(b)

Fig. 13.3 (a) A debounce circuit and (b) its waveforms.

Debounce circuit

An elementary example using this flip-flop is the **debounce** circuit shown in figure 13.3. Suppose a piece of electronics is to change state under the action of a mechanical switch. When this switch is moved from position S to position R, the contacts may make and break several times at R before settling to a good contact. It is desirable that the electronics should respond to the first contact and then remain stable, rather than switching back and forth as the switch makes and breaks. This is achieved by the flip-flop of figure 13.3(a), which is reset to $Q = 0$ by the first 1 signal on line R and remains in a fixed state until the switch is moved back to position S, when a 1 signal appears on line S and sets the flip-flop to $Q = 1$.

13.2   Clocks

In a large and complicated system like a computer, it is easy for small timing differences to appear between different parts of the circuitry. Signals propagate just below the speed of light, 30 cm ns^{-1}. Switching times of 1 or 2 ns are realistic in high-speed computers, so timing is a serious problem.

For this reason, it is common to synchronise the switching of gates by a **clock** or **strobe** pulse. To achieve this, it is desirable to reorganise the flip-flop as in figure 13.4 so that it uses NAND gates rather than NOR gates; then the clock pulse C can be a pre-requisite for the NAND gates A and B to respond. In the absence of a clock pulse, both gates A and B give high outputs, so both NAND gates A' and B' of the flip-flop itself are enabled and can respond to the feedback between Q and $\overline{Q}$. You may wish to work through the state table and excitation table, to follow the quiescent logic and the switching. The waveforms on the figure show switching from $Q(t) = 1$ to $Q(t+1) = 0$. The clock pulse C is made narrower than the R pulse and synchronises the switching, as indicated by the broken lines. Logic like this, regulated by a

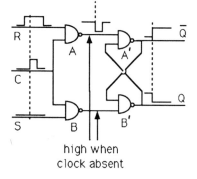

high when clock absent

Fig. 13.4. Clocked RS flip-flop.

clock pulse, is called **synchronous**. Logic operating without clock pulses is called **asynchronous**.

The RS flip-flop suffers the disadvantage that its output is not well defined if $S = R = 1$. A way of avoiding this contradictory situation is shown in figure 13.5. The S and R inputs are derived from a single input D using an inverter to create R from S. This is known as a **D flip-flop**, where D stands for data: the state of the flip-flop is dictated by the data fed to its input. If $D = 1$ it sets to 1 when activated by a clock pulse; if $D = 0$ it resets to 0. It is the basic unit of most memory elements: registers and random-access memory (RAM), and is available commercially in a wide variety of ICs. An example is the 74170 of appendix D, figure D.3.

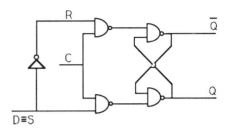

Fig. 13.5. The D flip-flop.

13.3 The JK Flip-flop

As discussed above, the RS flip-flop does not respond in a well-defined way to $R = S = 1$. A step towards solving this problem is given in figure 13.6(a). Here, the Q output is fed back to the input AND gate 1 so as to enable this gate when $Q = 1$; likewise, the $\overline{Q}$ output is fed back to the other input AND gate 2, enabling it when $Q = 0$. Because Q and $\overline{Q}$ are opposite logic states, only one of the AND gates can open and the switching action is well defined, as given in figure 13.6(b). If $Q = 0$, only the J input does anything; if $Q = 1$, only the K input responds. The upshot is that J alone acts as a set input, K alone acts as a reset, and if both inputs are present, the flip-flop changes state, or 'toggles', via whichever AND gate is currently open.

Unfortunately, there is a snag with figure 13.6(a), and it cannot be used as it stands. Suppose the clock pulse is long compared with the switching time of the flip-flop. In this case, as soon as the Q and $\overline{Q}$ outputs change state, the enabling of the input AND gates swops from J to K or vice versa. If $J = K = 1$ and the clock pulse is still there, the flip-flop will toggle back again to its original state. It may continue to do this several times until one of C, J or K inputs disappears; the final state is ill defined.

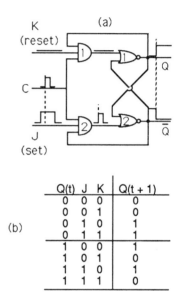

Q(t)	J	K	Q(t + 1)
0	0	0	0
0	0	1	0
0	1	0	1
0	1	1	1
1	0	0	1
1	0	1	0
1	1	0	1
1	1	1	0

Fig. 13.6 (a) A primitive JK flip-flop and (b) its excitation table.

Edge triggering
In figure 13.6(a), the switching action responds to the **level** of the input clock pulse C. An alternative is to make it respond to its **edge**. (This might be done with a CR network, but commercial units in fact adopt more complicated designs which lock out the input pulse once switching has started.) If the switching responds to the transition of the C pulse from low to high, the flip-flop is known as **positive edge triggering**; if it responds to the reverse transition from high to low, it is known as **negative edge triggering**. The manufacturer will specify that the logic signals J and

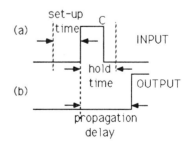

Fig. 13.7 (a) set-up and hold times for positive-edge triggering, (b) propagation delay.

K must be present for a minimum time called the **set-up time** (figure 13.7(a)) before the clock pulse arrives, and must persist for a time called the **hold time** after the appropriate edge of the clock pulse, positive or negative. There is then a **propagation delay** before the outputs switch. Several flip-flops may be involved in one piece of logic; an example is a four-bit counter containing four flip-flops, one for each bit. If the switching of one flip-flop depends on the state of another, the user must make sure that small timing errors will not cause the output of one flip-flop to change state before trigger pulses have finished triggering all flip-flops. For this reason, the propagation delay is deliberately made longer than the hold time. The delay is a compromise between this requirement and speed.

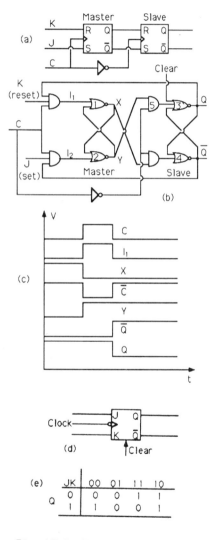

13.4   Master–slave Flip-flop

An alternative solution is shown in figures 13.8(a) and (b). In this arrangement, there are two flip-flops, called master and slave. The first master flip-flop responds to the clock pulse and may change state if J and K are set appropriately; the second slave flip-flop accepts the output of the master flip-flop only when the clock pulse disappears. Consequently, the feedback of Q and $\overline{Q}$ to the input AND gates can never arrive until the clock pulse has disappeared, and multiple toggling is locked out. The waveforms of figure 13.8(c) describe the response to an input $K = 1$ (reset) starting from $Q = 1$. In this case, the switching action of the slave is initiated at gate 5 by $\overline{C}$ going high with $Y = 1$. Exercise 7 shows how the flip-flop may be made purely from NAND gates.

Again the manufacturer will specify set-up and hold times, and the propagation delay must be long enough to accommodate timing jitter between flip-flops of one particular set of logic. Commercial chips usually have the additional feature, shown in (d), of an extra input which will clear the flip-flop to $Q = 0$ even without a clock pulse, i.e. asynchronously. This allows a register of flip-flops to be cleared initially to zero. The clear input bypasses the initial AND gates of figure 13.8(b) and is fed directly to NOR 3. The layout of the 7476 JK flip-flop is shown in figure D.3 of appendix D.

The triangle at the clock input to the flip-flop (figure 13.8(d)) indicates that it is edge sensitive; the bubble indicates that it triggers on the negative-going edge. If it were to trigger on the positive edge, as in figure 13.6, the bubble would be omitted.

A feature of the JK flip-flop is that with $S = R = 1$ it toggles from $Q = 1$ to 0 or vice versa. This makes it ideal for use in counters. A version is available with the J and K lines permanently connected together. If $J = K = 0$ it does nothing; if $J = K = 1$ it changes state. Such a device is called a **toggle (T) flip-flop**.

Fig. 13.8. The JK flip-flop,
(a) schematic, (b) gate layout,
(c) waveforms when J = K = 1
and Q is initially 1,
(d) symbolic form,
(e) Karnaugh map for Q(t + 1).

In following the action of the JK flip-flop, it will be helpful later to use its **state equation** or **characteristic equation**:

$$Q(t+1) = J.\overline{Q} + \overline{K}.Q.$$

This may be derived from the Karnaugh map of figure 13.8(e) and may be checked against the excitation table of figure 13.6(b). The characteristic equation of the T flip-flop is simply $Q(t+1) = \overline{T}.Q + T.\overline{Q}$.

13.5   A Scale-of-4 Counter

In order to appreciate how to use flip-flops, let us consider as an example how to make a counter which will count from 0 to 3 in binary and then reset. Suppose A represents its more significant digit and B the less significant one. The switching action required in response to clock pulses is shown in figures 13.9(a) and (b). Using set and reset signals applied to the A and B flip-flops, the required characteristic equations are:

$$S_B = \overline{B} \qquad\qquad R_B = \overline{S}_B$$

and either

$$S_A = \overline{A}.B \qquad\qquad R_A = A.B$$

or

$$S_A = A \oplus B \qquad\qquad R_A = \overline{S}_A.$$

The former changes A only when it is necessary to do so, i.e. it changes A from 0 to 1 or from 1 to 0; the latter sets all zeros and ones of the Karnaugh map every time and therefore includes some redundancy. An arrangement for achieving the latter with D flip-flops is shown in figure 13.9(c). Alternatively, because the JK flip-flop toggles when $J = K = 1$, the circuit of figure 13.9(d) achieves the same result without the XOR gate required in figure 13.9(c).

If clock pulses are fed in at a constant rate, the Q output of flip-flop B goes positive at half the rate of input clock pulses; $Q(A)$ goes positive at a quarter of the rate. Counters can therefore be used for frequency division.

In the circuits of figures 13.9(c) and (d), both digits A and B change in synchronism with the clock pulse. An alternative is shown in figure 13.9(e), where the clock pulse for A is provided by the positive edge given by $\overline{Q}_0$ switching to 1 (Q_0 switches at this instant to 0). In this case, A switches only after the propagation delay through the less significant digit B. If the counter has many digits, and is in the state ...1111, the carry propagates through each of the lower digits in turn, with a delay in each. This is called **ripple-through carry**. This mode of operation may be

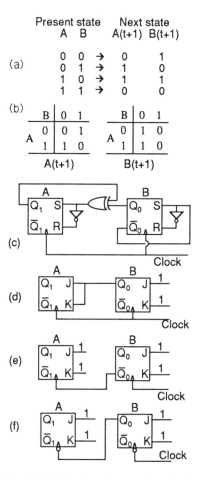

Fig. 13.9. Design of a scale-of-4 counter, (a) excitation table, (b) Karnaugh maps, (c) circuit using RS or D flip-flops, (d) circuit using JK or T flip-flops, (e) and (f) ripple-through carry.

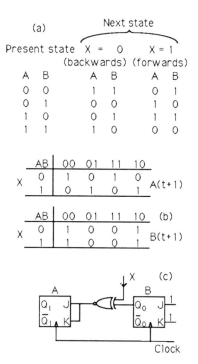

(a)   Next state

Present state		X = 0 (backwards)		X = 1 (forwards)	
A	B	A	B	A	B
0	0	1	1	0	1
0	1	0	0	1	0
1	0	0	1	1	1
1	1	1	0	0	0

AB	00	01	11	10	
X 0	1	0	1	0	A(t+1)
1	0	1	0	1	

AB	00	01	11	10	(b)
X 0	1	0	0	1	B(t+1)
1	1	1	0	0	

(c)

Clock

Fig. 13.10. Design of a bidirectional scale-of-4 counter: (a) excitation table, (b) Karnaugh maps and (c) circuit using JK flip-flops.

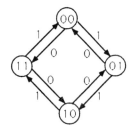

Fig. 13.11. State diagram for the bidirectional counter.

quite satisfactory at low speeds, but is undesirable in a high-speed counter, since input pulses may be arriving at a rate comparable with the rate of carry propagation, and at any instant the counter will not be giving an up-to-date reading; it would give a false result if it drives further logic. If the input pulse train is halted, the counter will catch up and give a correct reading after the carries have propagated through.

In figure 13.9(f), the flip-flop for A is drawn with a circle on its clock input. This indicates that it triggers on a **negative-going** edge; accordingly it is driven by Q_0 going from 1 to 0.

Bidirectional counter

Next, suppose the counter is to count forwards if a control signal X is 1 but backwards if $X = 0$. This makes a **bidirectional counter**. The excitation table and Karnaugh maps are given in figures 13.10(a) and (b), and the characteristic equations are

$$R_B = B \qquad\qquad S_B = \overline{R_B}$$
$$R_A = A \oplus B \oplus X \qquad S_A = \overline{R_A}$$

if the logic is set for every clock pulse or alternatively

$$S_A = \overline{A}.(\overline{X \oplus B}) \qquad\qquad R_A = A.(\overline{X \oplus B})$$

if it is set only when a change is required. These can be implemented with minor changes to figure 13.9(c) using RS or D flip-flops. Using JK flip-flops, the toggling action of A is opposite for $X = 1$ and $X = 0$. This can be achieved with the circuit of figure 13.10(c).

Four-bit ripple-through counters are available commercially as the IC 7493. This counter is actually made in the form of separate scale-of-2 and scale-of-8 counters, which can work independently or can be made into a scale-of-16 counter by feeding the output of one counter to the other. The 74193 shown in appendix D, figure D.4, is a four-bit synchronous up–down counter, i.e. all bits change in synchronism with the clock pulse.

Suppose you want to construct a BCD system. The scale-of-16 counter may be turned into a scale-of-10 counter by using combinational logic to recognise the binary state 1010 and generating a reset (and a carry to the next decade). The counter may be similarly modified to count up to any number $n < 16$.

13.6   State Diagrams

A useful way of representing the switching action of the bidirectional counter is shown in figure 13.11. The steady states are shown in the circles, and the arrows indicate the direction of the changes when $X = 0$ or 1. This is called a **state diagram**.

As a second example, consider the switching of traffic lights, discussed in the previous chapter. There are four transitions:

$$\text{red} \rightarrow \text{red} + \text{yellow} \rightarrow \text{green} \rightarrow \text{yellow} \rightarrow \text{red}.$$

Again there are four quiescent states, which can be labelled a, b, c and d; the state diagram, figure 13.12, is equivalent to the scale-of-4 counter if these four states are represented either by

$$a = 00 \qquad b = 11 \qquad c = 10 \qquad d = 01$$

or

$$a = 00 \qquad b = 01 \qquad c = 10 \qquad d = 11.$$

In either case, additional logic has to be provided to convert the output of the A and B flip-flops into a code driving the three colours of the traffic lights, for example $b \rightarrow \text{red} + \text{yellow}$.

An alternative arrangement is instead to code the colours of the traffic lights directly in the states a, b, c and d. For example, in figure 13.12(b) three digits ABC are used to code red, yellow and green. Now the excitation table and Karnaugh maps are shown in figures 13.13(a) and (b); X represents a don't care condition. There are now many possible characteristic equations to do the logic, but a simple choice would be that shown by the full(S) and broken (R) loops: either

$$S_A = \overline{A}.B \qquad\qquad R_A = A.B$$

or

$$S_A = A.\overline{B} + \overline{A}.B \qquad R_A = \overline{S}_A$$

then

$$S_B = \overline{B} \qquad\qquad R_B = \overline{S}_B = B$$
$$S_C = A.B \qquad\qquad R_C = \overline{S}_C.$$

Four of the eight possible logical combinations of A, B and C do not occur in the state diagram (figure 13.12(b)). What happens if the logic gets into one of these states? A power surge, for example, might temporarily set the combination yellow + green = 011. It is necessary to check that the system does not get stuck in the four unwanted combinations. figure 13.14(a) shows that the logic chosen here leads in every case back to one of the required combinations. A system which returns to the required sequence whatever the initial state is said to be **self-starting**.

However, to illustrate a point, it is possible to derive characteristic equations which would drive the lights round the unwanted loop of four 'illogical' states shown in figure 13.14(b), without ever

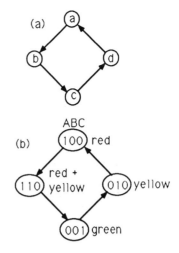

(a)

(b)

ABC

100 red

red + yellow (110)   010 yellow

001 green

Fig. 13.12 (a) State diagram for traffic lights and (b) a possible digital representation.

Present state			Next state		
A	B	C	A	B	C
1	0	0	1	1	0
1	1	0	0	0	1	(a)
0	0	1	0	1	0
0	1	0	1	0	0

(b)

A \ BC	00	01	11	10
0	X	0	X	1
1	1	X	X	0

$A(t+1)$

A \ BC	00	01	11	10
0	X	1	X	0
1	1	X	X	0

$B(t+1)$

A \ BC	00	01	11	10
0	X	0	X	0
1	0	X	X	1

$C(t+1)$

Fig. 13.13 (a) Excitation table and (b) Karnaugh maps for Fig. 13.12(b).

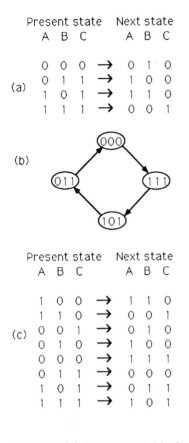

Present state | | | Next state | | |
A	B	C	A	B	C
0	0	0	0	1	0
0	1	1	1	0	0
1	0	1	1	1	0
1	1	1	0	0	1

(a)

(b)

Present state | | | Next state | | |
A	B	C	A	B	C
1	0	0	1	1	0
1	1	0	0	0	1
0	0	1	0	1	0
0	1	0	1	0	0
0	0	0	1	1	1
0	1	1	0	0	0
1	0	1	0	1	1
1	1	1	1	0	1

(c)

Fig 13.14 (a) Excitation table for faulty states, (b) loop of faulty states, (c) the excitation table for equations (13.1).

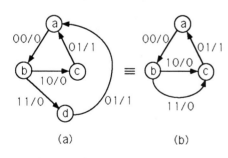

(a)   (b)

Fig. 13.15. Two equivalent trapping systems.

returning to the desired states. As an example, such a set of equations is

$$S_A = \overline{B}.\overline{C} + A.B.C + \overline{A}.B.\overline{C} \qquad R_A = \overline{S}_A$$
$$S_B = \overline{B} \qquad\qquad\qquad\qquad R_B = \overline{S}_B \qquad (13.1)$$
$$S_C = A.B + \overline{A}.\overline{B}.\overline{C} + A.C \qquad R_C = \overline{S}_C$$

as you can readily demonstrate by checking the excitation table they produce, figure 13.14(c). It clearly pays to think about the 'don't care' states, and in this example it might be sensible to arrange that all unwanted states lead to yellow at the next change. This leads to a unique Karnaugh map with no don't care states.

13.7   Trapping Sequences: Pattern Recognition

Another example is rather artificial in the simplified form given here, but is typical of a class of problems and illustrates several points. Consider a digital system which receives numbers at regular intervals governed by a clock. For simplicity, suppose these numbers are restricted to 0, 1, 2 or 3. A situation might arise where we want to recognise the sequences 0, 2, 1 or 0, 3, 1 and take some action. This is a **pattern recognition** problem. It might be used, for example, to trap faulty operation in a machine or a process. An analogous situation arises in a telephone system, where the operator must be alerted by an emergency code (999 in the UK); this involves more binary digits, but is otherwise the same problem.

The sequences may be recognised from the state diagram shown in figure 13.15. In (a), the first pair of digits on each arrow indicates the pair of binary input digits which take the system through the required patterns $a \rightarrow b \rightarrow c \rightarrow a$ or $a \rightarrow b \rightarrow d \rightarrow a$; any other pair of input digits restores the system to state a. The number after the stroke is 1 if the system is to produce an output.

The response of states c and d to any pair of input digits is identical. It is a general rule that in this case there is redundancy in the system, and one of the two states can be eliminated. This is done in (b) by eliminating state d and providing two alternative routes from b to c.

It is arbitrary how states a, b and c are coded. Suppose they are represented by flip-flops AB taking values $a = 00$, $b = 01$, $c = 11$. The excitation diagram and Karnaugh maps are shown in figure 13.16, where x and y represent the two input digits. Figure 13.16(a) is easily constructed by putting in entries corresponding to the four arrows of figure 13.15(b) and filling the remaining positions with zeros. There is an unused combination $AB = 10$, which should not occur. To deal with it, corresponding entries in the table may be set to 0, so that the system returns to state a.

The switching logic for A and B and the output C is:

$$A(t+1) = \overline{A}.B.x \tag{13.2}$$
$$B(t+1) = \overline{A}.B.x + \overline{A}.\overline{B}.\overline{x}.\overline{y} \tag{13.3}$$
$$C = A.B.\overline{x}.y.$$

It is straightforward to arrange this logic with RS or D flip-flops, e.g. $S_A = \overline{A}.B.x$, $R_A = \overline{S}_A$. Suppose, however, JK flip-flops are to be used instead. So far, these flip-flops have been used in an intuitive way in counters, and it is not immediately clear how to use their three modes of operation (set, reset and toggle) in the present problem. The solution is to use the characteristic equation of this flip-flop:

$$Q(t+1) = J.\overline{Q} + \overline{K}.Q.$$

Equation (13.3), for example, may be rewritten

$$B(t+1) = \overline{K}_B.B + J_B.\overline{B}$$
$$J_B = \overline{A}.\overline{x}.\overline{y} \tag{13.4}$$
$$\overline{K}_B = \overline{A}.x \quad \text{or} \quad K_B = A + \overline{x}. \tag{13.5}$$

Likewise, equation (13.2) may be rewritten

$$A(t+1) = J_A.\overline{A} + \overline{K}_A.A$$
$$J_A = B.x \tag{13.6}$$
$$\overline{K}_A = 0 \quad \text{or} \quad K_A = 1.$$

You may easily check using the Karnaugh maps that these equations give the correct transitions. For example, for the two entries in the top left-hand corner with $A = B = x = y = 0$, B toggles ($J_B = K_B = 1$) and A resets ($K_A = 1$, $J_A = 0$).

It is possible but very laborious to check visually by drawing the waveforms which describe the sequence of operations. This is shown in figure 13.17 using equations (13.4)–(13.6) and checks that A and B switch as required. Generally this is much harder than checking the algebra of the Karnaugh maps.

There is one important final point about constructing the logic of JK flip-flops not illustrated by this example. Sometimes it happens that the expression for $A(t+1)$, say, does not depend on A. For sake of example, the equation might be

$$A(t+1) = B.x \tag{13.7}$$

resembling equation (13.2) but not containing the $\overline{A}$. How can this be written in the form of the characteristic equation $Q(t+1) = J_A.\overline{A} + \overline{K}_A.A$? The answer is simply to use the fact that $A + \overline{A} = 1$ and write equation (13.7) as

$$A(t+1) = B.x.(A + \overline{A}).$$

(a)

Present state AB	x y	Next state AB	Output C
a = 00	0 0	01	0
	0 1	00	0
	1 1	00	0
	1 0	00	0
b = 01	0 0	00	0
	0 1	00	0
	1 1	11	0
	1 0	11	0
c = 11	0 0	00	0
	0 1	00	1
	1 1	00	0
	1 0	00	0

(b)

		AB = 00	01	11	10
	xy = 00	0	0	0	0
	01	0	0	0	0
A(t+1)	11	0	1	0	0
	10	0	1	0	0
	00	1	0	0	0
	01	0	0	0	0
B(t+1)	11	0	1	0	0
	10	0	1	0	0

Fig. 13.16 (a) Excitation table and (b) Karnaugh maps for the flip-flops in this trapping system.

Then
$$J_A = B.x$$
$$\overline{K}_A = B.x \qquad \text{or} \qquad K_A = \overline{B} + \overline{x} = \overline{J}_A.$$

13.8 Registers and Memory

The memory of a computer holds numbers (and instructions) in a large array of registers. One register consists of a group of flip-flops, each of which holds one bit of information (a 1 or a 0). Figure 13.18(a) illustrates a four-bit register made of edge-triggered D flip-flops. All digits of the register are loaded in parallel from lines D_3 to D_0 under the control of a clock pulse, which is allowed through an AND gate by an enable signal.

Registers are usually of 8, 16, 32 or 64 bits. One such register holds a **word** of information; one **byte** is 8 bits. Usually provision is made for a master clear signal which can reset the whole register to zero. A **memory** is an assembly of registers, each of which can be addressed individually. Figure 13.18(b) illustrates how this is done for a memory consisting of 16 registers. The address in this case is given by 4 binary digits, which 'point' to a particular register. The address is transmitted along 4 lines called the **address bus**; decoders driven by this bus enable individual registers via AND gates. The inputs D_3 to D_0 are connected in common to all registers and are called the **data bus** (figure 13.18(c)). To write a word to a particular register, the control unit of the computer puts a word on to the data bus and at the same time sends an address via the address bus. A write signal via the **control bus** then activates the actual write operation. The data bus is also used to read information from the register by means of AND gates which connect outputs D_3' to D_0' back to the data bus. The gates at the required address are opened by a **read** signal from the central processor. Nowadays, VLSI (very-large-scale integration) allows a memory of 256 Kbytes and its attendant buses to be laid down on a single chip. A memory of 65 536 words (usually referred to as 64K) requires 16 address bits; a memory of 1 048 576 words (1M) requires 20 address bits.

13.9 Computer Architecture

We now review briefly the way a computer is organised internally. This is shown schematically in figure 13.19. The heart of the computer is the control processor unit (CPU) which communicates with memory and input–output peripherals via address, data and control buses. The peripherals are often treated as simple extensions of memory.

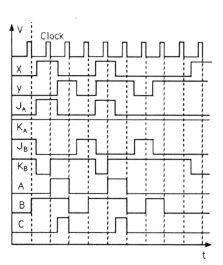

Fig. 13.17. Waveforms showing the sequence of operations in the trapping system of Fig. 13.16.

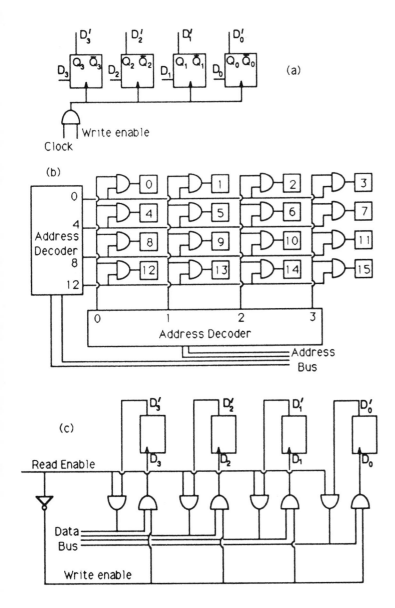

Fig. 13.18 (a) a 4-bit register and (b) a 16-word memory and its connections to the address bus, (c) connections to the data bus.

The computer operates under the direction of a program stored in memory. The program consists of a sequence of instructions. One instruction might be, for example, to add the contents of a particular memory register to the contents of the accumulator. In executing one such instruction, there are three basic steps controlled

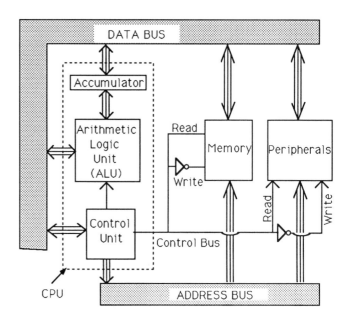

Fig. 13.19. Schematic organisation of a computer or microprocessor.

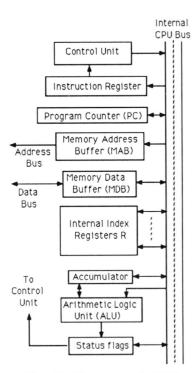

Fig. 13.20. A simple CPU

by a sequence of clock pulses. The control unit first FETCHES the instruction from the program by sending a read instruction over the control bus and an address via the address bus. The memory puts the instruction on the data bus, and it is read into the instruction register. There it is DECODED by the control unit. The third step is to EXECUTE the instruction.

The internal layout of the CPU varies from computer to computer, but might for example be as shown in figure 13.20. Data are transferred to and from the CPU via the **memory data buffer** (MDB). Addresses are put on to the address bus via the **memory address buffer** (MAB). A **program counter** (PC) keeps track of the address of the instruction to be executed.

The microprogram

The FETCH step would consist of the gating steps:

Cycle 1 Transfer the reading address from the program counter: PC out, memory address buffer in.

Cycle 2 Read the instruction: enable memory read, memory data buffer in.

Cycle 3 Transfer the instruction to the control unit: memory data buffer out, instruction register in.

Cycle 4 Increment PC by 1, ready for the next instruction.

The DECODE step sends control in the microprogram to an address chosen by the bits of the instruction code. The EXECUTE step depends on the instruction. Let us follow as an example the addition of the contents of memory register X to those of the accumulator; suppose the address of X is held in register $R1$ of the CPU. The steps would be:

Cycle 5   Address the memory: $R1$ out, memory address buffer in.

Cycle 6   Read from memory: memory read, memory data buffer in.

Cycle 7   Do the addition: memory data buffer out and add to the contents of the accumulator in the ALU.

At the end of cycle 7, the instruction is finished and control is restored to Cycle 1. At this point, the ALU automatically sets up a **status register** indicating what happened in this instruction. One bit indicates whether an overflow has occurred, another whether a carry is generated, a third whether the result is zero and a fourth the sign of the result. The next instruction might refer to these bits and jump to a new part of the program if one of them is set.

The microprogram given here is a set of instructions stored in a fast read-only memory. Each instruction consists of a pattern of bits. In the control unit, the instruction is fed to a set of decoders whose outputs drive individual operations. Each bit activates a particular operation, such as reading or writing to or from memory. A group or **field** of these bits tells the ALU what instruction to perform, for example, none, addition, form $\overline{A}$, $A.B$ and so on. Computers differ in the versatility of these instructions, the number of internal registers R, the number of bits, and some have more than one internal bus for speed. For details of individual microprocessors and computers, you need to refer to the manufacturer's literature and specialised texts.

13.10   Shift Registers

One of the fundamental operations needed for arithmetic is a shift register. Shifting the digits of a binary number one place left is equivalent to doubling it (see figure 13.21(a)); decimal 5 is 0101 in binary while 1010 is equal to decimal 10. Conversely, shifting the digits one place right halves the number. A register which will execute these steps is called a **shift register**.

Shift registers are made as commercial integrated circuits in many varieties. The simplest is capable only of accepting data serially at one end and transmitting it serially at the other. Such a shift register is useful for driving a low-speed output by feeding it one bit at a time. Conversely, it can accumulate serial input from a low-speed device. It can also function as a digital delay line. One

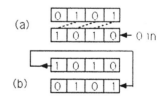

Fig. 13.21 (a) The shift operation, (b) the rotate operation.

IC, the 74195, is shown in figure D.4 of appendix C. Other varieties are capable of parallel input or output, or both. Bidirectional shift registers are capable of shifting left or right under the control of a select line. Provision is made for carries into or out of the register to right and left and several such units may be cascaded. In halving a number, the digit fed in from the extreme left needs to be zero if the number is positive but one if it is negative. Sometimes the digits carried out of the register are discarded. However, while doing arithmetic the logic will generally examine the digits at the left-hand end. When the bit moved out differs from the new leftmost bit, an **overflow** is indicated. To convince yourself of this, consider as an example an eight-bit register holding numbers from -128 to $+127$; work out what happens if you double 63, 127, -64 and -128 as examples.

Sometimes, while doing logical operations, it is desirable to transmit digits from the left-hand end to the right-hand end or vice versa. Such a shift register is called a **circular** shift register, and the operation is called **rotate**. It is illustrated in figure 13.21(b).

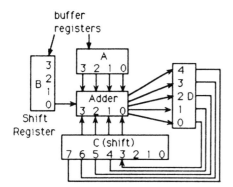

Fig. 13.22. A multiplier unit.

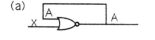

(a)

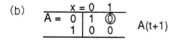

(b)

Fig. 13.23 (a) An unstable circuit and (b) its excitation table.

Multiplication

The shift register is a basic unit involved in the multiplication of two numbers. Figure 13.22 shows one possible scheme for achieving this, though there are many others. It is a challenging project for the student to set up such a hard-wired multiplier using units from appendix D.

Suppose a four-digit number A is to be multiplied by a four-digit number B, and suppose both are unsigned, for simplicity. Initially the accumulator C is cleared and the numbers A and B are loaded from memory into buffer registers; the B register is a shift register. At the first cycle, A is to be multiplied by the least significant bit of B; to achieve this, A and C_3–C_6 are added and loaded into a temporary register D. At cycle 2, the contents of D are loaded into C_3–C_7 if the least significant bit of B is 1, but otherwise this step is vetoed and the previous cycle was wasted. With master–slave units, we could dispense with D and write the sum directly back into C_3–C_7 in a single cycle controlled by B_0. At cycle 3, B and C are shifted one place right. This sequence is repeated with B_1–B_3 and stops at cycle 11, before shifting B and C right in the final iteration.

The cycles are controlled by a counter and each cycle is identified by a decoder. Suppose the output of this counter is coded $T1$–$T11$. Then the logical expression for loading the contents of D into accumulator C would be $(T2 + T5 + T8 + T11).B0$ and the shift instruction would be $(T3 + T6 + T9)$. In this way the logic can be hard-wired, rather than being controlled by a microprogram.

13.11* Asynchronous Logic

So far, attention has been concentrated on synchronous logic governed by clock pulses. It is necessary finally to consider the more difficult situations which arise when there is no clock pulse and the logic is **asynchronous**. The logic must now respond sensibly at whatever times input signals arrive and in whatever order. Two complications arise: (a) not all states are stable, and the system may run through a series of intermediate states before stopping in a stable state, (b) the sequence of states may depend critically upon the order in which digits change, giving rise to what is called a race situation. The first task is to decide which states are stable and which unstable.

Some circuits do not respond in a stable way to inputs, but instead oscillate between two states; alternatively, they may cycle continuously round a loop. The former behaviour is illustrated in figure 13.23(a). The output of the NOR gate is given in (b) in terms of its inputs A and x. For $x = 1$, there is a stable state, ringed in the table, with $A = 0$; if A is initially 1, it switches to $A = 0$ and thereafter is stable. However, if $x = 0$ there is no stable state, and the circuit oscillates between $A = 0$ and 1. (In passing, this is a simple way of making a pulse generator.)

As a second example, consider a pair of flip-flops A and B which respond to an input x according to the excitation table of figure 13.24(a). If $x = 0$, A and B remain unchanged; these stable states are ringed in the table. However, if $x = 1$, there is no stable state and the flip-flops loop continuously through the sequence $00 \rightarrow 01 \rightarrow 10 \rightarrow 11 \rightarrow 00$. This is called an **autonomous sequential circuit**. The lights on a pin-ball machine or disco run through a sequence like this. Figure 13.24(b) shows the Karnaugh maps for $A(t + 1)$ and $B(t + 1)$, and figure 13.24(c) the logic which will drive A and B flip-flops through such a sequence. Figure 13.24(d) shows the waveforms when $x = 1$.

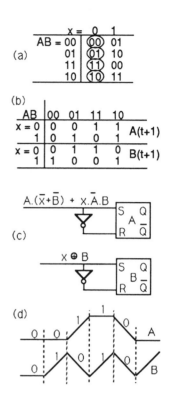

Fig. 13.24. A system which loops through AB = 00, 01, 10, 11, 00 for x = 1: (a) excitation table, (b) Karnaugh maps for A and B, (c) circuit and (d) waveform when x = 1.

Races

In this second example, it was assumed that A and B can switch states together: $AB = 01 \rightarrow 10$ and $11 \rightarrow 00$. In reality, one of the flip-flops will always switch faster than the other, though this might not be reproducible. It is necessary to consider the effects this will have. A situation where two (or more) states try to change simultaneously is known as a **race condition**. If the outcome is independent of which state switches first, it is called a **non-critical race**. If the outcome does depend on which state wins the race, it is called a **critical race**. In the example, suppose A switches significantly faster than B. The sequence with $x = 1$ will become $00 \rightarrow 01 \rightarrow 11$ (instead of 10) $\rightarrow 01$ (instead of 00) $\rightarrow 11$, etc. Conversely, if B wins the race the sequence is $00 \rightarrow 01 \rightarrow 00 \rightarrow 01$,

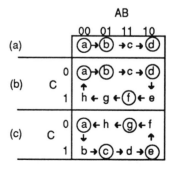

Fig. 13.25. Sequences for (a) AB = 00,01,10, (b) and (c) AB = 00,01,10,11.

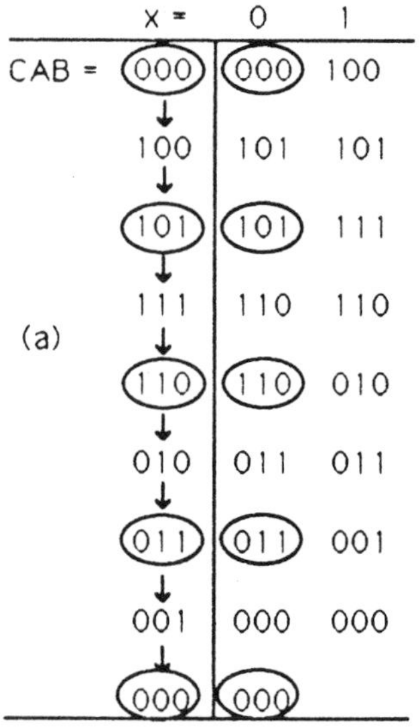

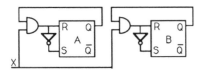

CAB =	x = 0	1
(000)	(000)	100
↓ 100	101	101
(101)	(101)	111
↓ 111	110	110
(110)	(110)	010
↓ 010	011	011
(011)	(011)	001
↓ 001	000	000
(000)	(000)	

(a)

AB =	00	01	11	10	
xC = 00	0	0	0	0	
01	1	1	1	1	C(t+1)
11	1	1	1	0	
10	1	0	0	0	
00	0	0	1	1	
01	0	0	1	1	A(t+1)
11	0	1	1	1	
10	0	0	0	1	
00	0	0	1	1	
01	1	1	0	0	B(t+1)
11	1	1	0	0	
10	0	0	1	1	

Fig. 13.26 (a) Excitation table for Figure 13.25(d); (b) the Karnaugh maps for this excitation table.

so there is a critical race in this example. In a well-behaved system, critical races must be avoided. Non-critical races are acceptable.

It may be possible to side-step this problem by making A and B with master–slave JK flip-flops and using the trigger signal as clock pulse. The master responds to the required logic changes and passes the result to the slave only on the trailing edge of the trigger pulse. The outputs of the slaves remain unchanged while the master is changing either or both of A and B. However, such logic relies on the trailing edge of the trigger pulse arriving at flip-flops A and B simultaneously. If the trigger pulse is sluggish, one flip-flop may respond faster than the other. Output A, say, might change before the trigger pulse has disappeared from the master of B, and this might allow time for B to respond to the new A value, giving the wrong result. With care, the right timing can be arranged, but it is better to make the logic proof against race situations. This necessarily makes it more complicated.

To illustrate how this may be achieved, consider a system following the familiar sequence $AB = 00 \rightarrow 01 \rightarrow 10 \rightarrow 11 \rightarrow 00$ in response to an input signal $x = 1$. Suppose successive states are labelled a, b, c, etc. Figure 13.25(a) (previous page) represents the transition from stable state a (ringed to indicate stability) to b (stable), then via an unstable state c to d (stable). This deals successfully with the sequence $00 \rightarrow 01 \rightarrow 10$. However, there is now no stable state for 11. This can only be achieved by the addition of another digit C. Figure 13.25(b) shows that this allows a stable state f with $C = 1$, $AB = 11$, reached via unstable state e. Finally, stable state a is reached via g and h. An alternative representation is shown in figure 13.25(c).

Figure 13.26(a) shows the excitation table corresponding to figure 13.25(c). Starting at the top left-hand corner, 000 is stable for $x = 0$, but switches to 100 when $x = 1$. This state is unstable and switches to 101 whether $x = 1$ or 0. You can follow the remaining transitions through the table. Figure 13.26(b) gives the Karnaugh maps for $C(t + 1)$, $A(t + 1)$ and $B(t + 1)$, from which the characteristic equations are derived:

$$C(t + 1) = C.(\overline{x} + B + \overline{A}) + x.\overline{A}.\overline{B}$$
$$A(t + 1) = A.(\overline{B} + \overline{x} + C) + x.C.B$$
$$B(t + 1) = C.\overline{A} + \overline{C}.A.$$

Manifestly, this is more complicated than the logic of a synchronous counter doing the same job.

13.12   Exercises

1.   Two RS flip-flops hold a two-bit binary number. Design a circuit to convert it to its 1's complement if an input signal x is 1. (Ans: figure 13.27.)

Fig. 13.27.

2. Two clocked *RS* flip-flops hold a two-bit binary positive number. Design a circuit so that a clock pulse x converts it to its 2's complement. (Ans: figure 13.28.)

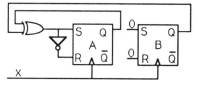

Fig. 13.28

3. (RHBNC 1989) Give the circuit of a counter of three *D* flip-flops which goes through the following sequence: $ABC = 000, 100, 110, 111, 011, 001, 000$. Draw the transition diagram of the unused states. (Ans: figures 13.29 and 13.30.)

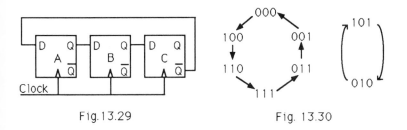

Fig.13.29   Fig. 13.30

4. (QMC 1988) A five clock-cycle sequence generator is required to produce the waveforms for signals *A*, *B* and *C* as shown in figure 13.31. Design the sequential circuit to use three negative edge-triggered *T* flip-flops designated *FFA*, *FFB* and *FFC*. Draw the state diagram of the solution using don't care states and show that if the input to *FFA* is $A + B.\overline{C}$ it is not self-starting. By modifying the toggle input of *FFA*, redesign the circuit to be self-starting and redraw the final state diagram. (Ans: figures 13.32 and 13.33; modified input to $FFA = B.\overline{C}$; figure 13.34.)

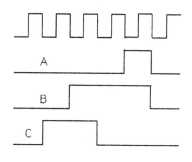

Fig. 13.31

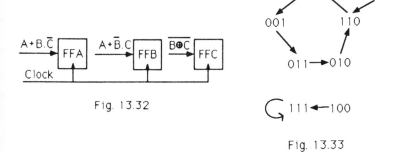

Fig. 13.32

Fig. 13.33 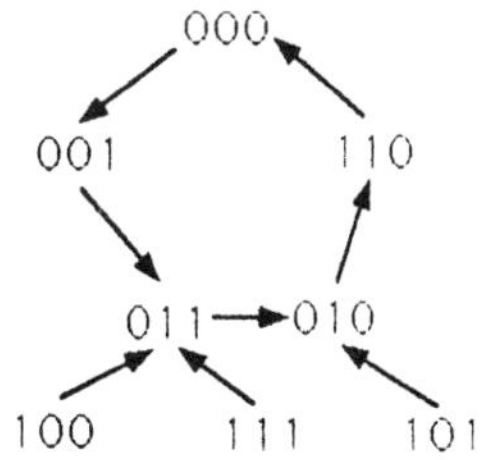   Fig. 13.34

5. (RHBNC 1988) Draw the circuit and give the state transition table for a *JK* flip-flop. What is the characteristic equation for the transition?

6**. Show that a *JK* flip-flop is obtained by changing all the gates of figure 13.9(b) to NAND gates.

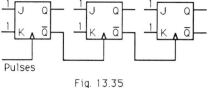

Fig. 13.35

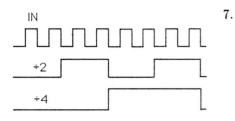

IN

÷2

÷4

Fig. 13.36

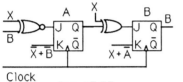

Clock

Fig. 13.37

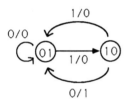

1/0

0/0

0/1

Fig. 13.38

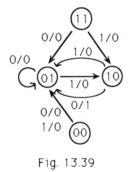

Fig. 13.39

7. (RHBNC 1987) Draw the connections for three JK negative edge-triggered master–slave flip-flops to act as a divide-by-eight ripple counter. Sketch the following waveforms so as to show the relative times of the transitions in each waveform: (a) at the input, (b) at the output, (c) at the intermediate divide-by-2 and divide-by-4 outputs. (Ans: figures 13.35 and 13.36.)

8. (RHBNC) Design from first principles a synchronous divide-by-three up/down counter using JK flip-flops and states 00, 01 and 10. (Ans: figure 13.37; $x = 1$ counting upwards, 0 counting downwards.)

9. (RHBNC 1989) Using JK flip-flops, design a synchronous code sequence generator circuit that gives the following code-words on successive clock pulses: $ABC = 000, 001, 011, 010,$ 110, 111, 101, 100, 000. (Ans: $J_A = B.\overline{C}$, $K_A = \overline{B}.\overline{C}$, $J_B = \overline{A}.C$, $K_B = A.C$, $J_C = A.B + \overline{A}.\overline{B}$, $K_C = \overline{J}_C$.)

10*. (QMC 1986) Design a clocked sequential circuit using positive edge-triggered master–slave JK flip-flops to follow the state diagram shown in figure 13.38. The external input and the output are x and y, respectively. Draw up the state table and derive the transition and excitation table required for the flip-flop inputs and the circuit output. Deduce what will happen if the unused states are accidentally entered. Redraw the state diagram including the unused states. Finally draw the logic diagram. How can you ensure that the output $y = 1$ only occurs for the case shown in figure 13.38? (Ans: $J_A = x.B$, $K_A = \overline{J}_A$, $J_B = 1$, $K_B = x$; $y = \overline{x}.A.\overline{B}$; figure 13.39; make $y = 0$ for unused states.)

11. (QMC 1987) Design a clocked sequential circuit using positive edge-triggered master–slave JK flip-flops to follow the state table given below.

Present state	Next state		Output, y	
	$x = 0$	$x = 1$	$x = 0$	$x = 1$
a	b	e	0	1
b	c	a	0	0
c	d	b	0	0
d	e	c	0	0
e	a	d	1	0

In this table x and y are the external input and output, respectively. If the states a to e are encoded in binary as 000 to 100 consecutively, draw the encoded state diagram and derive the transition and excitation table required for the flip-flop inputs and the circuit output. Finally draw the logic diagram.

(Ans: figure 13.40; $J_A = x.\overline{B}.\overline{C} + \overline{x}.B.C$, $K_A = x + \overline{B} + \overline{C}$,
$J_B = \overline{x}.C + x.A$, $K_B = \overline{x}.C + x.\overline{C}.\overline{A}$, $J_C = \overline{x}.\overline{A} + x.A + B$,
$K_C = \overline{x} + \overline{A}$, $y = \overline{B}.\overline{C}.(x.\overline{A} + \overline{x}.A).$)

12. Using JK flip-flops, design a synchronous decade counter which will count from 0 to 9 in binary and then reset. (Ans: figure 13.41.)

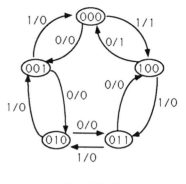

Fig. 13.40

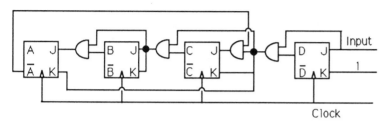

Fig. 13.41

13**. (RHBMC 1988) Draw the state transition diagram for the synchronous counter of figure 13.42. The flip-flops have no direct set or clear inputs. However, it is required to modify the circuit so that it can be reset on the next clock pulse when a given logic input $x = 1$. The circuit must also be self-starting and correcting by directing all states into the sequence which contains 0000. Show how to make the necessary modifications to the circuit. Show how the modulo of the counter may be multiplied by two using only one extra flip-flop, while retaining synchronous operation. (Ans: $0000 \rightarrow 1000 \rightarrow 1100 \rightarrow 1110 \rightarrow 1111 \rightarrow 0111 \rightarrow 0011 \rightarrow 0001 \rightarrow 0000$; $0010 \rightarrow 1001 \rightarrow 0100 \rightarrow 1010 \rightarrow 1101 \rightarrow 0110 \rightarrow 1011 \rightarrow 0101 \rightarrow 0010$; $J_A = \overline{x}.\overline{D}$, $K_A = \overline{J}_A$, $J_B = \overline{x}.A$, $K_B = \overline{J}_B$, $J_C = \overline{x}.B$, $K_C = \overline{J}_C$, $J_D = \overline{x}.C$, $K_D = \overline{J}_D$; many methods of correction, e.g. $x = A.\overline{B}.\overline{C}.D$; e.g. figure 13.43.)

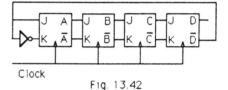

Fig. 13.42

14. (RHBNC 1989) Explain briefly, with the aid of a timing diagram, the sequence of signals in the address, data and control buses involved in writing a byte of data to computer memory.

15. A circuit has two inputs x_1 and x_2 and two output states y_1 and y_2. Its transition table is given in the following table. Are there any non-critical races, critical races or cycles?

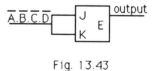

Fig. 13.43

x_1x_2	00	01	11	10
y_1y_2 00	01	00	11	00
01	01	11	11	10
11	11	01	11	10
10	11	00	11	10

(Ans: (a) Input 01, cycle $01 \to 11 \to 01$; (b) input 11, non-critical race $00 \to 11$; (c) input 1, critical race with $y_1 y_2 = 01$.)

16. (QMC 1972) Distinguish between synchronous, asynchronous and autonomous sequential circuits. Discuss the instabilities which can occur in sequential systems and indicate how they may be avoided. Illustrate your answer with examples of stable and unstable circuits.

17**. Design an asynchronous counter which will count through the sequence $00 \to 01 \to 11 \to 10 \to 00$ on the transitions $x = 1 \to x = 0$ of a control signal. (Ans: $J_A = x.B$, $K_A = \overline{x}.\overline{B}$, $J_B = \overline{x}.\overline{A}$, $K_B = \overline{x}.A$.)

18. Devise a scheme for making a 24-hour digital clock using a 50 Hz pulse generator, several counters 74LS193 shown in appendix D, figure D.4, and the 74LS47 BCD to seven-segment converter.

14

Resonance and Ringing

14.1 Introduction

Resonance in mechanical systems is a familiar phenomenon. If you pluck the string of a violin it vibrates at a frequency which depends on its length and tension. Likewise, rotating machinery is liable to resonate violently at certain speeds (for example, a washing machine). We shall find that a circuit containing an inductance L and capacitance C also oscillates with a natural frequency $\omega_0 = 1/\sqrt{LC}$; over a range of frequencies close to this, large currents or voltages can be excited in the circuit.

This chapter will go through two explicit examples of resonance; a fair amount of algebraic manipulation is necessary, but it is straightforward. These two cases may be regarded as worked examples of the ideas of impedance developed in Chapter 6. This work will pay a dividend later in the discussion of active filters (Chapter 16) and will also serve as a useful example for Fourier analysis in the next chapter.

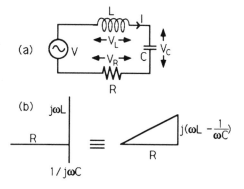

14.2 Resonance in a Series LCR Circuit

The classic example of a resonant circuit is shown in figure 14.1(a) and consists of L, C and R in series. As an exercise, this circuit will be discussed using (i) phasor diagrams, and (ii) complex numbers.

From Kirchhoff's voltage law,

$$V = V_R + V_C + V_L = RI + (1/C) \int I \, dt + L\frac{dI}{dt}. \qquad (14.1)$$

Suppose

$$I = I_0 e^{j\omega t} \qquad (14.2)$$
$$V = V_0 e^{j(\omega t - \phi)} \qquad (14.3)$$

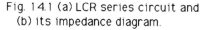

Fig. 14.1 (a) LCR series circuit and (b) its impedance diagram.

217

so that ϕ is the angle by which the current *leads* the applied voltage. Then

$$V = \left(R + \frac{1}{j\omega C} + j\omega L\right) I = ZI. \qquad (14.4)$$

(i) From the impedance diagram (figure 14.1(b)),

$$V_0 = I_0|Z| = I_0[R^2 + (\omega L - 1/\omega C)^2]^{1/2} \qquad (14.5)$$
$$\text{phase}(V) = \text{phase}(I) - \phi = \text{phase}(I) + \text{phase}(Z)$$

so

$$\phi = -\tan^{-1}(\omega L - 1/\omega C)/R. \qquad (14.6)$$

(ii) The same conclusion can be reached algebraically using complex numbers from (14.2)–(14.4):

$$V_0 e^{-j\phi}e^{j\omega t} = [R + j(\omega L - 1/\omega C)]I_0 e^{j\omega t}. \qquad (14.7)$$

Cancelling the factor $e^{j\omega t}$ and taking real and imaginary parts,

$$V_0 \cos\phi = RI_0 \qquad (14.8)$$
$$-V_0 \sin\phi = (\omega L - 1/\omega C)I_0. \qquad (14.9)$$

Squaring and adding (14.8) and (14.9),

$$V_0^2 = I_0^2[(\omega L - 1/\omega C)^2 + R^2]$$

in agreement with (14.5); and dividing (14.9) by (14.8):

$$\tan\phi = -(\omega L - 1/\omega C)/R$$

in agreement with (14.6).

Can you arrive at these same results using $V = V_0 \cos(\omega t - \phi)$ and $I = I_0 \cos \omega t$?

Bode plot
The fundamental results of these alternative approaches are contained in equations (14.5) and (14.6) for I_0 and ϕ. At this point, it is however necessary to mention a source of confusion over the sign of the phase. In mechanics and in atomic and nuclear physics it is conventional to plot instead the angle $\psi = -\phi$; this would correspond here to the angle by which current *lags* voltage; i.e. if $V = V_0 \exp(j\omega t)$, $I = I_0 \exp[j(\omega t - \psi)]$. In figure 14.2, the convention used in electrical engineering is followed and ϕ is displayed; the plot is made for a small value of R giving a narrow peak. The

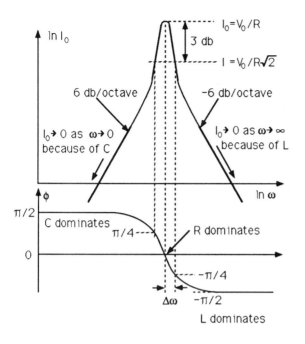

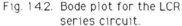

Fig. 14.2. Bode plot for the LCR
series circuit.

curve for I_0 is called a **resonance** curve. The frequency ω_0 of the
peak is given by

$$\omega_0 L = 1/\omega_0 C$$

or

$$\boxed{\omega_0 = 1/\sqrt{LC}.} \tag{14.10}$$

At this frequency, the reactances of L and C cancel, so $\phi = 0$
and I_0 peaks at the value V_0/R. As $\omega \to 0$, C dominates, and as
$\omega \to \infty$, L dominates.

Resonance width
The circuit may be used as a narrow-band filter (for example, in
tuning a radio) since it only passes significant current near $\omega = \omega_0$.
It is of interest to find a quantitative measure of the width of the
peak. It turns out that this is related to power considerations. The
average power dissipated is:

$$P = \frac{1}{2}I_0^2 \operatorname{Re}(Z) = \frac{\frac{1}{2}V_0^2 R}{(\omega L - 1/\omega C)^2 + R^2}. \tag{14.11}$$

On either side of the resonant peak, the power falls to half its
maximum value when $(\omega L - 1/\omega C) = \pm R$, i.e. $\tan\phi = \pm 1$ or
$\phi = \pm\pi/4$. At this point the current falls from its maximum value

by a factor $1/\sqrt{2}$. The half-width $\Delta\omega$ of the peak is obtained by setting $\omega = \omega_0 \pm \Delta\omega$. Then

$$(\omega_0 \pm \Delta\omega)L - \frac{1}{(\omega_0 \pm \Delta\omega)C} = \pm R. \tag{14.12}$$

To simplify this expression, the binomial expansion is used to make the approximation:

$$(\omega_0 \pm \Delta\omega)^{-1} = \frac{1}{\omega_0}\left(1 \pm \frac{\Delta\omega}{\omega_0}\right)^{-1} \simeq \frac{1}{\omega_0}\left(1 \mp \frac{\Delta\omega}{\omega_0}\right).$$

Then equation (14.12) reads

$$(\omega_0 \pm \Delta\omega)L - \frac{1}{\omega_0 C}\left(1 \mp \frac{\Delta\omega}{\omega_0}\right) = \pm R$$

and using (14.10)

$$R = \Delta\omega(L + 1/\omega_0^2 C) = \Delta\omega(L + L)$$

so finally
$$\Delta\omega = R/2L. \tag{14.13}$$

The width of the peak is proportional to R and hence to the power dissipated. The resonance peak is narrow for small R, and its height is V_0/R. The product of height × width is independent of R. A narrow, high resonance can be changed to a wide, low one by increasing R.

Q

A popular measure of the sharpness of the resonance curve is the **quality factor** Q. For resonances in general this is defined by

$$Q = 2\pi \times \frac{\text{maximum energy stored}}{\text{energy dissipated per cycle}}. \tag{14.14}$$

For the LCR series circuit,

$$\boxed{Q = \frac{2\pi \times \frac{1}{2}LI_0^2}{\frac{1}{2}I_0^2 R/f_0} = \frac{\omega_0 L}{R} = \frac{\omega_0}{\text{full width}}.} \tag{14.14a}$$

This latter form is easy to remember. The same result will emerge for parallel resonance, which is discussed later. A high-Q circuit gives a narrow resonance and small bandwidth.

The treatment here ignores the coupling of the circuit to electromagnetic waves. Any oscillatory circuit like this acts as an antenna and radiates radio waves. If the radiation is significant, it adds to the width of the curve (**power broadening**). To take account of this, an equivalent series resistance called **radiation resistance** should really be inserted into the circuit diagram:

$$R_{\text{radiation}} = 2 \times \text{power radiated}/I_0^2.$$

The voltages across C and L are obtained from $I/j\omega C$ and $(j\omega L)I$. At resonance, they have magnitudes $V_0/\omega_0 CR$ and $V_0\omega L/R$ which may be large. For example, suppose $R = 10\ \Omega$, $C = 0.1\ \mu$F and $L = 4$ mH. Then $\omega_0 = 5 \times 10^4$ rad/s and $V_C = V_L = 20V_0$, i.e. much larger than the applied voltage. On the phasor diagram, figure 14.1(b), V_C and V_L are individually large compared with V_R, but on resonance they cancel exactly. Off resonance, the difference between V_C and V_L swamps V_R if the resonance is narrow. Because V_C and V_L are individually large, this type of resonance is called **voltage resonance**. The narrower the resonance, the larger these voltages are; for a very narrow resonance, there is the danger that they actually damage components. What is happening is that a small applied voltage $V_0 e^{j\omega t}$ is exciting large resonant voltages in C and L. The next sections will examine more closely just how this comes about.

Incidentally, the name voltage resonance is potentially confusing, since the magnitude of the current I_0 also peaks at $\omega = \omega_0$; this is because voltages V_L and V_C have opposite signs, so they cancel, leaving the full applied voltage across R.

14.3 Transient in a CL Circuit

The resonant frequency $\omega_0 = 1/\sqrt{LC}$ depends only on C and L. This is a clue that these two elements should be considered alone (figure 14.3). Transient oscillations in this circuit will be discussed. The result which emerges is that the current performs simple harmonic oscillations. There is no damping because there is no resistance to dissipate energy.

For this circuit,

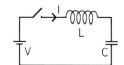

Fig. 14.3. CL resonant circuit.

$$V = L\frac{dI}{dt} + (I/C)\int I\,dt$$

or

$$C\frac{dV}{dt} = CL\frac{d^2I}{dt^2} + I.$$

Suppose the switch is closed at $t = 0$. Except at this instant, the left-hand side is zero, resulting in the equation

$$d^2I/dt^2 + I/CL = 0. \tag{14.15}$$

This is the equation of **simple harmonic motion**

$$I = I_0 \cos(\omega_0 t + \phi_1) \tag{14.16}$$

as is now demonstrated by explicit differentiation and back-substitution into (14.15):

$$dI/dt = -\omega_0 I_0 \sin(\omega_0 t + \phi_1)$$
$$d^2 I/dt^2 = -\omega_0^2 I_0 \cos(\omega_0 t + \phi_1) = -\omega_0^2 I.$$

This satisfies equation (14.15) with

$$\omega_0^2 = 1/CL.$$

So the current executes natural oscillations at angular frequency $\omega_0 = 1/\sqrt{CL}$. Energy is stored alternately as kinetic energy $\frac{1}{2}I^2 L$ in the form of current and as potential energy $\frac{1}{2}Q^2/C$ in the form of charge on the capacitor. As long as there is no resistance in the circuit, this oscillation continues indefinitely. The resonance in the LCR circuit is clearly associated with this oscillation; the AC voltage excites large oscillations in L and C if the applied frequency is close to that of natural oscillations.

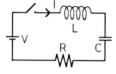

Fig. 14.4. A switched LCR series circuit.

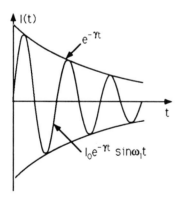

Fig. 14.5. Damped simple harmonic motion.

14.4   Transient in the Series LCR Circuit

After this preliminary discussion, let us now return to the LCR series circuit of figure 14.4. In this circuit the resistance dissipates energy, so when the switch is closed the current executes the damped oscillations shown in figure 14.5. This is called **ringing** and is a common result of a pulse in an electrical circuit.

The algebra is only a minor extension of the previous case. With the addition of resistance R,

$$V = L\frac{dI}{dt} + RI + (1/C)\int I\,dt$$

or

$$C\frac{dV}{dt} = CL\frac{d^2 I}{dt^2} + CR\frac{dI}{dt} + I \tag{14.17}$$

and, except at the instant when the switch is closed, the left-hand side is zero. The equation is rewritten in the standard form

$$\boxed{\frac{d^2 I}{dt^2} + 2\gamma\frac{dI}{dt} + \omega_0^2 I = 0} \tag{14.18}$$

where

$$\gamma = R/2L \tag{14.19}$$
$$\omega_0^2 = 1/CL. \tag{14.20}$$

This equation is easily solved by the trial substitution:

$$I = I_0 e^{st + \phi} \tag{14.21}$$

which gives

$$(s^2 + 2\gamma s + \omega_0^2)I = 0$$

or

$$s = -\gamma \pm (\gamma^2 - \omega_0^2)^{1/2} \tag{14.22}$$

$$s = -\gamma \pm j(\omega_0^2 - \gamma^2)^{1/2}. \tag{14.22a}$$

The latter expression describes oscillations if the quantity

$$\omega_1^2 = \omega_0^2 - \gamma^2 \tag{14.23}$$

is positive, i.e. for small R. As $R \to 0$, $\gamma \to 0$ and $\omega_1 \to \omega_0$ and the results of the previous section reappear.

Damped SHM
The two alternative values of s give the general solution

$$I = e^{-\gamma t}(A e^{j\omega_1 t} + B e^{-j\omega_1 t}). \tag{14.24}$$

For I to be real, A and B must be complex with $B = A^*$. Then

$$I = 2e^{-\gamma t}(\text{Re}A \cos\omega_1 t - \text{Im}A \sin\omega_1 t)$$

which may be written in the alternative forms

$$I = e^{-\gamma t}(A' \cos\omega_1 t + B' \sin\omega_1 t) \tag{14.24a}$$

or

$$I = I_0 e^{-\gamma t} \sin(\omega_1 t + \phi) \tag{14.24b}$$

where

$$I_0 \cos\phi = B' \qquad I_0 \sin\phi = A'.$$

The two constants A' and B' are to be fixed by the initial conditions. If $I = 0$ when $t = 0$, then $\phi = 0$ and

$$I = I_0 e^{-\gamma t} \sin\omega_1 t. \tag{14.24c}$$

This equation describes damped simple harmonic motion. The envelope of the oscillation in figure 14.5 is exponentially damped by the factor $e^{-\gamma t}$. As expected, the damping constant $\gamma = 2R/L$ is proportional to R. The damping reduces the frequency of natural oscillation from ω_0 to $\omega_1 = (\omega_0^2 - \gamma^2)^{1/2}$.

Can you see the relation between I_0 and V_0 in (14.24c)? Hint: what is dI/dt just after the switch is closed?

There is an intimate relation between the transient behaviour of figure 14.5 and the resonance curve of figure 14.2. An AC signal close to frequency ω_1 finds it easy to drive the *LCR* circuit. In fact, the driving voltage is supplying energy at exactly the rate at which the circuit dissipates it. The oscillations are largest for small damping. The damping constant $\gamma = R/2L$ of the transient is equal to the half-width of the resonance curve. This is no accident: Chapter 15 reveals that it is a general consequence of Fourier's theorem.

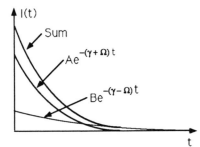

Fig 14.6. Overdamped response.

Overdamping

If the damping is large, ω_1^2 may become negative, in which case there is no oscillation and the values of s given by (14.22) are appropriate. Then

$$I = I_0 e^{-\gamma t}(A e^{-\Omega t} + B e^{\Omega t}) \qquad (14.25)$$

where

$$\Omega = (\gamma^2 - \omega_0^2)^{1/2}. \qquad (14.26)$$

The current is now the sum of two falling exponentials (figure 14.6). The circuit is said to be **overdamped**. At large enough times, the more slowly falling exponential dominates.

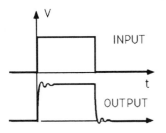

Fig. 14.7. Ringing response to a step function input.

Critical damping

In electronics, and particularly in servo-systems, the circuit should generally follow the input as faithfully as possible. A step in the input should be followed with a step in the output. Overshooting and ringing (figure 14.7), are to be avoided. Conversely, we want to avoid the sluggish response which is a consequence of overdamping (figure 14.6). The optimum response is **critically damped**, corresponding to $\gamma = \omega_0$ and $\Omega = 0$.

14.5 Parallel LCR

A second very common circuit displaying resonance is the parallel LC arrangement shown in figure 14.8. It is called a **tuned circuit**. We shall find that V_{out} follows a resonance curve very similar to figure 14.2; and if a voltage step is applied, the current executes damped oscillations very similar to the series LCR circuit. The equations can be manipulated into a form identical to those of the previous sections with minor changes of parameters.

When V and R are replaced with their Norton equivalent circuit, as in figure 14.9, the resistance appears in parallel with L and C. If there is a resonant current in L and C, some current is diverted to R and dissipates energy, damping the oscillation.

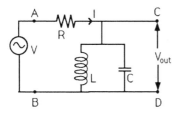

Fig. 14.8. Parallel LC circuit.

Consider first the response to an AC voltage. The LC arrangement has impedance Z_{CL} where

$$1/Z_{CL} = j\omega C + 1/j\omega L = (1 - \omega^2 CL)/j\omega L.$$

From figure 14.8, $V = IZ$ with

$$Z = R + j\omega L/(1 - \omega^2 CL).$$

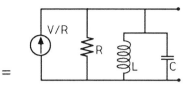

Fig. 14.9. Norton equivalent of Fig. 14.8.

The phasor diagram is shown in figure 14.10. If $V = V_0[\exp j(\omega t - \phi)]$ and $I = I_0 \exp(j\omega t)$,

$$I_0 = V_0 / \left[R^2 + \left(\frac{\omega L}{1 - \omega^2 CL} \right)^2 \right]^{1/2} \quad (14.27)$$

$$\tan \phi = \omega L/[R(\omega^2 CL - 1)]. \quad (14.28)$$

There is a resonant peak in V_{out} of figure 14.8 very similar to figure 14.2, centred at $\omega = 1/\sqrt{LC}$, where the impedances $j\omega L$ and $1/j\omega C$ balance. At resonance, $Z_{CL} \to \infty$ and I_0 drops to 0; however, large balancing AC currents I_1 and I_2 flow through L and C individually. For this reason, the resonance is called **current resonance**. On resonance, $I_0 \to 0$, so there is no voltage drop across R and the full applied voltage $V_0 \exp(j\omega t)$ appears across the tuned circuit. Off resonance, this voltage falls by a factor $1/\sqrt{2}$ when $\tan \phi = \pm 1$, i.e.

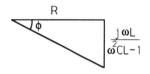

Fig. 14.10. Phasor diagram for Fig. 14.8.

$$\pm \omega L/R = \omega^2 CL - 1.$$

Using the same approximations as in the LCR case, this occurs for $\omega = \omega_0 \pm \Delta\omega$ where

$$\Delta\omega \simeq 1/2CR. \quad (14.29)$$

On resonance, $|I_L| = V_0/\omega L$ and $|I_C| = \omega C V_0$. The Q of the circuit is

$$A = \frac{2\pi \times \frac{1}{2} I_L^2 L}{(\frac{1}{2} V_0^2/R)/f_0} = \omega_0 \frac{R}{\omega_0^2 L} = \omega_0 RC = \frac{\omega_0}{2\Delta\omega} = \frac{\omega_0}{\text{full width}}$$

as for the LCR series circuit.

Coil resistance

In practice, the inductor will have a small resistance r. So far this has been neglected in the interests of keeping algebra to a minimum. There is a useful trick which allows it to be included approximately if it is small. The admittance of the parallel CL combination is

$$Y_{CL} = j\omega C + \frac{1}{j\omega L + r} = j\omega C + \frac{1}{j\omega L(1 + r/j\omega L)}. \quad (14.30)$$

Using the binomial theorem to expand $(1 + r/j\omega L)$,

$$Y_{CL} \simeq j\omega C + \frac{1 - r/j\omega L}{j\omega L} = j\omega C + \frac{1}{j\omega L} + \frac{r}{\omega^2 L^2}. \quad (14.30a)$$

This is the same admittance as is given by a parallel combination of C with a pure inductance L and a resistor $R' = \omega^2 L^2 / r = L/Cr$. The resistor R' may be combined in parallel with R of figure 14.9 to make R'', and thereafter the previous algebra goes through with R'' replacing R. On resonance, the impedance of the tuned circuit is R', and is proportional to $1/r$. Because of R', the impedance of the tuned circuit is not infinite on resonance, so V_{out} of figure 14.8 is less than V. This allows an easy measurement of R'.

Considering energy dissipation, why does the series LCR circuit give a broad peak if R is large while the parallel LCR circuit gives a narrow peak?

Any circuit contains a small amount of stray capacitance and stray inductance. Why don't all circuits resonate at high frequency?

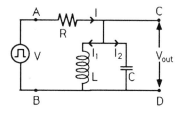

Fig. 14.11. Square waves applied to the parallel LC circuit.

Transients

It will now be demonstrated that transients in the parallel LC circuit of figure 14.11 obey the same equation for damped simple harmonic motion as for the series LCR circuit, except for a different value of γ. Because there is the same voltage across L and C

$$L \frac{dI_1}{dt} = (1/C) \int I_2 \, dt$$

so

$$I_2 = CL \frac{d^2 I_1}{dt^2}.$$

Then

$$V = RI + L \frac{dI_1}{dt} = R(I_1 + I_2) + L \frac{dI_1}{dt}$$

or

$$V/CLR = \frac{d^2 I_1}{dt^2} + (1/CR) \frac{dI_1}{dt} + \omega_0^2 I_1. \tag{14.31}$$

This is the same equation as (14.18) except that (a) $\gamma = 1/2CR$, and (b) the constant term on the left-hand side adds a DC current V/R; it is obvious that this must be so after the transient oscillations have died away. Notice that again the damping constant γ is equal to the half-width $\Delta\omega$ of the resonance curve, given by equation (14.29).

It is instructive to demonstrate both resonance and transient oscillations for yourself on the oscilloscope using the circuits of figures 14.11 and 14.8. The input signal V_{AB} may be displayed on one trace and the signal V_{CD} across the tuned circuit on the second. Suitable circuit parameters are $R = 10^5 \ \Omega$, $C = 0.01 \ \mu$F and $L = 4$ mH, giving $\omega_0 \simeq 1.6 \times 10^5$ rad/s or $f_0 \simeq 25$ kHz. If you omit C, the circuit will resonate at a higher frequency due to stray capacitance across the coil. You will probably find a discrepancy with the calculated width of the resonance curve (and R'). At high frequencies, current is concentrated in the surface of the wire by the skin effect (see textbooks on electromagnetic theory) and the resistance r of the coil increases significantly above its DC value.

If you switch to square waves and reduce the frequency of the generator to about 100 Hz, you can examine the transients. By varying the resistance R or by putting a variable 4 kΩ resistor in parallel with the tuned circuit, you can vary the damping and demonstrate that oscillations disappear when $\gamma = \omega_0$. It is also interesting to insert a soft iron core into the inductor. This increases the value of L, therefore reducing ω_0 and increasing the impedance R' of the tuned circuit on resonance, and hence increasing the magnitude of the output signal.

14.6* Poles and Zeros

As you vary the frequency of the generator in figure 14.8 through resonance, the output signal peaks and the phase ϕ moves rapidly through 90°. The algebra will now be recast slightly so as to display better the origin of this rapid phase variation. An understanding of this point will be helpful in a wide range of advanced phenomena where damped oscillations occur, for example, in servosystems.

If $V = V_0 e^{j\omega t}$, equation (14.17) gives

$$Fe^{j\omega t} = \frac{d^2 I}{dt^2} + 2\gamma \frac{dI}{dt} + \omega_0^2 I \tag{14.32}$$

where $F = j\omega V_0/L$. Equation (14.31) gives a similar result except for a slight change in F. Suppose $I = I_1 e^{j\omega t}$ where I_1 is complex and includes the phase dependence. Substituting this into (14.32),

$$F = I_1(-\omega^2 + 2j\gamma\omega + \omega_0^2).$$

The right-hand side can be factorised into $-I_1(\omega - \omega_A)(\omega - \omega_B)$ where $j\omega_A = s_A$ and $j\omega_B = s_B$ are the two solutions (14.22). After these manipulations,

$$I_1 = -\frac{F}{(\omega - \omega_1 - j\gamma)(\omega + \omega_1 - j\gamma)}. \tag{14.33}$$

The magnitude of I_1 peaks around the resonant frequency $\omega = \omega_1$ because the first term in the denominator is small. The second term does not vary rapidly and can be approximated in the vicinity of the resonance by $(\omega + \omega_1 - j\gamma) \simeq 2\omega_1$. Then

$$I_1 \simeq -\frac{F/2\omega_1}{\omega - \omega_1 - j\gamma}. \tag{14.34}$$

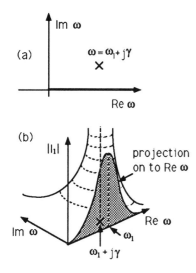

Fig. 14.12 (a) location of the pole in the complex ω plane, (b) 3-D plot of |I_1| for complex values of ω.

The denominator goes to zero when $\omega = \omega_1 + j\gamma$. This is called a **pole** or **singularity** of I_1. The value of ω at the pole is complex, so it is not a physically realisable frequency. Physically ω can vary only along the real axis of figure 14.12(a) from $\omega = 0$ to ∞. However, as a piece of mathematics the function $|I_1|$ can be plotted for both real and imaginary values of ω. (This is called the complex ω plane.) What happens (figure 14.12(b)) is that $I_1 \to \infty$ for $\omega = \omega_1 + j\gamma$. The function has a surface shaped like a volcano centred at the pole. In the vicinity of this (complex) frequency, the behaviour of I_1 is dominated by this singularity. What is measured physically is the slice along the real ω axis. The resonance is narrow and high if the pole lies close to the real axis, i.e. if $j\gamma$ is small. Near the resonance, the complex value of I_1 depends on 1/(complex distance from the pole). This gives instantly the two decisive features of a resonance: a peak in the magnitude and a rapid phase variation. The phase variation goes hand in hand with the resonant peak and (except in contrived situations) one cannot arise without the other.

For parallel resonance, the current goes to zero as $\omega \to \omega_0$. For series resonance, the current peaks at $\omega = \omega_0'$. It may be shown that the impedance of *any* circuit may be factorised into the form

$$Z = \frac{(\omega - \omega_0)(\omega - \omega_1)\dots(\omega - \omega_n)}{(\omega - \omega_A)(\omega - \omega_B)\dots(\omega - \omega_N)}$$

where values of ω_0–ω_n and ω_A–ω_N are complex. Voltage resonances occur for values of ω near the zeros ω_0–ω_n and current resonances for ω close to the poles ω_A–ω_N.

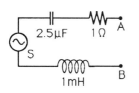

Fig. 14.13

14.7   Exercises

1.   (QMC 1970) In figure 14.13, S is a constant voltage generator of variable frequency and zero internal impedance. For what range of frequencies can this circuit be tuned by placing a variable capacitance across the terminals AB? Sketch a curve showing the amplitude of the voltage across the 1 Ω resistor as a function of frequency if a 1 μF capacitor is connected across AB. What is the resonant frequency and the half-width $\Delta\omega$ of the resonance curve? (Ans: $\omega > 2 \times 10^4$ rad/s; $\omega_0 = 3.8 \times 10^4$ rad/s; $\Delta\omega = 500$ rad/s.)

2.   Define the Q of a series LCR circuit. If $Q = 100$, $R = 50$ Ω and the resonant frequency ω_0 is 20×10^3 rad/s, what are the values of L, C and the half-width $\Delta\omega$ of the resonance? (Ans: $\Delta\omega = 10^2$ rad/s, $L = 0.25$ H, $C = 0.01$ μF.)

3*.   (QMC 1971) Calculate the value of the variable capacitor C' when figure 14.14(a) has been tuned to voltage resonance and

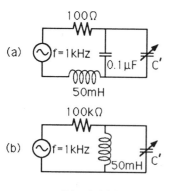

Fig. 14.14

figure 14.14(b) has been tuned to current resonance. Calculate also the impedance seen by the AC source in each case (a) at resonance, (b) when C' is 10% higher than the resonant value. (Ans: 0.4 μF, 0.5 μF, 100 Ω, ∞, $100 + j23.3$ Ω, $10^5 - j3137$ Ω.)

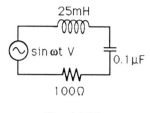

Fig. 14.15

4. (QMC 1981) Find the current and the voltage across each element of figure 14.15 at resonance and sketch how current varies with ω over the range 0.5 to 1.5 times the resonant frequency. (Ans: 10 mA, 50 V, 50 V, $\omega_0 = 2 \times 10^4$ rad/s, $\Delta\omega = 2 \times 10^3$ rad/s.)

5. Derive the transfer function of the Wien filter shown in figure 14.16 and hence show that it is a band-pass filter with maximum transmission at frequency $\omega = 1/CR$. Deduce the value of the transfer function at this frequency. (Ans: $j\omega CR/(1 - \omega^2 C^2 R^2 + 3j\omega CR)$; 1/3.)

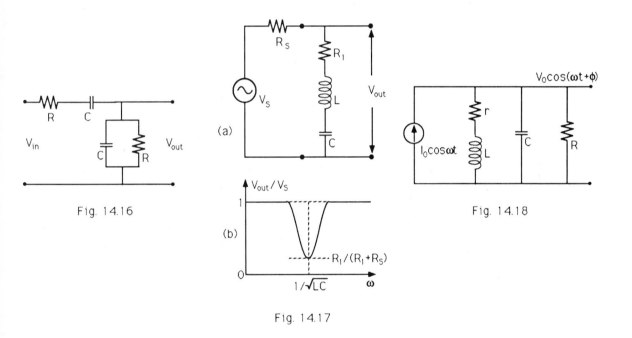

Fig. 14.16

(a)

(b)

Fig. 14.17

Fig. 14.18

6. In figure 14.17(a), the AC source has output impedance $R_S \gg R_1$. Find $|V_{out}/V_S|$ as a function of angular frequency ω and plot it. What use is this circuit? (Ans: $[R_1^2 + (\omega L - 1/\omega C)^2]^{1/2}/[(R_1 + R_S)^2 + (\omega L - 1/\omega C)^2]^{1/2}$, figure 14.17(b), rejects a narrow band of frequencies around $\omega = 1/\sqrt{LC}$.)

7. Figure 14.18 shows a tuned circuit with a lossy inductor. (a) Show that at resonance (when voltage and current are in phase), $\omega_0^2 = (1/LC) - r^2/L^2$. (b) The circuit is used as

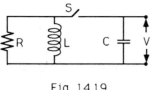

Fig. 14.19

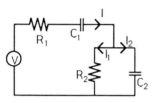

Fig 14.20

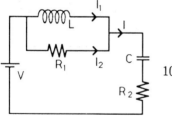

Fig 14.21

an aerial tuned circuit in a radio receiver with $C = 300\ \mu\text{F}$, $R = 20\ \text{k}\Omega$, $L = 85\ \mu\text{H}$ and $r = 16\ \Omega$. A current of 1 A is obtained from the aerial. The Q of the circuit is given by $1/Q = 1/(\omega_0 CR) + r/(\omega_0 L)$. Calculate (i) the resonant frequency ω_0, (ii) the bandwidth and (iii) the current in the resistor at resonance. (Ans: 6.26×10^6 rad/s, 3.55×10^5 rad/s, $0.47\ \mu\text{A}$.)

8*. In figure 14.19, the capacitor is charged to voltage V_0 with the switch open. Find the differential equation for the voltage V across the capacitor if the switch is closed. What is this voltage and dV/dt immediately after the switch is closed? What are the frequency and decay time of the oscillations of V if $L = 4$ mH, $C = 0.5\ \mu\text{F}$, and $R = 20\ \text{k}\Omega$? Find a formula for $V(t)$ if the switch is closed at time $t = 0$. (Ans: $CL\,d^2V/dt^2 + L/R\,dV/dt + V = 0$; $V = V_0$, $dV/dt = -V_0/RC$; $\omega_1 = 2.24 \times 10^4$ rad/s, $\tau = 0.02$ s; $V_0 \exp[-(t/2CR)][\cos\omega_1 t - (1/2\omega_1 CR)\sin\omega_1 t]$, where $\omega_1^2 = [(1/CL) - (1/4C^2R^2)]^{1/2}$.)

9*. For the circuit of figure 14.20, show that I_2 obeys the differential equation

$$C_1 C_2 R_2 \frac{d^2V}{dt^2} = C_1 C_2 R_1 R_2 \frac{d^2 I_2}{dt^2}$$
$$ + (C_1 R_1 + C_1 R_2 + C_2 R_2)\frac{dI_2}{dt} + I_2.$$

Show that this circuit does not show ringing (i.e. natural oscillations) for any values of C_1, C_2, R_1 and R_2.

10*. Find the equation governing transients in I_1 in the circuit of figure 14.21. If $\tau_1 = L/R_1$ and $\tau_2 = CR_2$, show that (a) $\gamma_1 = (1+\tau_1/\tau_2)/2(\tau_1+L/R_2)$ and (b) the circuit rings if $4L\tau_2/R_2 > (\tau_1 - \tau_2)^2$. (Ans: $C\,dV/dt = (CL + \tau_1\tau_2)\,d^2I_1/dt^2 + (\tau_1 + \tau_2)\,dI_1/dt + I_1$.)

15

Fourier's Theorem

15.1 Introduction

Up to now, we have dealt largely with two special cases:

(a) direct current, where the applied voltage is constant, and
(b) alternating current of a single frequency: $V_{\text{applied}} = V_0 \cos \omega t$.

Our methods need to be extended to deal with the possibility of several frequencies being superimposed and with voltages of any shape $V(t)$.

Suppose initially that an applied voltage is repeated at a fixed frequency ω_0, as in figure 15.1, but that the waveform is complicated. A musical sound is an example of such a waveform. Different musical instruments playing the same note produce waveforms of different shapes. Fourier's theorem states that $V(t)$ can be written as a series of terms at frequency ω_0 and multiples (harmonics) of it:

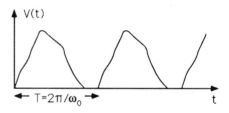

Fig. 15.1. A repeating waveform.

$$V(t) = a_0 + a_1 \sin(\omega_0 t + \phi_1) + a_2 \sin(2\omega_0 t + \phi_2) + \ldots \quad (15.1)$$

$$= \sum_{n=0}^{\infty} a_n \sin(n\omega_0 t + \phi_n).$$

Consider an example when $V(t)$ is a square wave (figure 15.2). Then the Fourier series is

$$V(t) = \frac{4V_0}{\pi}\left(\sin \omega_0 t + \tfrac{1}{3}\sin 3\omega_0 t + \tfrac{1}{5}\sin 5\omega_0 t + \ldots\right). \quad (15.2)$$

The way successive harmonics build up the square wave is illustrated in figure 15.2.

Why are even harmonics absent in this case?

Suppose this square wave is applied to a circuit with impedance $Z(\omega)$. The current $I(\omega)$ may be found for each separate harmonic

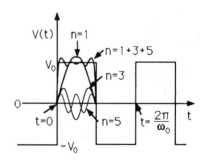

Fig. 15.2. A square wave and its Fourier components.

from $I(\omega) = V(\omega)/Z(\omega)$; then $I(t)$ may be built up by adding these components using the superposition principle. The important conclusion is that the general case for $V(t)$ can be solved with a superposition of simple AC solutions with suitable coefficients.

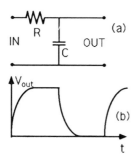

Fig. 15.3 (a) A low-pass filter, (b) $V_{\text{out}}(t)$.

15.2   A Square Wave Applied to a CR Filter

Suppose square waves are applied to the input of the filter circuit of figure 15.3(a) at a frequency ω_0 well below the cut-off frequency. The fundamental and low harmonics will be transmitted faithfully, but high harmonics above the cut-off will be attenuated (and modified in phase). Consequently, the output will have rounded shoulders (figure 15.3(b)). Later we will follow algebraically how the harmonics build up the shape of $V_{\text{out}}(t)$.

An oscilloscope itself has some input capacitance C_{in} and therefore a bandwidth limited to $\omega \leq 1/C_{\text{in}}R_{\text{in}}$. Even if perfectly square voltage pulses could be applied to its input, the time constant observed on the screen would be limited to

$$\tau = C_{\text{in}}R_{\text{in}}. \tag{15.3}$$

Conversely, a high-pass filter cuts off low frequencies. If the cut-off frequency is less than that of the square waves, only direct current is rejected; direct current arises from $n = 0$ in equation (15.1).

If the waveform shown in figure 15.3(b) is applied to the terminals of figure 15.3(a) labelled OUT, will a square wave be observed at the terminals labelled IN?

15.3   How to Find Fourier Coefficients

Another way of writing equation (15.1) is

$$V(t) = b_0 + b_1 \cos \omega_0 t + b_2 \cos 2\omega_0 t + \dots$$
$$+ a_1 \sin \omega_0 t + a_2 \sin 2\omega_0 t + \dots$$
$$= \sum_{n=0}^{\infty} [a_n \sin(n\omega_0 t) + b_n \cos(n\omega_0 t)].$$

If both sides are multiplied by $\sin(m\omega_0 t)$ and integrated over one complete cycle of the waveform from $t = 0$ to $T = 2\pi/\omega_0$,

$$\int_0^{2\pi/\omega_0} V(t) \sin(m\omega_0 t)\, \mathrm{d}t = \sum_{n=0}^{\infty} \int_0^{2\pi/\omega_0} [a_n \sin(n\omega_0 t)$$
$$+ b_n \cos(n\omega_0 t)] \sin(m\omega_0 t)\, \mathrm{d}t$$

it turns out that all but one of the terms on the right-hand side is zero. To see this, remember that

$$2\sin(n\omega_0 t)\sin(m\omega_0 t) = \cos[(n-m)\omega_0 t] - \cos[(n+m)\omega_0 t]$$
$$2\cos(n\omega_0 t)\sin(m\omega_0 t) = \sin[(n+m)\omega_0 t] - \sin[(n-m)\omega_0 t].$$

When the cosines and sines on the right-hand side are integrated over a complete cycle they give zero, except for the cosine term with $n - m = 0$. Thus

$$a_n = (\omega_0/\pi) \int_0^T V(t)\sin(n\omega_0 t)\,dt \qquad (15.4)$$

or

$$a_n = (1/\pi) \int_0^{2\pi} V(t)\sin(nx)\,dx \qquad (15.4a)$$

where

$$x = \omega_0 t.$$

The latter form (15.4a) is the one which is convenient to use in most applications. Likewise

$$b_n = (1/\pi) \int_0^{2\pi} V(t)\cos(nx)\,dx \qquad (15.4b)$$

except that the DC term is obviously given by the mean value of $V(t)$:

$$b_0 = (1/2\pi) \int_0^{2\pi} V(t)\,dt. \qquad (15.4c)$$

A simple application of (15.4a) is to derive the Fourier coefficients in equation (15.2).

Can you understand base-line shift of Chapter 3 in terms of b_0?

In words, what equations (15.4) are saying is that:

$a_n = 2 \times$ *mean value of* $V(t)\sin(n\omega_0 t)$ *over a cycle*
$b_n = 2 \times$ *mean value of* $V(t)\cos(n\omega_0 t)$ *over a cycle (for* $n \geq 1)$
$b_0 =$ *mean value of* $V(t)$ *over a cycle.*

The origin of the factor of two difference between b_0 and remaining b_n is the same as the factor of two difference between DC power $I_0 V_0$ and AC power $\frac{1}{2}I_0 V_0$.

Why are subharmonics not present in the Fourier series?

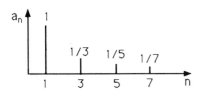

Fig. 15.4. Fourier or frequency spectrum of the square waves.

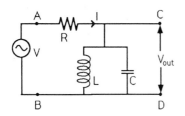

Fig. 14.8. Parallel LC circuit.

Fourier spectrum

When these coefficients are plotted against n, as in figure 15.4, the histogram is called the **spectrum** of the pulse. Electronic **spectrum analysers** are available commercially to plot out the Fourier components of a waveform; the distinctive character of any musical instrument depends on its Fourier spectrum. **Synthesisers** do the reverse: they reproduce the sound of an instrument by reconstructing the waveform using oscillators which generate the fundamental and its harmonics with appropriate coefficients and phase relationships.

A nice demonstration of Fourier decomposition is to apply square waves to the resonant circuit of figure 14.8. If you adjust their frequency successively close to $1/2$, $1/3$, $1/4$, ... of the resonant frequency, the circuit acts as a narrow pass filter and at these frequencies you will observe on the oscilloscope maxima at the output of the filter. The outputs are sine waves at the resonant frequency, i.e. harmonics of the square wave transmitted by the filter. Their magnitudes measure the spectrum shown in figure 15.4. As you vary the frequency away from an exact sub-harmonic of the resonant frequency, you will observe very complicated output waveforms. What is happening is that harmonics are undergoing phase shifts because of the phase variation away from resonance.

If $V(t)$ is symmetrical about $t = 0$, i.e. if $V(t) = V(-t)$, the Fourier series contains only cosine terms and all a_n are zero. If it is symmetrical above and below the horizontal axis, even harmonics are absent, as in equation (15.2). (Why?) If $V(t)$ is antisymmetric about $t = 0$, i.e. $V(t) = -V(-t)$, it contains only sine terms and all b_n are zero.

15.4   The Heterodyne Principle

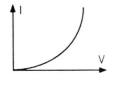

Fig. 15.5. A non-linear characteristic.

Diodes have characteristic curves which are non-linear, like that in figure 15.5. It is instructive to follow through the consequences of this non-linear relation between I and V. To keep the algebra simple, suppose

$$I = aV + bV^2.$$

Next, suppose V is the sum of two AC signals of different frequency:

$$V = V_0 \cos \omega_0 t + V_1 \cos \omega_1 t.$$

Then

$$I = a(V_0 \cos \omega_0 t + V_1 \cos \omega_1 t)$$
$$+ b(V_0^2 \cos^2 \omega_0 t + V_1^2 \cos^2 \omega_1 t + 2V_0 V_1 \cos \omega_0 t \cos \omega_1 t).$$

The second term may be rewritten

$$I = \tfrac{1}{2} b \{ V_0^2 + V_0^2 \cos 2\omega_0 t$$
$$+ V_1^2 + V_1^2 \cos 2\omega_1 t + 2V_0 V_1 [\cos(\omega_0 - \omega_1)t + \cos(\omega_0 + \omega_1)t] \}.$$

In this second term there are outputs at the beat frequencies:

(a) $\omega_0 - \omega_1$
(b) $\omega_0 + \omega_1$
(c) $\omega_0 - \omega_0 = 0$
(d) $\omega_0 + \omega_0 = 2\omega_0$
(e) $\omega_1 - \omega_1 = 0$
(f) $\omega_1 + \omega_1 = 2\omega_1$.

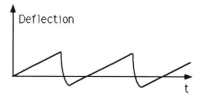

Fig. 15.6. Deflection of a violin string with time.

From a *single* input $V_0 \sin \omega_0 t$ we get (c) and (d), i.e. a DC term and the first harmonic. The spectrum is distorted by this harmonic. A good audio amplifier is linear and gives very little harmonic distortion. A low-fi radio produces music which is still recognisable because it contains the fundamental frequency ω_0, but it sounds bad because the harmonic spectrum is distorted.

If the characteristic curve contains a term cV^3, similar algebra shows that a term appears in the output at frequency $3\omega_0$, and so on for higher powers. A very non-linear characteristic curve generates many harmonics.

A musical instrument gives the complicated waveform of figure 15.1 just because the output depends non-linearly on the input driving force. When a bow is pulled (scraped) across the string of a violin, the string moves in a series of jerks (figure 15.6) as it sticks to the bow; different bowing gives a different pattern of jerks, and hence a difference in spectrum or timbre. Distortion of harmonic content in recording or reproduction confuses one instrument with another.

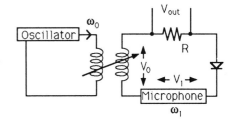

Fig. 15.7. Schematic diagram for broadcasting.

15.5 Broadcasting

Suppose ω_0 is large (radio frequencies 10^7 rad/s) and ω_1 is small (audio frequencies $\sim 2 \times 10^3$ rad/s for middle C). This might be achieved with the circuit of figure 15.7. If the output is fed through a filter passing only frequencies close to ω_0 (as in figure 15.8), the only components which get through the filter are

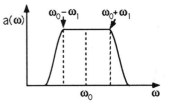

Fig. 15.8. The spectrum passed by a narrow-band filter.

$$I = aV_0 \cos \omega_0 t + bV_0 V_1[\cos(\omega_0 - \omega_1)t + \cos(\omega_0 + \omega_1)t]. \quad (15.5)$$

The signals at frequencies $(\omega_0 - \omega_1)$ and $(\omega_0 + \omega_1)$ are called **sidebands**. The expression may be manipulated further:

$$I = aV_0 \cos \omega_0 t + 2bV_0 V_1 \cos \omega_0 t \cos \omega_1 t \quad (15.6)$$
$$= aV_0 \cos \omega_0 t[1 + (2b/a)V_1 \cos \omega_1 t].$$

Graphically this has the form shown in figure 15.9. The ratio $2bV_1/a$ is called the **modulation depth** or **modulation index**. The signal contains information about the audio signal of angular frequency ω_1, but the Fourier components of equation (15.5) have

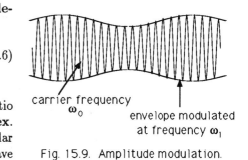

carrier frequency ω_0

envelope modulated at frequency ω_1

Fig. 15.9. Amplitude modulation.

been shifted to $(\omega_0 - \omega_1)$ and $(\omega_0 + \omega_1)$. This arrangement is called **amplitude modulation (AM)**. It was the first form used for radio broadcasting at frequencies around $f = 1$ MHz, $\lambda = 300$ m. At frequencies within a factor of five or so of this, radio waves bounce backwards and forwards between the Earth and the ionosphere readily; this allows propagation of radio waves over large distances round the Earth. Much shorter or longer wavelengths penetrate the ionosphere.

Suppose the audio signal covers the range up to $\omega_1 = 10^5$ rad/s. The filter of figure 15.8 needs to have a bandwidth twice this, i.e. from, say, 5.9×10^6 to 6.1×10^6 rad/s. Then one station uses 3% of the available frequency range. Because there are vastly more than 30 stations, a medium-wave receiver picks up beats between the carrier frequencies of different stations, leading to a steady background whine at frequency $\omega_0 - \omega_0'$; this is all too familiar.

Equation (15.5) shows that both sidebands carry the same information. It is possible to suppress one of them and halve the required bandwidth. This is called **single-sideband amplitude modulation**.

15.6   Phase and Frequency Modulation (FM)

An alternative to modulating the amplitude is to keep the amplitude fixed and to vary the phase ϕ. Suppose $\phi = A_\mathrm{p} \cos \omega_1 t$, where ω_1 is an audio frequency. A **phase-modulated** signal of unit amplitude is then

$$I = \cos(\omega_0 t + A_\mathrm{p} \cos \omega_1 t)$$
$$= \cos(\omega_0 t) \cos(A_\mathrm{p} \cos \omega_1 t) - \sin(\omega_0 t) \sin(A_\mathrm{p} \cos \omega_1 t).$$

If A_p is small, $\cos(A_\mathrm{p} \cos \omega_1 t) \simeq 1$ and $\sin(A_\mathrm{p} \cos \omega_1 t) \simeq A_\mathrm{p} \cos \omega_1 t$, so

$$I \simeq \cos \omega_0 t - A_\mathrm{p} \sin \omega_0 t \cos \omega_1 t \qquad (15.7)$$
$$= \cos \omega_0 t - \tfrac{1}{2} A_\mathrm{p} [\sin(\omega_0 + \omega_1)t + \sin(\omega_0 - \omega_1)t].$$

Again, two new Fourier components are generated, but the sidebands differ in phase by 90° from the carrier.

Frequency modulation
Technically, it is easier to modulate the frequency than the phase. Suppose

$$\omega = \omega_0 + B(t) \qquad (15.8)$$
$$= \omega_0 + A_\mathrm{f} \cos \omega_1 t.$$

The quantity A_f is called the **frequency deviation**; it gives the range of variation of the frequency. The phase of the resulting signal is

$$\phi = \int \omega \, dt = \omega_0 t + \int B(t') \, dt'$$

$$= \omega_0 t + \int A_f \cos \omega_1 t \, dt = \omega_0 t + (A_f/\omega_1) \sin \omega_1 t.$$

The quantity A_f/ω_1 is called the **modulation index** for frequency modulation. Algebra similar to that given above results in

$$I \simeq \cos \omega_0 t - (A_f/\omega_1) \sin \omega_0 t \sin \omega_1 t \qquad (15.9)$$
$$= \cos \omega_0 t + (\tfrac{1}{2} A_f/\omega_1) \left[\cos(\omega_0 + \omega_1)t - \cos(\omega_0 - \omega_1)t \right].$$

The Fourier spectrum is like that in amplitude modulation, except that one sideband is reversed in sign with respect to the carrier. The waveform and its spectrum are shown in figure 15.10.

So far, modulation by a pure cosine term has been considered. More generally the modulating amplitude in equation (15.8) has a complicated time dependence $B(t)$; in the small-angle approximation used so far,

$$I = \cos \omega_0 t - A(t) \sin \omega_0 t$$

where

$$A(t) = \int_0^t B(t') \, dt'.$$

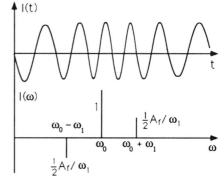

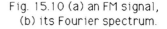

Fig. 15.10 (a) an FM signal, (b) its Fourier spectrum.

The signal $A(t)$ can be Fourier analysed, and the second term in equation (15.9) is replaced by its Fourier spectrum. However, it is not essential to restrict the algebra to the approximation that $A(t)$ is small compared to 1. Physically, the wave shown in figure 15.10(a) is perfectly well defined for large $A(t)$, but evaluation of the corresponding expression to equation (15.9), and hence obtaining the bandwidth, is more elaborate. This is explored in exercises 5 and 6.

If you cut off the top and bottom of the amplitude in figure 15.10(a) at 50% of the peaks and feed the resulting waveform through a filter centred at ω_0, what are the consequences?

15.7 Frequency Multiplexing

By international agreement, the bandwidth for telephone conversations extends from 300 to 3400 Hz. Most of the spectrum produced by the voice is within this range, and although the removal of high

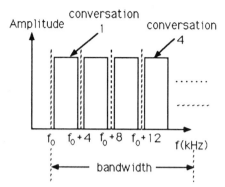

Fig.15 11. Frequency multiplexing.

frequencies reduces the quality of the sound the result is perfectly adequate for conversation.

This bandwidth is much less than is available in modern communications. Many separate telephone conversations can be stacked at intervals of 4 kHz in the bandwidth of the link, as indicated in figure 15.11. Individual conversations are modulated on to carrier frequencies spaced by 4 kHz, using a single sideband and suppressing the carrier itself. At the other end of the line, they are recovered by demodulation, i.e. beating against signals at the individual carrier frequencies.

This is called **frequency multiplexing**. The gap of 900 Hz between each conversation and the next is needed for two reasons. Firstly, the filter which isolates one conversation from its neighbours is not perfectly sharp but has rounded edges. Secondly, control information (e.g. on synchronisation and routing) can be transmitted within the gap, normally at 3825 Hz above the carrier frequency.

15.8   Time Division Multiplexing

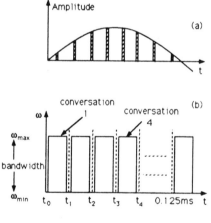

Fig. 15.12 (a) sampling a waveform,
(b) time division multiplexing.

An alternative to sharing the line in frequency intervals is to share it in time. Suppose the signal from one conversation is sampled at regular intervals (figure 15.12(a)). A slice of the signal could be despatched down the link and at the other end a suitable smoothing scheme could rejoin the slices. If the sampling is done at a frequency well above 4 kHz and if the bandwidth of the smoothing circuit is limited to 4 kHz, the interpolating curve is limited to 4 kHz and there will be no distortion below this frequency. There is a theorem, due to Shannon, that the signal must be sampled at least twice per cycle, so in practice the sampling is done at 8 kHz.

Slices of individual conversations are transmitted sequentially (figure 15.12(b)), using the full bandwidth ω_{max} of the link. Each conversation uses a time interval Δt slightly less than given by $\omega_{max}\Delta t = 1$, so that they do not spread into one another. The input to the link dials through a number of conversations, returning to the first at the sampling frequency of 8 kHz, i.e. after 0.125 ms. This is called **time multiplexing**.

Pulse code modulation

In fact, current practice is not to send the slices themselves but to digitise them. The possible range of amplitudes is broken up into 256 intervals. Thus the amplitude is converted to an eight-bit binary number and it is this number which is transmitted. This is called **pulse code modulation**. Necessarily, there is some truncation in this digitisation, and the result can be in error by up to half an interval. The range of signals from smallest to largest which can be digitised is known as the **dynamic range**.

Companding

The digitisation is not done linearly. Small signals are allocated more intervals, while the large amplitude range is treated coarsely (figure 15.13). This is done by *comp*ressing the signal at the input with a non-linear amplifier, then digitising linearly, and then exp*anding* it again at the other end of the line using the inverse transformation. This scheme is called **companding**. Different transformations are employed in the USA and in Europe.

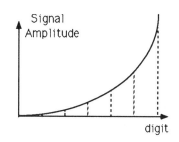
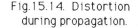

Fig. 15.13. Digitisation profile.

Noise considerations

There are eight bits per digitisation and sampling is at 8 kHz, so each conversation requires that bits are transmitted at 64 kHz. Isn't this uneconomical compared with the bandwidth of 4 kHz used in frequency multiplexing? The difference arises because frequency multiplexing produces a sideband whose spectrum can vary continuously in amplitude, whereas digital transmission limits the possibilities to 0 or 1. The penalty of the former scheme is that any noise lying at the frequency of the signal will add coherently to it, changing its amplitude or phase from A to $A + \Delta A$. The signal is corrupted. If the attenuation of the signal along the line is frequency dependent, this will also corrupt it, though some recovery is possible by compensation. The digital signal, however, is not corrupted unless the noise is so large as to convert 0 to 1 or vice versa. The digital signal will distort progressively along the line from an initial square wave of figure 15.14(a) to a rounded shape (b), because of the bandwidth of the channel. However, it can be regenerated at a repeater station before the distortion causes confusion between 0 and 1.

Let us examine this in a little more detail. Suppose the frequency-multiplexed signal were digitised at the downstream end of the link. If, for simplicity, it is digitised into n equally spaced intervals, the ratio of noise to interval size depends on $n\Delta A$, where ΔA is the amplitude of the noise. In contrast, for pulse code modulation the noise is ΔA in each of n channels. The total noise power is the same in both cases but is differently distributed. The amplitude of the noise distribution follows a Gaussian curve (figure 15.15), peaking at zero and falling as $\exp(-A^2)$ for large amplitude A. Because of the factor n multiplying ΔA, the frequency multiplexed signal is more likely to be corrupted. For TV or telephone signals this is not very important, but for transmission of data from one computer to another it is crucial.

Compact discs act like time division multiplexers. They encode the digitised amplitude as sixteen bits at a sampling rate of

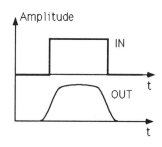

Fig.15.14. Distortion during propagation.

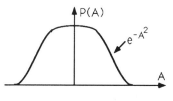
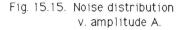

Fig. 15.15. Noise distribution v. amplitude A.

41.1 kHz. Like satellite links, they include the feature of checking the digitised signal against corruption by forming regular sum checks on the recorded signal. Sufficient redundant information is encoded that if a sum check fails the mistake can usually be corrected. Primitive schemes for such error recovery were discussed in Chapter 12.

Intelsat V satellite links carry 12 000 voice channels and two TV channels (each with a bandwidth of 8 MHz). Intelsat VI will carry 33 000 voice channels and four TV channels.

15.9   Fourier Series using Complex Exponentials

Yet another way of writing the Fourier series is in terms of complex exponentials. This is actually simpler algebraically than using sines and cosines. Remember that

$$\cos n\omega_0 t = \tfrac{1}{2}(e^{jn\omega_0 t} + e^{-jn\omega_0 t})$$

$$\sin n\omega_0 t = (1/2j)(e^{jn\omega_0 t} - e^{-jn\omega_0 t}).$$

It follows that $V(t)$ may be expressed as a series of complex exponentials, but now the series runs from $n = -\infty$ to $+\infty$:

$$V(t) = \sum_{n=-\infty}^{\infty} c_n e^{jn\omega_0 t}. \tag{15.10}$$

The Fourier coefficients are easily obtained by multiplying both sides by $\exp(-jn\omega_0 t)$ and integrating over a cycle:

$$c_n = \frac{\omega_0}{2\pi} \int_{-\pi/\omega_0}^{\pi/\omega_0} V(t)e^{-jn\omega_0 t} \, dt$$

or

$$\boxed{c_n = \frac{1}{2\pi} \int_{t=-T/2}^{T/2} V(t)e^{-jn\omega_0 t} \, d(\omega_0 t).} \tag{15.11}$$

In words, c_n is the mean value of $V(t)\exp(-jn\omega_0 t)$ over a cycle. In general, c_n are complex and convey information about both magnitude and phase; two spectra like figure 15.4 are required, one displaying the magnitude and the other the phase of every harmonic.

What is the relation between a_n and b_n of equations (15.4) and c_n?

15.10   Fourier Transforms

The waveform in figure 15.1 was periodic, i.e. it repeated itself after one period T. The idea of Fourier analysis is now extended to a single waveform which does *not* repeat. This leads to ideas of fundamental importance throughout electrical engineering, and most of physics. Heisenberg's famous uncertainty principle is an example.

The requisite formulae can be derived from those we already have by letting $T \to \infty$, in which case the fundamental frequency $\omega_0 \to 0$. The spectrum of frequencies becomes continuous. There are some mathematical conditions concerning discontinuities which must be obeyed for the following algebra to be valid; these are called the Dirichlet conditions. For any *physical* system, these discontinuities do not occur and the conditions are obeyed.

Suppose we set $\omega_0 = 2\pi/T = \mathrm{d}\omega$, an infinitesimally small quantity, and $n\omega_0 = \omega$; then from the first form of equation (15.11)

$$c_n \to \frac{\mathrm{d}\omega}{2\pi} \int_{-\infty}^{\infty} V(t)e^{-\mathrm{j}\omega t}\, \mathrm{d}t = c(\omega)\,\mathrm{d}\omega/\sqrt{2\pi} \qquad (15.12)$$

where

$$c(\omega) = \sqrt{\frac{1}{2\pi}} \int_{-\infty}^{\infty} V(t)e^{-\mathrm{j}\omega t}\, \mathrm{d}t \qquad (15.13)$$

and from (15.10)

$$V(t) = \sqrt{\frac{1}{2\pi}} \int_{-\infty}^{\infty} c(\omega)e^{\mathrm{j}\omega t}\, \mathrm{d}\omega. \qquad (15.14)$$

The factor $1/\sqrt{2\pi}$ was introduced to make (15.13) and (15.14) symmetrical. The distribution $c(\omega)$ is the spectral representation of $V(t)$ and is mathematically completely equivalent to it. It is called the **Fourier transform** of $V(t)$. We shall see shortly that it is both familiar from previous chapters and very useful. Conversely, $V(t)$ is the Fourier transform of $c(\omega)$.

There is a further convention which may at first sight appear confusing. It is conventional to refer to $V(\omega)$ instead of $c(\omega)$, i.e. equation (15.13) is written

$$\boxed{V(\omega) = \sqrt{\frac{1}{2\pi}} \int_{-\infty}^{\infty} V(t)e^{-\mathrm{j}\omega t}\, \mathrm{d}t.} \qquad (15.13a)$$

The reason for this notation is that the same information is embodied in $V(\omega)$ as in $V(t)$, though in a different form: $V(t)$ gives the dependence on t and is said to be **in the time domain**, while $V(\omega)$ gives the dependence on ω and is said to be **in the frequency domain**. The symmetrical pair of relations (15.14) and (15.13a) is easy to remember; only the derivation is tricky.

What are Fourier transforms physically? An example may help. The next sections will demonstrate that if the current $I(t)$ is the transient (time) response of a circuit to an impulse, then $I(\omega)$ is its frequency response to an AC signal as a function of ω. This gives the clue to why we found an intimate relation between time constant and bandwidth in Chapters 3, 4 and 14. Exponential decay or ringing (as a function of t) is the Fourier transform of the Bode plot (dependence on ω) and vice versa. In order to build up some feeling for the meaning of Fourier transforms, the algebra is now followed through two examples in detail.

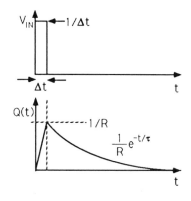

15.11 Fourier Analysis of an Exponential Decay

Firstly, what is an impulse? It is a kick. Mathematically it is an input $\delta(t)$ (figure 15.16), lasting for an infinitesimally short time Δt, with a height $1/\Delta t$ so that integrated over time it gives one:

$$\int_{-\infty}^{\infty} \delta(t) \, dt = 1.$$

Suppose, for example, a voltage $\delta(t)$ is applied to the CR circuit of figure 15.3(a). While the voltage is applied, current $\delta(t)/R$ flows. The charge deposited on the capacitor is $\int I \, dt = (1/R) \int \delta(t) \, dt = 1/R$. Subsequently, the capacitor discharges again with time constant τ:

$$Q(t) = (1/R)e^{-t/\tau}$$
$$I(t) = dQ/dt = -(1/\tau R)e^{-t/\tau} = I_0 e^{-t/\tau} \qquad (15.15)$$

Fig. 15.16 (a) An impulse, (b) the charge developed on a capacitor.

where $\tau = RC$. The Fourier transform of equation (15.15) is

$$I(\omega) = \sqrt{\frac{1}{2\pi}} \int_0^{\infty} I_0 e^{-t/\tau} e^{-j\omega t} \, dt = \frac{I_0}{\sqrt{2\pi}} \left[\frac{e^{-t/\tau} e^{-j\omega t}}{-1/\tau - j\omega} \right]_0^{\infty}.$$

The decaying exponential is zero as $t \to \infty$, so

$$I(\omega) = \frac{I_0}{\sqrt{2\pi}} \frac{1}{1/\tau + j\omega}. \qquad (15.16)$$

Apart from a factor $j\omega\sqrt{2\pi}$, (which will be accounted for later) this is just the AC response for a CR filter, as will now be shown:

$$V = ZI = \left(R + \frac{1}{j\omega C} \right) I = \frac{1 + j\omega CR}{j\omega C} I$$

$$I = \frac{j\omega CR(V/R)}{1 + j\omega CR} = \frac{I_0 j\omega \tau}{1 + j\omega \tau} = \frac{I_0 j\omega}{1/\tau + j\omega}.$$

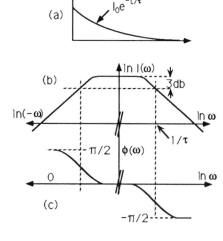

Fig. 15.17 (a) An exponentially falling pulse and (b) and (c) its Fourier transforms.

Thus, apart from the factor $j\omega\sqrt{2\pi}$, the Bode plot is the same as the Fourier transform $I(\omega)$ shown in figures 15.17(b) and (c). This example also demonstrates the relation (15.3) between τ and band-width; from equation (15.16), $I(\omega)$ falls to $1/\sqrt{2}$ of its maximum value when

$$\boxed{\omega\tau = 1.}$$
(15.3a)

15.12 Fourier Analysis of a Damped Oscillator

As a second example, consider the response of a damped oscillator to an impulse at $t = 0$:

$$
\begin{aligned}
I(t) &= I_0 e^{-t/\tau} \sin\omega_1 t \qquad \text{for } t > 0 \\
&= (I_0/2\mathrm{j})e^{-t/\tau}(e^{\mathrm{j}\omega_1 t} - e^{-\mathrm{j}\omega_1 t}) \\
&= (I_0/2\mathrm{j})(e^{s_1 t} - e^{s_2 t})
\end{aligned}
$$
(15.17)

where

$$
\begin{aligned}
s_1 &= (-1/\tau) + \mathrm{j}\omega_1 \\
s_2 &= (-1/\tau) - \mathrm{j}\omega_1.
\end{aligned}
$$

Then

$$
\begin{aligned}
I(\omega) &= \sqrt{\frac{1}{2\pi}}\frac{I_0}{2\mathrm{j}} \int_0^\infty (e^{s_1 t} - e^{s_2 t})e^{-\mathrm{j}\omega t}\, dt \\
&= \frac{-I_0}{2\mathrm{j}}\sqrt{\frac{1}{2\pi}}\left(\frac{1}{\mathrm{j}(\omega_1 - \omega + \mathrm{j}/\tau)} - \frac{1}{\mathrm{j}(-\omega_1 - \omega + \mathrm{j}/\tau)}\right) \\
&= \frac{I_0}{\sqrt{8\pi}}\left(\frac{1}{\omega_1 - \omega + \mathrm{j}/\tau} + \frac{1}{\omega_1 + \omega - \mathrm{j}/\tau}\right) \\
&= \frac{I_0\omega_1}{\sqrt{2\pi}}\frac{1}{\omega - \omega_1 - \mathrm{j}/\tau}\frac{1}{\omega + \omega_1 - \mathrm{j}/\tau}.
\end{aligned}
$$
(15.18)

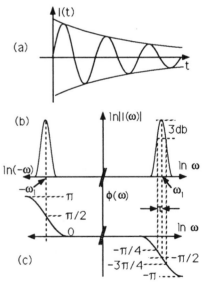

Again, apart from numerical factors, this is just the AC response of the oscillator, equation (14.33), with poles at $\omega = \omega_1 + \mathrm{j}/\tau$ and $\omega = -\omega_1 + \mathrm{j}/\tau$. The waveform $I(t)$ and its Fourier transform are shown in figure 15.18. There are peaks in the magnitude of the spectrum around ω_1 and $-\omega_1$ falling by 3 db when ω moves off the peak by $1/\tau$.

The origin of these results is that the impedance Z depends only on ω and not on time, so if

$$V(t) = Z(\omega)I(t)$$

then

$$\boxed{V(\omega) = Z(\omega)I(\omega).}$$
(15.19)

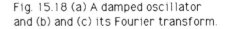

Fig. 15.18 (a) A damped oscillator and (b) and (c) its Fourier transform.

This connection will now be traced explicitly for the Fourier transform of the exponential decay in the CR filter circuit, and the missing factor $\mathrm{j}\omega\sqrt{2\pi}$ will be located.

If $V(t)$ is an impulse of magnitude $1/\Delta t$ from $t = 0$ to Δt,

$$V(\omega) = \sqrt{\frac{1}{2\pi}} \int_0^{\Delta t} \frac{1}{\Delta t} e^{j\omega t}\, dt \rightarrow \sqrt{\frac{1}{2\pi}} \quad\text{as}\quad \Delta t \rightarrow 0$$

since the exponential is 1 for $t = 0$. During the impulse, the current which flows is

$$I_1 = \frac{1}{R\Delta t} \quad\text{from}\quad t = 0 \text{ to } \Delta t.$$

Charge $Q = I\Delta t = 1/R$ builds up on the capacitor and the subsequent discharge current is

$$I_2 = \frac{dQ}{dt} = -\frac{1}{\tau R} e^{-t/\tau} \quad\text{from}\quad t = \Delta t \text{ to } \infty$$

so, from equation (15.16),

$$I_2(\omega) = -\frac{1}{R\sqrt{2\pi}} \frac{1}{1 + j\omega\tau}.$$

Also

$$I_1(\omega) = \frac{1}{R\sqrt{2\pi}}.$$

This was the component which was omitted earlier. Adding I_1 and I_2,

$$I(\omega) = \frac{1}{R\sqrt{2\pi}} \frac{j\omega\tau}{1 + j\omega\tau}. \tag{15.20}$$

Now

$$Z(\omega) = R(1 + 1/j\omega\tau) = R(1 + j\omega\tau)/j\omega\tau$$

so

$$Z(\omega)I(\omega) = \sqrt{1/2\pi}$$

which agrees with $V(\omega)$. Similar algebra applies for the damped oscillator.

15.13   The Perfect Filter

A desirable objective is to make an ideal filter with a flat response from zero frequency up to ω_0 and a sharp cut-off at this frequency (figure 15.19(a)). The spectrum transmitted by this circuit will be $h(\omega)g(\omega)$, where $g(\omega)$ is the Fourier transform of the input signal. Unfortunately it turns out to be theoretically impossible. If $h(\omega)$ is to have the shape of figure 15.19(a),

$$h(t) = \sqrt{\frac{1}{2\pi}} \int_0^{\omega_0} e^{j\omega t}\, d\omega = \sqrt{\frac{1}{2\pi}} \frac{1}{jt} \left(e^{j\omega_0 t} - 1\right)$$

$$= \sqrt{\frac{1}{2\pi}} e^{\frac{1}{2} j\omega_0 t} \frac{\sin(\frac{1}{2}\omega_0 t)}{\frac{1}{2}t}. \tag{15.21}$$

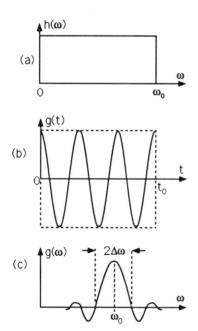

Fig. 15.19 (a) the ideal low-pass filter, (b) a cosine wave of finite duration and (c) its spectrum near $\omega = \omega_0$.

This requires $h(t)$ to be non-zero for $t < 0$, i.e. the circuit must respond *before* the impulse arrives. This defies common sense, or the so-called **causality principle** that effect follows cause: $h(t) = 0$ for $t < 0$. So it is a physical impossibility to make the perfect filter of figure 15.19(a).

The sync function

A second situation often arises where a sine or cosine wave is applied to a circuit for a finite length of time from $t = 0$ to t_0, as in figure 15.19(b):

$$g(t) = \cos\omega_0 t \qquad \text{for} \qquad t = 0 \text{ to } t_0.$$

Then using (15.21) for algebraic manipulations,

$$g(\omega) = \tfrac{1}{2}\sqrt{1/2\pi} \int_0^{t_0} \left(e^{j\omega_0 t} + e^{-j\omega_0 t}\right) e^{-j\omega t} \, dt$$

$$= \sqrt{\frac{1}{2\pi}} \left(\exp[j(\omega_0 - \omega)t_0/2] \frac{\sin \tfrac{1}{2}(\omega_0 - \omega)t_0}{\omega_0 - \omega} \right.$$

$$\left. - \exp[-j(\omega_0 + \omega)t_0/2] \frac{\sin \tfrac{1}{2}(\omega_0 + \omega)t_0}{\omega_0 + \omega} \right).$$

This spectrum is called the **sync function**. It is peaked around $\omega = \omega_0$ and $\omega = -\omega_0$ and in each case has a shape like that in figure 15.19(c). The half-width $\Delta\omega$ of the peak is given by

$$\tfrac{1}{2}\Delta\omega t_0 = \pi \qquad \text{or} \qquad \Delta\omega = 2\pi/t_0. \tag{15.22}$$

A narrow spectrum requires large t_0.

The uncertainty principle

When an atom radiates a light wave, it gets interrupted every now and then by collisions, which alter the phase of the wave in a random way. If the average length of the wave train is $t_0 = \Delta t$, we can substitute $E = h\nu = h\omega/2\pi$ for the energy of the light. From (15.22),

$$\Delta E \Delta t = h \tag{15.23}$$

where h is Planck's constant. This is Heisenberg's uncertainty principle: the energy spread $2\Delta E$ is dictated by the duration Δt of the signal and cannot be less than the value given by equation (15.23). Sometimes Δt and ΔE are defined in terms of RMS values, in which case small but unimportant extra numerical factors appear in equation (15.23).

15.14   Exercises

(Exercises 1 and 2 give some practice in evaluating Fourier coefficients. The algebra is often tricky and sometimes lengthy. Once you have the idea, you may subsequently refer to tables for standard results.)

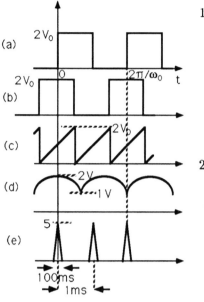

(a)

(b)

(c)

(d)

(e)

100 ms

1 ms

Fig. 15.20

1*.   Equation (15.2) is the Fourier series for the square waves of figure 15.2, where the signal alternates between $+V_0$ and $-V_0$; verify this result by evaluating the integrals (15.4*a*) explicitly for $n = 1$, 2, 3, 4 and 5. If instead the square waves are between 0 and 2 V, as in figure 15.20(a), what change is there in the Fourier series? Find the Fourier series for the square waves of figure 15.20(b) by redefining $t = 0$. (Ans: a DC term V_0 is added; $V_0 + (4V_0/\pi)[\cos \omega_0 t - (1/3)\cos 3\omega_0 t + (1/5)\cos 5\omega_0 t - \ldots]$.)

2**.   Show that the waveforms in figure 15.20(c), (d) and (e) are given for one cycle $x = -\pi$ to $+\pi$ about $t = 0$ by:

(c) $V = V_0 + V_0 x/\pi$, hence

$$V = V_0 + (2V_0/\pi)(\sin \omega_0 t - \tfrac{1}{2}\sin 2\omega_0 t + \tfrac{1}{3}\sin 3\omega_0 t - \ldots)$$

(d) $V = 1 + \cos(\tfrac{1}{2}x)$, hence

$$V = 1 + 2/\pi + (4/3\pi)\cos \omega_0 t - (4/15\pi)\cos 2\omega_0 t$$
$$+ (4/35\pi)\cos 3\omega_0 t - \ldots$$

where $\omega_0 = \pi/2$ rad/s.

(e) $V = 5 - 50|x|/\pi$ for $|x| < \pi/10$, hence

$$V = \frac{1}{4} + \frac{100}{\pi^2}\sum_1^\infty \frac{1}{n^2}\left(1 - \cos\frac{n\pi}{10}\right)\cos n\omega_0 t$$

where $\omega_0 = 2\pi \times 10^3$ rad/s.

3.   An RF signal of 100 mA at 300 m wavelength is mixed with an audio signal of frequency 1 kHz and magnitude 1 mA by a device having a characteristic $V = AI + BI^2$. The output is passed through a filter with a passband flat between 0.9 and 1.1 MHz and zero elsewhere. What is the output of the filter? What would be the effect of a term CI^3 in the characteristic of the device? (Ans: $V = 0.1A\cos \omega_1 t + 10^{-4}B[\cos(\omega_1 - \omega_2)t + \cos(\omega_1 + \omega_2)t]$ where $\omega_1 = 2\pi \times 10^6$ and $\omega_2 = 2\pi \times 10^3$ rad/s; $1.5 \times 10^{-7}C[\cos \omega_1 t + \tfrac{1}{2}\cos(\omega_1 - 2\omega_2)t + \tfrac{1}{2}\cos(\omega_1 + 2\omega_2)t] + 0.5 \times 10^{-3}\cos \omega_1 t$.)

4.   (QMC 1968) Describe how non-linear systems are used in amplitude modulation, demodulation, and frequency changing.

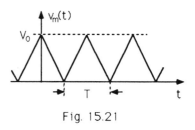

5. (QMC 1987) With the aid of time and frequency diagrams, describe frequency modulation. Derive the frequency modulation formula given below and explain each of the terms: $V_{fm}t = A_C \cos(\omega_C t + M \sin \omega_M t)$. An FM transmitter has a bandwidth of 120 kHz. If the carrier frequency is 100 MHz, what is the maximum frequency of the modulating signal if the modulation index M is 5? (Ans: $f_M = 12$ kHz.)

Fig. 15.21

6. A modulation signal, consisting of a series of triangular pulses as shown in figure 15.21, is used to produce the FM signal $v_C(t) = V_C \cos[\omega_C t + A v_m(t)]$, where ω_C is the carrier frequency and A a constant. Describe the FM signal $v_C(t)$ and its frequency spectrum. Compare the bandwidths of the modulation signal and the FM signal assuming $T = 1$ μs, $A = 1$ radian per V for the cases: (a) $V_0 = 1$ V, (b) $V_0 = 100$ V. (Ans: signals at $\omega = \omega_C \pm 2AV_0/T$; (a) fundamental at $\omega_0 = 2\pi/T$ + harmonics, where $\omega_0 = 2\pi \times 10^6$ rad/s; FM bandwidth $\Delta\omega = 4 \times 10^6$ rad/s, (b) $\omega_0 = 2\pi \times 10^6$ rad/s and $\Delta\omega = 4 \times 10^8$ rad/s.)

7. If the Fourier transform of a function $x(t)$ is $X(f)$, show that (i) the transform of $x(t) \cos(2\pi f_C t)$ is $\frac{1}{2}[X(f + f_C) + X(f - f_C)]$ and (ii) the transform of $x'(t)$ is $j\omega X(f)$, where $\omega = 2\pi f$.

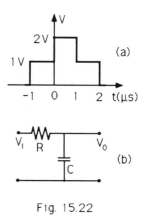

8. (Adapted from QMC 1971) A signal $V = V_0 \exp(j\omega_0 t)$ lasts from $t = -\frac{1}{2}\tau$ to $+\frac{1}{2}\tau$ and is zero outside this time interval. Show that its frequency spectrum is $(2V_0/x\sqrt{2\pi}) \sin(\frac{1}{2}x\tau)$ where $x = \omega - \omega_0$. Show that if it is centred at time t_1 it becomes $V(x) \exp(-jxt_1)$. Hence find the spectrum of the pulse shown in figure 15.22(a). Write down the spectrum of the pulse after it has passed through the filter circuit shown in figure 15.22(b). (Ans: $V'(x) = (8/x\sqrt{2\pi}) \exp(-\frac{1}{2}jx\tau) \cos^2(\frac{1}{2}x\tau) \sin(\frac{1}{2}x\tau)$ where $\tau = 1$ μs; $V''(x) = V'(x)/(1 + j\omega CR)$.)

Fig. 15.22

16

Active Filters

16.1 Types of Filter

In the previous chapter, it became apparent that there is a need for filters with well defined cut-offs, separating one band of frequencies from another. In this chapter ideas of how to achieve this are introduced. It is an extensive subject, and the professional will need to consult further specialist texts.

In outline, there are five basic types of filter. The first four are illustrated in figure 16.1. The low-pass and high-pass filters are similar to the simple CR filters of Chapter 4 except that they cut off more sharply. The band-pass filter selects a specific range of frequencies and the band-reject or notch filter eliminates this range. There is a fifth type, called the all-pass filter; it transmits all frequencies with constant amplitude, but introduces a phase variation with frequency (Chapter 7).

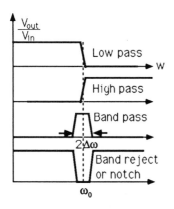

Fig. 16.1. Four filter types.

16.2 Quadratic Filters

In Chapter 14, resonant circuits were considered; these are of the band-pass variety. Here the discussion will be generalised to other circuits obeying similar equations. This will allow replacement of the inconvenient inductance by other more convenient elements. In turn, this introduces the freedom to tune freely the resonant frequency and the width of the resonance curve, i.e. the bandwidth.

Let us start from a familiar example, the series LCR circuit of figure 16.2. This obeys the equation

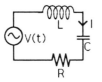

Fig. 16.2. Series LCR circuit.

$$L\frac{dI}{dt} + RI + (1/C)\int I\,dt = V(t)$$

$$\frac{d^2 I}{dt^2} + (R/L)\frac{dI}{dt} + \frac{I}{CL} = (1/L)\frac{dV}{dt}.$$

In standard form, this reads

$$\boxed{\frac{d^2 I}{dt^2} + 2\gamma\frac{dI}{dt} + \omega_0^2 I = F(t).}$$

(16.1)

Recall that for small damping ω_0 is the resonant frequency and the half-width of the resonant peak is, from equations (14.13) and (14.19),

$$\boxed{\Delta\omega = \gamma.}\qquad(16.2)$$

Another useful form, from equation (14.14a), is

$$\boxed{2\gamma = 2\Delta\omega = \omega_0/Q.}\qquad(16.3)$$

The properties of equation (16.1) will now be explored as a function of angular frequency ω. However, it is conventional to do this using the variable

$$\boxed{s = j\omega.}\qquad(16.4)$$

This is to avoid getting tangled up with j and j^2 where these are not of crucial importance. We shall mostly be concerned with the *magnitude* of the transfer function, which depends on $|s|$; only where the phase is relevant is it necessary to return to $j\omega$. With this convention, the impedances of capacitors and inductors are

$$Z_C = 1/sC\qquad(16.5)$$
$$Z_L = sL.\qquad(16.6)$$

Suppose $V(t) = V_0\exp{(j\omega t)} = V_0\exp(st)$ and $I(t) = I_0\exp(st)$, where I_0 is complex, incorporating the phase of the current relative to the voltage. Equation (16.1) gives

$$(s^2 + 2\gamma s + \omega_0^2)I_0 = sV_0/L$$

$$I_0 = (sV_0/L)/(s^2 + 2\gamma s + \omega_0^2).\qquad(16.7)$$

Then the voltages across L, R and C are

$$V_L = s^2 V_0/(s^2 + 2\gamma s + \omega_0^2)\qquad(16.8)$$
$$V_R = 2\gamma s V_0/(s^2 + 2\gamma s + \omega_0^2)\qquad(16.9)$$
$$V_C = \omega_0^2 V_0/(s^2 + 2\gamma s + \omega_0^2).\qquad(16.10)$$

Low-pass filter

Figure 16.3 displays the Bode plot for the magnitude of V_C; remembering that $s = j\omega$, this is given by

$$|V_C| = \frac{\omega_0^2 V_0}{[(-\omega^2 + \omega_0^2)^2 + (2\gamma\omega)^2]^{1/2}}.\qquad(16.11)$$

For small ω, $|V_C| \simeq V_0$. For $\omega \gg \omega_0$, $|V_C| \simeq V_0\omega_0^2/\omega^2$. These limiting cases are described by the two straight lines meeting at

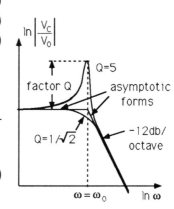

Fig. 16.3. Bode plot for V_C.

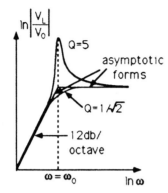

Fig. 16.4. Bode plot for V_L.

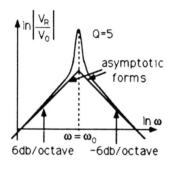

Fig. 16.5. Bode plot for V_R.

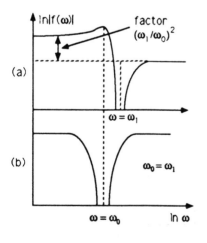

Fig. 16.6. Bode plots of notch filters.

$\omega = \omega_0$. At resonance, the height of the peak above this intersection is determined by

$$|V_C| = \omega_0 V_0 / 2\gamma = V_0 Q$$

using equation (16.3), i.e. a factor Q in magnitude. For $Q = 1/\sqrt{2}$, V_C gives a low-pass filter with cut-off frequency ω_0 and 12 db/octave rolloff at high frequencies.

High-pass filter

The behaviour of $|V_L|$ against ω is shown in figure 16.4. Again it is approximated by two straight lines meeting at $\omega = \omega_0$ and a peak at this frequency a factor Q above the intersection point. For $Q = 1/\sqrt{2}$, V_L gives a high-pass filter with cut-off frequency ω_0 and 12 db/octave rolloff at low frequencies.

Band-pass filter

The behaviour of $|V_R|$ is shown in figure 16.5. It acts as a band-pass filter. Again the height of the peak is a factor Q above the intersection point of the two straight lines approximating the low and high frequency behaviour.

Band-reject filter

To see how the band-reject filter arises, consider the transfer function

$$\frac{V_{\text{out}}}{V_{\text{in}}} = f(s) = \frac{s^2 + \omega_1^2}{s^2 + 2\gamma s + \omega_0^2}. \tag{16.12}$$

Its magnitude is given by

$$|f(s)| = \frac{|\omega_1^2 - \omega^2|}{[(\omega_0^2 - \omega^2)^2 + (2\gamma\omega)^2]^{1/2}}. \tag{16.13}$$

Its behaviour for low Q is sketched in figure 16.6. At $\omega = \omega_1$, $|f| = 0$, making a band-reject filter at this frequency. If $\omega_1 = \omega_0$, figure 16.6(b), the transmission is flat except for a dip at this frequency.

All-pass filter

The fifth variety of filter has the transfer function

$$\frac{V_{\text{out}}}{V_{\text{in}}} = f'(s) = \frac{s^2 - 2\gamma s + \omega_0^2}{s^2 + 2\gamma s + \omega_0^2} \tag{16.14}$$

and

$$|f'(s)|^2 = \frac{(\omega_0^2 - \omega^2)^2 + (2\gamma\omega)^2}{(\omega_0^2 - \omega^2)^2 + (2\gamma\omega)^2} = 1.$$

An example of a circuit giving this type of transfer function was considered as a worked example in Chapter 7; it gives

$$V_{\text{out}} = V_{\text{in}} e^{-2j\phi}$$
$$\phi = \tan^{-1} 2\gamma\omega/(\omega_0^2 - \omega^2).$$

The transmission is constant in magnitude, but the phase lags by an angle 2ϕ, changing through $360°$ with frequency (figure 16.7). This is called an all-pass filter.

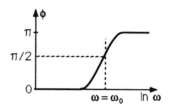

Fig. 16.7. Phase variation in the all-pass filter.

Summary

Any circuit producing the s dependence of equations (16.8)–(16.10), (16.12) and (16.14) is a suitable prototype for the required filters. It does not yet have a sharp cut-off against frequency, but it will become clear shortly how this can be improved.

16.3   The Biquad Active Filter

There are many varieties of feedback circuit which will produce the required quadratic s dependence. Generally they involve two capacitors to give two powers of s in the algebra. Several examples are given in the exercises at the end of the chapter. These examples generally suffer the disadvantage that component values interact in the expressions for gain, ω_0 and bandwidth, so that these quantities cannot be adjusted independently. Here one popular circuit which overcomes this problem is discussed; it suffers the minor expense of using four operational amplifiers. It is called the **biquad filter**, since it gives rise to a transfer function which is the ratio of two quadratic forms:

$$f = K \frac{s^2 + sa + b}{s^2 + sc + d} \tag{16.15}$$

where values of K, a, b, c and d depend on component values.

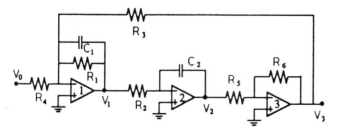

Fig. 16.8. The biquad active filter.

The circuit is given in figure 16.8 and is quite straightforward to analyse. From the currents at the negative terminals of amplifiers 1, 2 and 3 in turn:

$$\frac{V_0}{R_4} = -\frac{V_3}{R_3} - V_1\left(\frac{1}{R_1} + sC_1\right) \tag{16.16}$$

$$V_1/R_2 = -sC_2V_2 \tag{16.17}$$

$$V_2/R_5 = -V_3/R_6. \tag{16.18}$$

From (16.17) and (16.18),

$$V_1 = (sC_2R_2R_5/R_6)V_3. \tag{16.19}$$

Substituting in (16.16),

$$\frac{V_0}{V_3} = -R_4\left(\frac{1}{R_3} + \frac{sC_2R_2R_5}{R_1R_6}(1 + sC_1R_1)\right)$$

$$= \frac{-R_4}{R_1R_3R_6}(R_1R_6 + sC_2R_2R_3R_5 + s^2C_1C_2R_1R_2R_3R_5)$$

so

$$\frac{V_3}{V_0} = \frac{-R_6/(R_2R_4R_5C_1C_2)}{s^2 + s/(R_1C_1) + R_6/(R_2R_3R_5C_1C_2)}. \tag{16.20}$$

Thus V_3 behaves as a low-pass quadratic filter with gain $-R_3/R_4$ at low frequency,

$$\omega_0^2 = R_6/(R_2R_3R_5C_1C_2) \tag{16.21}$$

and bandwidth

$$2\Delta\omega = 2\gamma = 1/C_1R_1. \tag{16.22}$$

Independent control over the gain is given by R_4, which does not figure in ω_0^2 or the bandwidth; the cut-off frequency may be controlled by R_2 (or R_6/R_5) and the bandwidth by R_1.

Next let us solve for V_1 from equation (16.19):

$$\frac{V_1}{V_0} = \frac{-s/(C_1R_4)}{s^2 + s/(R_1C_1) + \omega_0^2}. \tag{16.23}$$

Thus V_1 behaves as a quadratic band-pass filter, whose gain on resonance can again be controlled by R_4.

The remaining filters may be synthesised by feeding V_0, V_1 and V_3 via resistors to an analogue adder (figure 16.9). The result is

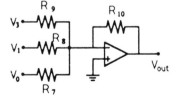

Fig. 16.9. Synthesis of f(s).

$$f(s) = \frac{V_{\text{out}}}{V_{\text{in}}}$$

$$= -(R_{10}/R_7)[s^2 + s/(R_1C_1) + \omega_0^2 - s(R_7/R_8C_1R_4)$$

$$- \omega_0^2(R_7R_3/R_9R_4)][s^2 + s/(R_1C_1) + \omega_0^2]^{-1}$$

$$= -(R_{10}/R_7)[s^2 + (s/C_1)(1/R_1 - R_7/R_4R_8)$$

$$+ \omega_0^2(1 - R_3R_7/R_4R_9)][s^2 + s/(R_1C_1) + \omega_0^2]^{-1}$$

from which the values of K, a and b of equation (16.15) can be read off. The values of a and b may be controlled by R_8 and R_9 to produce high-pass, notch or all-pass filters.

16.4 Butterworth Filters

The quadratic filter is a convenient starting point for making more sharply edged filters, but to achieve a rapid fall-off it is necessary to use higher powers of s. This may be achieved by cascading several quadratic filters. Again attention will be concentrated on one type, the Butterworth filter. Once you have the idea how this one works, it is a matter of detail to pursue other varieties in advanced texts.

Low-pass filter
Consider first a low-pass filter. The others may be derived from it by simple algebraic transformations. The idea is to construct a filter having transfer function $f(s)$ such that

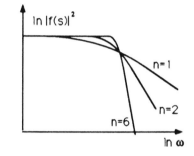

$$|f(s)|^2 = \frac{1}{1 + (\omega/\omega_0)^{2n}} \qquad (16.24)$$

$$|f(s)|^2 = \frac{1}{1 + (-s^2/\omega_0^2)^n}. \qquad (16.24a)$$

The transmission falls to $\frac{1}{2}$ at $\omega = \omega_0$ and thereafter falls extremely rapidly for large n (figure 16.10).

Fig. 16.10. Performance of Butterworth filters with $n = 1$, 2 and 6.

Zeros
In order to construct a filter having the form of equation (16.24a), some mathematical manipulations are necessary. The denominator is factorised in powers of s into the form $(s - \alpha)(s - \beta)(s - \delta) \ldots$. Each of the terms in this expression gives a zero at $s = \alpha$ or β or δ, etc, where from equation (16.24a)

$$(-s^2/\omega_0^2)^n = -1 = e^{j\pi(2m-1)}$$

with m an integer. At these zeros,

$$s^2 = -\omega_0^2 e^{j\pi(2m-1)/n} = \omega_0^2 e^{j\pi} e^{j\pi(2m-1)/n}$$

$$s = \omega_0 e^{j\pi/2} e^{j\pi(2m-1)/2n}. \qquad (16.25)$$

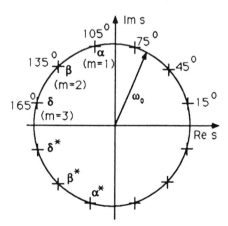

As an example, for $n = 6$ the zeros occur at values of s shown in figure 16.11; they lie on a circle for complex s of radius ω_0 and spaced by $30°$ in argument.

Fig. 16.11. Values of s for the zeros of the denominator of eqn. (16.24a)

f(s)

We are now ready to write down an expression for $f(s)$. In doing this it simplifies the algebra to make use of the left–right symmetry of figure 16.11 about the imaginary axis. It is convenient to group together the zeros of the left-hand half of the figure into one term and the zeros of the right-hand half into a second term. For example, with $n = 2$, setting $S = s/\omega_0$, the left-hand zeros give

$$f^{-1}(S) = \left(S + \frac{1}{\sqrt{2}} - \frac{j}{\sqrt{2}}\right)\left(S + \frac{1}{\sqrt{2}} + \frac{j}{\sqrt{2}}\right) = S^2 + \sqrt{2}S + 1.$$

(16.26)

To make use of the left–right symmetry, remember that $S = j\omega/\omega_0$, so changing the sign of $\mathrm{Re}(s)$ is equivalent to taking the complex conjugate of $f^{-1}(s)$, i.e.

$$[f^{-1}(S)]^* = S^2 - \sqrt{2}S + 1.$$

This second term is repesented by the zeros of the right-hand half of figure 16.11. You can easily check that $|f(S)|^2 = f(S)f^*(S) = 1/(1 + S^4)$ as required. For other values of n, examples are

$$f^{-1}(S) = 1 + S \quad \text{for} \quad n = 1 \tag{16.26a}$$

$$f^{-1}(S) = (S^2 + 0.76536S + 1)(S^2 + 1.84776S + 1) \text{ for } n = 4$$
(16.26b)

$$f^{-1}(S) = (S^2+0.5176S+1)(S^2+2S+1)(S^2+1.93185+1) \text{ for } n = 6.$$
(16.26c)

Tables of these functions are available for any n value.

How may $f(s)$ be generated for $n = 6$, as an example? Each of the brackets in (16.26c) can be formed with one biquad filter circuit with parameters generating the required coefficients. For example, in the first term $2\gamma = 0.5176\omega_0$; in the second, $2\gamma = 2\omega_0$, and so on. With the signal propagated through three such filters in succession, $f(s)$ is formed. For the user, the task is reduced to looking up the required expansions (16.26) and choosing component values appropriate to the required value of ω_0.

Other filters

This deals with the low-pass filter. What about the other types of filter? To generate a high-pass filter, each bracket in (16.26) needs to be transformed by the substitution $s \rightarrow \omega_0^2/s$. The effect is to change the quadratic low-pass function (16.10) to

$$f(s) \rightarrow \frac{\omega_0^2}{\omega_0^4/s^2 + 2\gamma\omega_0^2/s + \omega_0^2} = \frac{s^2}{\omega_0^2 + 2\gamma s + s^2}$$

which is the high-pass filter function (16.8). This substitution is made in every term of equation (16.26) and in every corresponding biquad filter.

To generate a band-pass filter with $n = 2$, the trick is to substitute into the low-pass filter expression for $n = 1$ the transformation

$$s \to \frac{\omega_0}{2\gamma}\left(\frac{s}{\omega_0} + \frac{\omega_0}{s}\right). \tag{16.27}$$

Then $f(s)$ changes from $1/(s+1)$ to

$$f(s) = \left[1 + \frac{\omega_0}{2\gamma}\left(\frac{s}{\omega_0} + \frac{\omega_0}{s}\right)\right]^{-1} = \frac{2\gamma s}{\omega_0^2 + 2\gamma s + s^2}$$

which is the required band-pass filter equation (16.9). Finally, to generate the band-reject filter with $n = 2$, we substitute into $n = 1$ with the inverse of (16.27):

$$s \to \frac{2\gamma}{\omega_0}\left(\frac{s}{\omega_0} + \frac{\omega_0}{s}\right)^{-1}$$

$$f(s) \to \left[1 + \frac{2\gamma}{\omega_0}\left(\frac{s}{\omega_0} + \frac{\omega_0}{s}\right)^{-1}\right]^{-1} = \frac{(s/\omega_0) + (\omega_0/s)}{(s/\omega_0) + (\omega_0/s) + 2\gamma/\omega_0}$$

$$= \frac{s^2 + \omega_0^2}{s^2 + \omega_0^2 + 2\gamma s} \tag{16.28}$$

as required by equation (16.12) if $\omega_0 = \omega_1$.

As an example, to make a notch filter with $n = 8$, every term in (16.26*b*) is transformed in this way and then four biquad filters are used to generate the four quadratic terms of the form of (16.28). This is tedious but straightforward. If you want to vary the centre frequency or bandwidth of the filter, the resistor values controlling ω_0 and γ in each of the four filters have to be varied in unison.

16.5   Chebychev and Other Filters

The choice (16.24) is only one of many functions which fall rapidly at $\omega = \omega_0$. You can invent your own. However, you will find that if $|f(\omega)|^2$ falls continuously away from 1 with increasing ω, the Butterworth filter has the rapidest fall-off beyond $\omega = \omega_0$. To see this, expand it by the binomial expansion for small ω:

$$|f(s)|^2 \simeq 1 - (\omega/\omega_0)^{2n} + 0(\omega/\omega_0)^{4n}.$$

The first $(2n - 1)$ coefficients of ω are zero. For this reason, the behaviour of the Butterworth filter is described as 'optimally flat'.

Chebychev filters
It is possible to achieve a faster fall-off at $\omega = \omega_0$, but only at the expense of some ripple for $\omega < \omega_0$, as shown in figure 16.12. If this

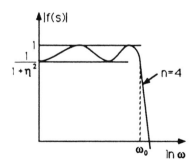

Fig. 16.12. The Chebychev filter for $n = 4$.

ripple is confined to the range 1 to $1/(1+\eta^2)$, it can be shown that the fastest falloff at $\omega = \omega_0$ is given by

$$|f(s)|^2 = \frac{1}{1 + \eta^2 C_n^2(\omega/\omega_0)}$$

where the $C_n(x)$ are Chebychev polynomials,

$$C_1(x) = x$$
$$C_2(x) = 2x^2 - 1$$
$$C_4(x) = 1 - 8x^2 + 8x^4$$
$$C_6(x) = 32x^6 - 48x^4 + 18x^2 - 1$$

and so on. In all cases $C_n(1) = 1$. figure 16.12 shows as an example the case where $n = 4$. Beyond $\omega = \omega_0$, the transfer function falls distinctly faster than the Butterworth filter with the same n value.

Elliptic filters

Beyond $\omega = \omega_0$, the transfer function of the Chebychev filter falls continuously. If the user is prepared to accept that it flattens off as in figure 16.13 to a second ripple, an even faster fall-off can be achieved at $\omega = \omega_0$. The frequency region below $\omega = \omega_0$ is known as the **passband** and the region of the second ripple is called the **stopband**. The theoretical optimum for this type of behaviour is called the elliptic or Cauer filter and is described by elliptic functions.

Bessel filters

Near $\omega = \omega_0$, all of these filters generate very rapidly varying phases $\phi(\omega)$. In consequence, if you feed a square pulse through any of them, it emerges with strong transients and overshoots at the leading edge and trailing edge. A type of filter which minimises such transients is the **Bessel filter**. It has the smoothest possible transient response. The price to be paid is that the roll-off at $\omega = \omega_0$ is less rapid than any of the other filters.

Summary

All the user has to do to make any of these various filters (Butterworth, Chebychev, Cauer, Bessel, high-pass, low-pass, notch, etc) is to look up tables of the functions $f(s)$ and then choose component values for the biquad filter giving the required coefficients. There are many alternatives to the biquad filter. A point to watch in practice is how sensitive the performance of the filter is to errors in the component values (due, for example, to temperature variations).

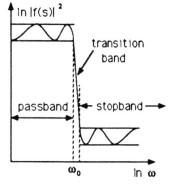

Fig. 16.13. The response of the elliptic filter.

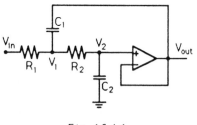

Fig. 16.14.

16.6   Exercises

1. Show from equation (16.9) that V_R on resonance is a factor Q above the intersection of the two straight lines of figure 16.5.

2. Figure 16.14 gives the Sallen–Key low-pass filter. Show that

$$V_{out}/V_{in} = 1/[s^2 C_1 C_2 R_1 R_2 + sC_2(R_1 + R_2) + 1].$$

What are ω_0^2 and the bandwidth and Q? (Ans: $1/C_1 C_2 R_1 R_2$, $(R_1 + R_2)/C_1 R_1 R_2$, $(R_1 + R_2)(C_2/C_1 R_1 R_2)^{1/2}$.)

3. Figure 16.15 gives the Sallen–Key high-pass filter. Show that

$$V_{out}/V_{in} = s^2/(s^2 + 2\gamma s + \omega_0^2)$$

and find 2γ and ω_0^2. (Ans: as exercise 2.)

4. In figure 16.16, Y_1–Y_5 are admittances. Show that

$$\frac{V_{out}}{V_{in}} = \frac{(R_6 + R_7)Y_1 Y_3/R_7}{Y_5(Y_1 + Y_2 + Y_3 + Y_4) + Y_3(Y_1 + Y_4 - Y_2 R_6/R_7)}.$$

How are components Y_i to be chosen to make a band-pass filter? (Ans: R_1, R_2, C_3, C_4, R_5.)

5. In figure 16.17, $Y_1 - Y_5$ are admittances. Show that

$$V_{out}/V_{in} = -Y_1 Y_3/[Y_5(Y_1 + Y_2 + Y_4) + Y_3(Y_4 + Y_5)].$$

What components would you choose for a band-pass filter? (Ans: R_1, R_2, C_3, C_4 and R_5.)

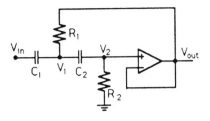

Fig. 16.15.

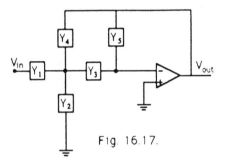

Fig. 16.16.

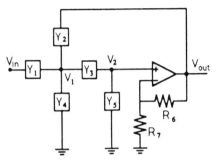

Fig. 16.17.

17

Equivalent Circuits for Diodes and Transistors

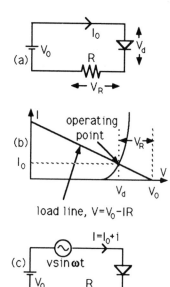

Fig. 17.1 (a) Diode circuit,
(b) determination of the
operating point, (c) the
effect of a small AC voltage.

17.1 Introduction

Transistors and the diode were introduced in Chapters 9 and 10 and their use in simple circuits was outlined qualitatively. In this chapter, they are treated quantitatively. It turns out that for small signals they may be represented by Thevenin or Norton equivalent circuits. This makes it easy to analyse their performance with simple algebra. The idea is introduced with the diode, and repeats some of the discussion of section 1.11.

17.2 AC Equivalent Circuit for the Diode

Suppose a DC voltage V_0 is applied to a resistor and diode in series (figure 17.1(a)). From Kirchhoff's voltage law, the voltage V_d across the diode is given by

$$V_0 = I_0 R + V_d(I_0). \qquad (17.1)$$

The DC resistance of the diode, V_d/I_0, varies with V_d. It does not obey Ohm's law. Using the diode as a switch, this is very obvious: its resistance is small for forward bias and very large for reverse bias.

The values of V_d and I_0 may be determined as in figure 17.1(b) from the intersection of the straight line $V_0 - IR$ with the diode curve. This intersection at $I = I_0$ is called the **operating point**. The straight line is called the **load line**, since it gives the DC voltage across the diode after allowing for the potential drop across the load R.

Next consider figure 17.1(c), where a small AC voltage $v\sin\omega t$ is added to the circuit. The source voltage swings up and down between the two lines of figure 17.2. The equilibrium point for the diode moves up and down the characteristic curve, along the

258

shaded part. The current changes by a small amount $\pm i$ and the voltage across the diode by $\pm i\rho$, where ρ is the slope of the diode curve dV_d/dI at $I = I_0$. Then

$$V_0 + v\sin\omega t = (I_0 + i)R + V_d(I_0) + i\rho. \qquad (17.2)$$

Subtracting (17.1),

$$v\sin\omega t = iR + i\rho. \qquad (17.3)$$

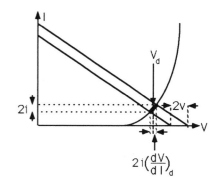

Equation (17.3) may be represented by means of the **AC equivalent circuit**, shown in figure 17.3. The circuit describes the AC part of the behaviour. The DC part has been subtracted out in the difference between equations (17.1) and (17.2). The quantity ρ is called the slope resistance:

Fig. 17.2. Effect of $v\sin\omega t$.

$$\boxed{\rho = (dV_d/dI)_{I=I_0}.}$$

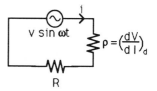

Effectively, the behaviour of the circuit has been separated into a DC *condition*, given by equation (17.1), and an AC *condition*, given by equation (17.3). This works even for power dissipation:

Fig. 17.3. Diode AC equivalent circuit for small signals.

$$P = \overline{VI} = \overline{(V_0 + v\sin\omega t)(I_0 + i\sin\omega t)} = V_0 I_0 + \tfrac{1}{2}vi.$$

The cross-terms involving $\sin\omega t$ average to zero.

The conclusion from figure 17.3 is that, as far as AC signals are concerned, we can

(a)   omit the battery (short circuit it), and

(b)   replace the diode by an equivalent resistance ρ.

If the diode characteristic is very steep, is ρ large or small? Can you understand the equivalent circuit direct from figure 17.2 in this case?

17.3   Slope Resistance of a pn Diode

In Chapter 9, an algebraic expression was derived for the current in the diode:

$$I = I_0\left[\exp\left(\frac{eV}{kT}\right) - 1\right]. \qquad (9.15)$$

Then

$$\frac{dI}{dV} = I_0\frac{e}{kT}\exp\left(\frac{eV}{kT}\right) \simeq \frac{eI}{kT}$$

when V is well away from zero and the 1 in (9.15) can be neglected. Hence the slope resistance of the diode is given by

$$\boxed{\rho = \frac{dV}{dI} \simeq \frac{kT}{e}\frac{1}{I} \simeq \frac{1}{40}\frac{1}{I(\text{A})} = \frac{25\ \Omega}{I(\text{mA})}.} \qquad (17.4)$$

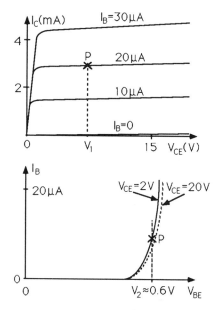

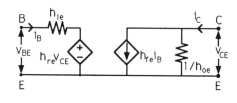

Fig. 17.4. Characteristic of
the bipolar transistor.

Fig. 17.5. The hybrid equivalent
circuit for the transistor when
the emitter is common to input
and output.

This is a useful rule of thumb. It applies also to the diode action between base and emitter in a bipolar transistor.

Of course, if the AC signal v is not infinitesimally small, ρ varies as the diode goes up and down the characteristic. Even in this case an average value of ρ in figure 17.3 may give a useful approximation.

From equation (17.4), what change in V do you expect as I varies from 1 to 10 mA? Use $dV = \rho\, dI$.

17.4   H Equivalent Circuit for the Bipolar Transistor

As a reminder, the typical characteristic curves for a bipolar transistor are reproduced in figure 17.4 from Chapter 9. It is possible, but laborious, to follow the behaviour of a transistor graphically, direct from these curves. However, when dealing with small AC signals it is easier to replace the transistor by an equivalent circuit, as for the diode. It will emerge that the equivalent circuit may be drawn as in figure 17.5. The input side, between base and emitter, is a Thevenin equivalent, where h_{ie} is the slope resistance of the base–emitter diode; usually the EMF $h_{re}v_{CE}$ is negligible. The output side is a Norton equivalent: a current source $h_{fe}i_B$ with a parallel resistance. Once this replacement has been made, circuit analysis is greatly simplified. Even when the AC signals are large and the linear approximation breaks down, the equivalent circuit gives a rough approximation to what is happening.

The circuit of figure 17.5 is a hybrid of Thevenin and Norton forms and the circuit parameters are denoted by the letter h for hybrid. The subscript e denotes the fact that the emitter is common to input and output circuits; i denotes input and o stands for output.

The derivation of the equivalent circuit is slightly more complicated than for the diode because collector current I_C now depends on two variables, I_B and V_{CE} (figure 17.4(a)). So does V_{BE} (figure 17.4(b)). A static operating point P is defined by DC voltages applied between emitter and collector ($V_{CE} = V_1$ on figure 17.4) and between emitter and base ($V_{BE} = V_2$). Then the dependence on small variations of both I_B and V_{CE} gives the equivalent circuit.

Output side

Consider first the output (collector) circuit. We can get from point P of figure 17.6(a) to any neighbouring point Q by first moving along one of the curves with I_B fixed, and then changing I_B keeping V_{CE} fixed. In the first step, as V_{CE} varies, figure 17.6(b) shows that

$$\delta I_C = \left(\frac{\partial I_C}{\partial V_{CE}}\right)\delta V_{CE}.$$

The ∂ notation indicates that only these two variables I_C and V_{CE} are changing; I_B is held constant. It is called a partial differential because only two variables change and the third is held constant. The slope of the line in figure 17.6(b) at the point P is the required differential.

Secondly, varying I_B, the change δI_B in figure 17.6(c) gives

$$\delta I_C = \left(\frac{\partial I_C}{\partial I_B}\right)\delta I_B.$$

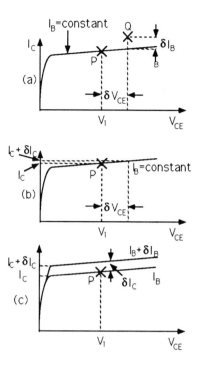

This time, V_{CE} is held constant. These are the only two changes to be made; varying V_{BE} reflects itself in changing I_B and V_{CE}, since it depends only on these two variables.

You may measure $\partial I_C/\partial I_B$ by observing the variation of I_C with I_B keeping V_{CE} constant. Likewise, you may measure $\partial I_C/\partial V_{CE}$ by varying V_{CE}, but you need to be careful to compensate for any small change in I_B because of the sensitivity of I_C to it; this is not easy to do accurately.

Putting these two terms together,

$$\delta I_C = \left(\frac{\partial I_C}{\partial I_B}\right)\delta I_B + \left(\frac{\partial I_C}{\partial V_{CE}}\right)\delta V_{CE}. \qquad (17.5)$$

Fig. 17.6. Changes to I_C.

The notation may be abbreviated by using small letters $v_{CE} = \delta V_{CE}$, $i_C = \delta I_C$, and $i_B = \delta I_B$ for the changes,

$$\boxed{i_C = h_{fe}i_B + h_{oe}v_{CE}.} \qquad (17.6)$$

Here $h_{fe} = (\partial I_C/\partial I_B)$ and $h_{oe} = (\partial I_C/\partial V_{CE})$.

This achieves the objective of deriving an equivalent circuit for the output side of the transistor. The Norton equivalent circuit of figure 17.5 is described by equation (17.6) where i_C, i_B and v_{CE} are small changes in these currents and voltages. The current i_C is made up of two components, (a) $h_{fe}i_B$ from the current source, and (b) $h_{oe}v_{CE}$ through the impedance $1/h_{oe}$. The manufacturer of the transistor will supply graphs of h_{fe} and h_{oe}. However, as already indicated in Chapter 9, $I_C \simeq \beta I_B$, so $h_{fe} = \partial I_C/\partial I_B \simeq \beta$. The f of h_{fe} stands for 'forward', denoting that a change in I_B propagates forward to a change in I_C.

The transistor characteristic of figure 17.4(a) indicates rather little dependence on V_{CE} above the knee. This implies that $h_{oe} = \partial I_C/\partial V_{CE}$ is small and $1/h_{oe}$ is large. A typical value is 20 kΩ. In the majority of electronic circuits, this is considerably greater than the load placed across the output terminals. If so, $1/h_{oe}$ can be neglected and the equivalent circuit of the output becomes very simple: just a current source βi_B amplifying the base current i_B. Typically, h_{fe} is 50 to 300.

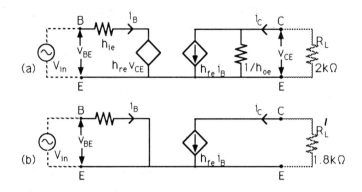

(a)

(b)

Fig. 17.8 (a) The complete equivalent circuit
in the common emitter configuration,
(b) a common simplification.

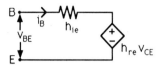

Fig. 17.7. Equivalent circuit for
the bipolar transistor input.

Input side

The input circuit can be treated similarly, expressing changes in V_{BE} in terms of δI_B and δV_{CE}:

$$\delta V_{BE} = \left(\frac{\partial V_{BE}}{\partial I_B}\right) \delta I_B + \left(\frac{\partial V_{BE}}{\partial V_{CE}}\right) \delta V_{CE} \qquad (17.7)$$

or

$$\boxed{v_{BE} = h_{ie} i_B + h_{re} v_{CE}} \qquad (17.8)$$

where $h_{ie} = (\partial V_{BE}/\partial I_B)$ and $h_{re} = (\partial V_{BE}/\partial V_{CE})$ are again partial differentials. In the former, V_{CE} is held constant, and in the latter I_B.

The equivalent circuit for equation (17.8) is shown in figure 17.7. It is common to refer loosely to h_{ie} as the input impedance. Strictly speaking, this is incorrect, since the input impedance is

$$r_{in} = v_{BE}/i_B = (h_{ie} i_B + h_{re} v_{CE})/i_B = h_{ie} + h_{re} v_{CE}/i_B.$$

For the common emitter configuration the distinction is a fine one, since the characteristics of figure 17.4 depend very little on V_{CE}, so h_{re} is very frequently negligible. In this case, the equivalent circuit of the input becomes very simple: just an input resistance h_{ie}. In the parameter h_{re}, the r denotes 'reverse', because h_{re} is input voltage divided by output: $\partial V_{BE}/\partial V_{CE}$. This parameter depends on several delicate effects in the transistor and can be positive or negative and can even change sign with V_{CE}. Typical values are 3–10×10^{-4}. The effect of h_{re} is to introduce some feedback from the output v_{CE} to the input.

The complete equivalent circuit of the bipolar transistor in the common emitter configuration is shown in figure 17.8(a). Frequently it can be simplified to figure 17.8(b) when $h_{re} \simeq 0$ and h_{oe} is small.

Let us now follow through the amplification of a small AC input. Suppose as an example that $v_{BE} = 10\sin\omega t$ mV and suppose that $h_{ie} = 1$ kΩ, $h_{re} = 5 \times 10^{-4}$, $h_{fe} = 100$ and $1/h_{oe} = 20$ kΩ. As a first approximation, $h_{re}v_{CE}$ may be neglected. Then the base current changes by

$$\delta I_B = i_B = v_{in}/h_{ie} = 10\sin\omega t \ \mu A. \qquad (17.9)$$

The collector current changes by

$$\delta I_C = i_C = h_{fe}i_B = h_{fe}v_{in}/h_{ie} = \sin\omega t \ mA. \qquad (17.10)$$

The transistor is behaving as a current amplifier with gain $h_{fe} = 100$ and input impedance $h_{ie} = 1$ kΩ. If the load placed across the output is $R_L = 2$ kΩ, this resistor appears in parallel with $1/h_{oe}$; together, they make a load of $R_L' = 1.8$ kΩ. The output voltage is

$$\delta V_{CE} = v_{CE} = -R_L'i_C = -1.8\sin\omega t \ V$$

and the voltage gain of the circuit (called the common emitter amplifier) is

$$G_V = -R_L'h_{fe}/h_{ie} = -180.$$

The minus sign indicates that the transistor inverts the input signal: if v_{BE} is positive, v_{CE} is negative, because of the direction of current flow through the load R_L' (from E to C). Superposing waveforms on the characteristic curves, as in figure 17.9, illustrates this inversion explicitly.

Finally, let us return to the effect of $h_{re}v_{CE}$. Its magnitude is $\sim 5 \times 10^{-4}v_{CE} = -0.9\sin\omega t$ mV. Its sign is such that in figure 17.8(a) the top of the generator $h_{re}v_{CE}$ is negative with respect to E when v_{BE} is positive. This increases the current i_B through h_{ie} by 9%. In turn this increases i_c and v_{CE} by 9%. However, this correction is likely to be within the tolerances with which h_{ie} is known. It will be demonstrated that h_{ie} varies quite rapidly with operating point. This is a second reason for not bothering too much about the effects of h_{re}.

Common base and common collector configurations of the transistor are also used commonly. A similar analysis leads to h equivalent circuits identical in appearance to the one for the common emitter, except that h parameters are labelled with a different subscript, for example, h_{ib} for the input resistance in the common base configuration. Referring forward a bit, figure 17.11(a) shows the h equivalent circuit for the common base configuration.

The manufacturer of a transistor normally supplies h parameters for these other configurations. However, this is not strictly necessary, since the performance of the transistor can be modelled in a way which allows the new h parameters to be derived from the ones we already have; let us now do this.

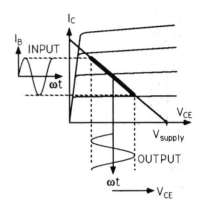

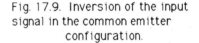

Fig. 17.9. Inversion of the input signal in the common emitter configuration.

17.5   The hybrid-π Equivalent Circuit

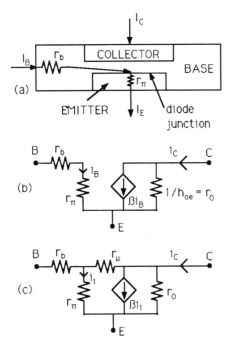

Fig. 17.10 (a) the bipolar transistor, (b) the hybrid-π equivalent circuit in simplified form, (c) including r_μ.

Figure 17.10(b) shows an equivalent circuit derived from the physical characteristics of the bipolar transistor (figure 17.10(a)). The resistance r_π is obtained from the slope resistance of the pn junction between base and emitter:

$$r_\pi = \frac{\mathrm{d}V_{\mathrm{BE}}}{\mathrm{d}I_{\mathrm{B}}} = \frac{\mathrm{d}I_{\mathrm{E}}}{\mathrm{d}I_{\mathrm{B}}}\frac{\mathrm{d}V_{\mathrm{BE}}}{\mathrm{d}I_{\mathrm{E}}} = \frac{(\beta+1)25\ \Omega}{I_{\mathrm{E}}(\mathrm{mA})}. \tag{17.11}$$

How much does r_π change as I_{E} goes from 1 to 10 mA if $\beta = 100$?

Strictly, it is necessary to add a further resistance r_{b} allowing for the resistance of the terminal connection and the ohmic resistance of the base region, though r_{b} is frequently negligible ($\leq 50\ \Omega$). Figure 17.10(b) gives

$$v_{\mathrm{BE}} = (r_{\mathrm{b}} + r_\pi)i_{\mathrm{B}}.$$

Comparing this with equation (17.8),

$$h_{\mathrm{ie}} = r_{\mathrm{b}} + r_\pi \tag{17.12}$$

$$h_{\mathrm{re}} = 0. \tag{17.13}$$

If $I_{\mathrm{E}} = 2$ mA and $\beta \simeq 100$, h_{ie} and r_π are of the order of 1 kΩ. This is a typical figure for the input resistance of the common emitter amplifier. However, if I_{E} varies over the cycle of an AC signal by a significant amount, r_π changes, and consequently so do h_{ie} and the voltage gain. The output signal is no longer a true replica of the input. Thus the rudimentary common emitter amplifier suffers from distortion. A way of overcoming this is to add a resistance R_{E} of the order of 100 Ω directly in series with the emitter, to swamp the junction resistance of the diode. This has the desirable effect of increasing the input impedance to $r_{\mathrm{b}} + r_\pi + (\beta+1)R_{\mathrm{E}} \simeq 10^4\ \Omega$. The factor $(\beta+1)$ arises because current i_{B} from the base circuit and βi_{B} from the collector circuit both flow through R_{E} (neglecting the small current through $1/h_{\mathrm{oe}}$).

From equation (17.13), it is clear that figure 17.10(b) does not allow for the small dependence of the input characteristic on V_{CE}. To accommodate this, a further resistor r_μ may be added, as in figure 17.10(c), between C and B. In practice this resistor is very large (typically 1 MΩ) and makes the algebra messy. Exercise 8 investigates this algebra. Both h_{re} and r_μ will be ignored here for simplicity. If r_μ is omitted, the output circuit between collector and emitter is identical to the h equivalent circuit of the common emitter configuration. Thus $r_0 = 1/h_{\mathrm{oe}} \simeq 20$ kΩ.

This model works for *all* configurations: common emitter, common base and common collector. Using this model, relations may be found between the h parameters in the different configurations.

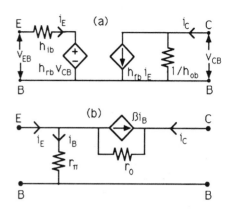

Fig. 17.11. The common base amplifier (a) h equivalent, (b) hybrid-π equivalent.

17.6* Common Base Amplifier

The common base amplifier is rather awkward because feedback from output to input influences its properties rather strongly. The starting point is a derivation of its h parameters, using the hybrid-π circuit in its simplified form (figure 17.11(b)). In this circuit the base B is common to input and output; the input signal is applied between emitter and base and the output is taken between collector and base. Its h equivalent circuit is shown in figure 17.11(a). It is conventional to define the signs of i_C and i_E so that they both flow into the circuit, as in the h equivalent circuit of figure 17.8 for the common emitter configuration. Note that for i_E and i_B this is the opposite convention to figure 17.10(a). Then

$$i_C = h_{fb}i_E + h_{ob}v_{CB} \tag{17.14}$$
$$v_{EB} = h_{ib}i_E + h_{rb}v_{CB}. \tag{17.15}$$

It is a matter of straightforward but tedious algebra to relate the parameters of the h equivalent and hybrid-π circuits. To keep the algebra as simple as possible, r_b of figure 17.10(b) is neglected. It is then simply necessary to express i_C and v_{EB} for the hybrid-π in terms of i_E and v_{CB} using figure 17.11(b):

$$i_B = i_E + i_C \tag{17.16}$$

$$v_{CB} = r_\pi i_B + (i_C + \beta i_B)r_0 = (r_\pi + \beta r_0)(i_E + i_C) + i_C r_0$$

$$v_{CB} = (r_\pi + \beta r_0)i_E + [r_\pi + (\beta + 1)r_0]i_C \tag{17.14a}$$

$$v_{EB} = r_\pi i_B = r_\pi \left(i_E + \frac{v_{CB}}{r_\pi + (\beta + 1)r_0} - \frac{(r_\pi + \beta r_0)i_E}{r_\pi + (\beta + 1)r_0} \right)$$

$$v_{EB} = \frac{r_\pi r_0 i_E}{r_\pi + (\beta + 1)r_0} + \frac{r_\pi v_{CB}}{r_\pi + (\beta + 1)r_0}. \tag{17.15a}$$

Comparing equation (17.14) with (17.14a), and likewise (17.15) with (17.15a),

$$h_{fb} = - (r_\pi + \beta r_0)/[r_\pi + (\beta + 1)r_0] \simeq -1 \tag{17.17}$$
$$h_{ob} = 1/[r_\pi + (\beta + 1)r_0] \tag{17.18}$$
$$h_{ib} = r_\pi r_0/[r_\pi + (\beta + 1)r_0] \simeq r_\pi/(\beta + 1) \tag{17.19}$$
$$h_{rb} = r_\pi/[r_\pi + (\beta + 1)r_0] \simeq r_\pi/(\beta + 1)r_0. \tag{17.20}$$

The approximations arise because $(\beta+1)r_0 \gg r_\pi$. Equation (17.17) is obvious, since most of the emitter current flows to the collector.

In the formula for h_{ib}, the factor $(\beta + 1)$ dividing r_π reflects the fact that most of the input current i_E does not flow to the base but to the collector. The input resistance h_{ib} is of the order of a few ohms. This low value is a distinctive feature of the common base amplifier. It makes it less popular than the common emitter amplifier for ordinary applications; another feature making it less popular is the current gain of -1. But it does have some special applications in integrated circuits and oscillators.

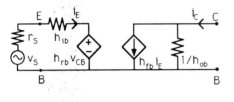

Fig. 17.12. Finding r_{out} for the common base amplifier.

*Output impedance

There is an important subtlety attached to h_{ob} and h_{rb}. One might think that (a) h_{rb} is negligible and (b) h_{ob} is the output impedance of the amplifier. Neither is true under many circumstances and this is relevant to how the common base amplifier is used in integrated circuits. To elucidate this point, let us find the output impedance for the circuit of figure 17.12 from the formula

$$r_{out} = \frac{v_{out}(\text{open circuit})}{i_{out}(\text{short circuit})}.$$

On open circuit,

$$v_{CB} = -h_{fb}i_E/h_{ob} = -h_{fb}(v_S - h_{rb}v_{CB})/[h_{ob}(r_S + h_{ib})]$$

so

$$v_{CB}[h_{ob}(r_S + h_{ib}) - h_{fb}h_{rb}] = -h_{fb}v_S.$$

On short circuit,

$$i_{out} = -h_{fb}v_S/(r_S + h_{ib}).$$

Then

$$r_{out} = \frac{r_S + h_{ib}}{h_{ob}(r_S + h_{ib}) - h_{fb}h_{rb}}$$

or

$$1/r_{out} = h_{ob} - h_{fb}h_{rb}/(r_S + h_{ib}). \tag{17.21}$$

The second term on the right-hand side arises via feedback of v_{CB} from output to input. It is positive since h_{fb} is negative and is by no means negligible. Indeed, if $r_S = 0$, substitution of (17.17)–(17.20) into (17.21) leads to the result

$$r_{out} = r_0$$

while $1/h_{ob} \simeq (\beta+1)r_0$, a vastly different result. (The former result may be confirmed direct from the hybrid-π equivalent circuit; if r_μ is included, r_{out} becomes the parallel combination of r_0 and r_μ.) However, when r_S is large, equation (17.21) shows that $r_{out} \simeq 1/h_{ob}$. In this limit, the common base amplifier has a high output impedance $r_{\text{out}} \simeq (\beta + 1)r_0 \simeq 2 \times 10^6 \ \Omega$. Under these conditions, it is an excellent constant current generator and this is its primary use.

*Input impedance

Likewise the input impedance is affected by $h_{\text{rb}}v_{\text{CB}}$. From figure 17.12,

$$v_{\text{CB}} = -h_{\text{fb}}i_{\text{E}}R_{\text{L}}'$$

where R_{L}' is the parallel combination of $1/h_{\text{ob}}$ with the load resistance across CB. Then

$$v_{\text{in}} = i_{\text{E}}h_{\text{ib}} + h_{\text{rb}}v_{\text{CB}}$$
$$= i_{\text{E}}(h_{\text{ib}} - h_{\text{rb}}h_{\text{fb}}R_{\text{L}}')$$
$$r_{\text{in}} = v_{\text{in}}/i_{\text{E}} = h_{\text{ib}} - h_{\text{rb}}h_{\text{fb}}R_{\text{L}}'. \qquad (17.22)$$

Note that h_{fb} is negative, so the second term is positive. If $R_{\text{L}}' = 2 \ \text{k}\Omega$, and $h_{\text{rb}} = 3 \times 10^{-4}$, the second term is 0.6 Ω, which is normally small compared with h_{ib}. However, if R_{L} is very large and $R_{\text{L}}' \simeq 1/h_{\text{ob}} = 2 \times 10^6 \ \Omega$, the second term dominates completely.

Summary

The transistor in the common base configuration is best treated with figure 17.13, where explicit values of h parameters are inserted in terms of hybrid-π parameters. The current gain is 1. The input impedance is a few ohms, provided that the load resistor $R_{\text{L}} \ll r_0 \simeq 2 \times 10^4 \ \Omega$. However, if R_{L} is made very large, the input impedance increases according to equation (17.22). The output impedance is likewise strongly influenced by feedback to the input side; it can be as high as $(\beta+1)r_0$ if the source resistance r_{S} is large ($\gg r_\pi$). In practice, making use of this high output impedance requires use of a load with a resistance greater than $(\beta + 1)r_0$; it may take the form of either a tuned circuit or a constant current source. It is not possible to achieve simultaneously very low input impedance (R_{L} small) and large output voltage (R_{L} large). The voltage gain G_{V} is $R_{\text{L}}'/r_{\text{in}}$, so if R_{L} is large ($\gg r_0$) and the second term of equation (17.22) dominates over the first, $G_{\text{V}} \simeq 1/h_{\text{rb}} = (\beta + 1)r_0/r_\pi$, the same as the common emitter amplifier.

17.7   The FET

The derivation of an equivalent circuit for the FET (of any variety) follows the same procedure as for the bipolar transistor. As a reminder, typical characteristic curves are repeated in figure 17.14. DC voltages applied at the gate and drain define an operating point

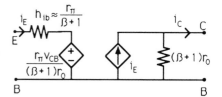

Fig. 17.13. The common base amplifier with common approximations.

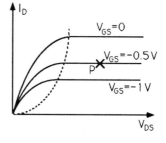

Fig. 17.14. Characteristic curves for the FET.

P, and the equivalent circuit describes small variations around this point. For small changes in V_{GS} or V_{DS},

$$\delta I_D = \left(\frac{\partial I_D}{\partial V_{GS}}\right)\delta V_{GS} + \left(\frac{\partial I_D}{\partial V_{DS}}\right)\delta V_{DS} \qquad (17.23)$$

or

$$\boxed{i_D = g_m v_G + (1/r_d)v_D} \qquad (17.24)$$

where i and v stand for the small changes in current or voltage. The quantity $g_m = \partial I_D/\partial V_{GS}$ is called the **mutual conductance**; a typical value is 2–3 mA V^{-1}. The quantity $r_d = \partial I_D/\partial V_{DS}$ is called the drain resistance and is typically 20–100 kΩ.

The right-hand half of figure 17.15(a) shows the equivalent circuit corresponding to equation (17.24). On the left-hand side, the input between gate and source is a reverse-biased diode with a very high resistance $r_{in}(\geq 10^6\ \Omega$, often *very* much greater). Usually this resistance is so high compared with other resistances in the circuit that the input of the FET can be treated as an open circuit (figure 17.15(b)).

The equivalent circuit is very simple. It is that of a current source controlled by the input voltage. It has high input resistance and voltage gain $-g_m R_L'$, where R_L' is the parallel combination of the load R_L with r_d. Very frequently, r_d is much larger than R_L and can be neglected. The input current is very small, so the current gain is extremely high, and so is the power gain.

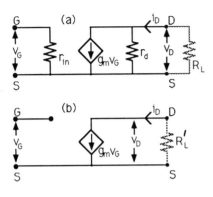

Fig. 17.15 (a) FET equivalent circuit, (b) a common simplification.

17.8   Exercises

1. A transistor in the common emitter configuration has $h_{ie} = 2$ kΩ, $h_{fe} = 120$, $h_{re} = 2 \times 10^{-4}$ and $1/h_{oe} = 15$ kΩ. It is fed by an AC signal $V_{in} = 3\cos\omega t$ mV and drives a load resistor $R_L = 2.2$ kΩ. Find the AC base current i_B, the AC collector current i_C and the AC current through the load resistor, first neglecting the effect of h_{re} and then including it as a perturbation. What are the current and voltage gains through R_L? What are the input and output impedances of the circuit? (Ans: with $h_{re} = 0$, $i_B = 1.5\ \mu$A, $i_C = 0.18$ mA, $i_L = 0.157$ mA, $v_{CE} = -0.345$ V, $h_{re}v_{CE} = -0.069$ mV; including h_{re}, $i_B = 1.53\ \mu$A, $i_C = 0.184$ mA, $i_L = 0.161$ mA, current gain $= -105$, voltage gain $= -118$, $r_{in} = 1.96$ kΩ, $r_{out} = 1.92$ kΩ.)

2. Appendix D gives the characteristic curves for (a) the BC182 npn transistor, and (b) the BC315L pnp transistor. Find values of h_{ie}, $1/h_{oe}$ and h_{fe} for operating points respectively (a) $V_{CE} = 5$ V, $I_{BE} = 20\ \mu$A, (b) $V_{CE} = -5$ V, $I_{BE} = -20\ \mu$A. Assume $h_{re} = 0$. If an AC voltage $\cos\omega t$ mV is

applied to each transistor, what is the AC base current, the AC emitter current and the AC voltage between collector and emitter if the load resistor is infinite? What is the maximum available voltage gain and current gain in each case? (Ans: (a) $\sim$1250 Ω, 20 kΩ, 350, (b) $\sim$2200 Ω, 15 kΩ, 180; (a) 0.8 $\cos\omega t$ μA, $-0.28\cos\omega t$ mA, $-5.6\cos\omega t$ V, (b) 0.45 $\cos\omega t$ μA, $-80\cos\omega t$ μA, $-1.2\cos\omega t$ V; (a) -5600, -350, (b) -1200, -180.)

3*. Appendix D gives h parameters for the BC107A transistor in the common emitter configuration. If the operating point is $I_C = 2$mA, $V_{CE} = 5$V, find the maximum available current and voltage gains for small AC signals if the load resistance is chosen suitably. What is the sign and magnitude of $h_{re}v_{CE}$? If, on the other hand, the load resistance is 5 kΩ, what is the magnitude of the current gain, the voltage gain and $h_{re}v_{CE}$? (Ans: -220, -12600, $-1.9v_{in}$ increasing; -202, -380, $-0.057v_{in}$.)

4. A transistor operating with $I_E = 2.5$ mA has $r_b = 100$ Ω, $\beta = 210$ and $r_0 = \infty$. What is r_π? The transistor is used in a common emitter amplifier fed by an AC voltage 5 $\cos\omega t$ mV, and drives a 1.8 kΩ load. Draw the equivalent circuit and find (a) i_B, (b) the output voltage across the load. What is the input impedance of the amplifier, the current gain and the voltage gain? (Ans: 2 kΩ, (a) 2.38 $\cos\omega t$ μA, (b) $-0.90\cos\omega t$ V; 2.1 kΩ, -210, -180.)

5. A transistor operating in the common emitter configuration with current gain $\beta = 200$ has an input impedance of 4 kΩ. Draw the hybrid-π equivalent circuit. Set $r_0 = \infty$ initially. If the input voltage is 4 mV, what is the output voltage across a load resistor of 2.2 kΩ? Using these values, what would be the current through an actual r_0 of 20 kΩ? *With this value of r_0, what is the true value of the voltage across the load resistor? (Ans: -440 mV, $22i_{in}$, -396 mV.)

6. In figure 17.16, a transistor is shown in its common emitter configuration with a resistor R_E in series with the emitter. Take the operating conditions to be $r_\pi = 4$ kΩ, and $\beta = 200$. Relate v_C and v_E to v_{in} for $R_E =$ (a) 50 Ω, (b) 1 kΩ, considering separately the cases (i) $r_0 = \infty$, *(b) $r_0 = 20$ kΩ. (Ans: (ai) $v_C = -47v_{in}$, $v_E = 0.715v_{in}$, (bi) $v_C = -3.22v_{in}$, $v_E = 0.980v_{in}$, (aii) $v_C = -44.8v_{in}$, $v_E = 0.683v_{in}$, (bii) $v_C = -3.20v_{in}$, $v_E = 0.976v_{in}$.)

7. In figure 17.17, a transistor is shown in its common base configuration. Take $r_\pi = 4$ kΩ and $\beta = 200$. Make the assumption initially that r_0 is infinite and find the input impedance, the current gain and the voltage across the load resistor R_L.

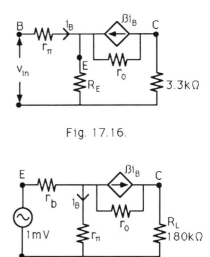

Fig. 17.16.

Fig. 17.17.

What changes are there to these quantities when the true value $r_0 = 20$ kΩ is used? (Ans: 20 Ω, 0.995, 9.0 V; V_L drops to 0.90 V, $r_{in} \to 190$ Ω, current gain $\to 0.952$.)

8. From figure 17.10(c), express i_1 in terms of v_{CE}, v_{BE} and i_B, hence v_{BE} in terms of i_B and v_{CE}. Show that $h_{re} = r_\pi/(r_\pi + r_\mu)$ and $h_{ie} = r_B + r_\pi/(1 + r_\pi/r_\mu)$. Express i_C in terms of i_B and v_{CE} and hence show that $h_{fe} = \beta - r_\pi(\beta + 1)/(r_\pi + r_\mu)$ and $h_{ob} = (1/r_0) + (\beta + 1)/(r_\pi + r_\mu)$. If $r_\mu = 1$ MΩ, $r_\pi = 1$ kΩ and $r_B = 100$ Ω, and $\beta = 100$, what are the magnitudes of h_{re}, h_{ie} and h_{fe}? If $1/h_{ob} = 6.67$ kΩ, what is r_0? (Ans: 10^{-3}, 1.1 kΩ, 99.8, $r_0 = 20$ kΩ.)

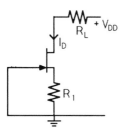

Fig. 17.18.

9. The circuit of figure 17.18 acts as an approximately constant source of current, insensitive to the magnitude of R_L. The voltage across the transistor is $V_{DS} = V_{DD} - (R_L + R_1)I_D$. By considering small changes in V_{GS}, V_{DS} and I_D, show that $\delta I_D = -I_D \delta R_L/[R_L + R_1 + r_d(1 + g_m R_1)]$. If $R_L = 3.3$ kΩ, $R_1 = 390$ Ω, $g_m = 3 \times 10^{-3}$ mho and $r_d = 40$ kΩ, what is the fractional change in I_D if R_L is halved? (Ans: +0.018.)

18

Transistor Amplifiers

18.1 Introduction

The circuit diagram of an integrated circuit is a bewildering array of transistors. Resistors have been mostly or entirely replaced by transistors and the circuit is often DC coupled to eliminate capacitors. Circuit design is frequently influenced by the technicalities of how to lay out the transistors on the chip. Nonetheless, recognisable building blocks of traditional amplifier circuits are there. In this chapter we examine some basic amplifier designs and outline some of the steps involved in converting them to integrated circuit form. You can try out these simple amplifiers with discrete transistors, so as to get a feeling for their capabilities.

18.2 Common Emitter Amplifier

The previous chapter showed that the bipolar transistor in the common emitter configuration amplifies both current and voltage. What remains is to set the operating point P of figure 17.4 to a suitable value. Chapter 9 argued towards an arrangement like that of figure 18.1. This circuit will now be discussed in detail, starting with the DC operating point.

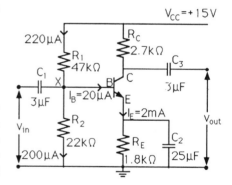

Fig. 18.1. Common emitter amplifier

Biasing
Suppose the supply voltage $V_{CC} = 15$ V. Next suppose the output voltage is to have the latitude to swing over a range of ± 5.5 V. (The reason for not aiming for ± 7.5 V will appear shortly.) This choice implies setting the DC voltage of the collector at 9.5 V. It can then swing up to the supply voltage and down to 4 V. If the latter takes the collector down to the knee of the characteristic curves ($V_{CE} \simeq 0.4$ V), the DC value of the emitter voltage is $V_E \simeq 3.6$ V and the base voltage is $V_B \simeq 4.2$ V. The DC voltage V_{CE} across the transistor is then 5.9 V. There is obviously some freedom of

choice in these conditions and other considerations might lead to a somewhat different operating point.

The choice of resistor values goes as follows. The potentiometer R_1 and R_2 fixes the DC value of the base voltage V_B, and hence V_E. The value of R_E then defines the DC current I_E through the emitter and hence $I_C = \beta I_E/(\beta+1)$ and $I_B = I_E/(\beta+1)$. It is this requirement (that R_E sets I_E) which determines that the DC value of V_E be non-zero and prevents the output voltage from swinging all the way down to zero. The load resistor R_C defines the collector voltage, hence the DC voltage V_{CE} across the transistor. Values of R_1 and R_2 are chosen so that 90% of the current through R_1 flows through R_2 and only 10% into the base of the transistor. As we shall see, this protects the operating point against variations in β from transistor to transistor. The large capacitor C_2 holds the DC voltage of the emitter nearly constant, but bypasses R_E for AC signals; for the latter, the emitter is effectively earthed and the input signal is developed between emitter and base; the output signal is then developed between emitter and collector. Hence this is the common emitter configuration for AC signals.

Suppose I_E is chosen to be 2 mA. Then $V_E = 3.6$ V requires $R_E = 1.8$ kΩ. Then $R_C = 2.7$ kΩ gives $V_C = 9.6$ V. Suppose next that the transistor has a current gain of $\beta = 100$, so that $I_B = 20\mu$A and $I_1 = 220$ μA. If the difference in current through R_1 and R_2 is ignored and if V_B is to be 4.2 V, $R_2 \simeq 19$ kΩ and $R_1 \simeq 49$ kΩ. To allow for the fact that I_B flows through R_1 but not R_2, R_2 is rounded up to 22 kΩ and R_1 down to 47 kΩ, the nearest available common values.

What change is there in operating point if all resistors are halved?

In view of the tolerances on resistors and the uncertainty in β, the arithmetic given above is usually adequate. Suppose, however, we wish to check the sensitivity to β by doing the arithmetic exactly. We have

$$V_E = V_B - 0.6 = R_E I_B(\beta + 1)$$
$$(V_{CC} - V_B)/R_1 = V_B/R_2 + I_B = V_B/R_2 + (V_B - 0.6)/R_E(\beta + 1)$$

so

$$V_B \left(\frac{1}{R_1} + \frac{1}{R_2} + \frac{1}{R_E(\beta + 1)} \right) = \frac{V_{CC}}{R_1} + \frac{0.6}{R_E(\beta + 1)}.$$

Inserting the chosen values, $V_B = 4.46$ V, $I_E = 2.14$ mA and $V_C = 9.21$ V. If β changes to 50, $V_B = 4.19$ V, $I_E = 1.99$ mA and $V_C = 9.62$ V. These changes are within 10% tolerances, so the potentiometer R_1 and R_2 is successful in making the operating point insensitive to β.

Suppose, however, R_2 were omitted completely and V_B were set via the potential drop through R_1 due to base current I_B:

$$V_B = V_{CC} - R_1 I_B.$$

Now V_B becomes directly sensitive to I_B hence β, so this is a poor choice.

18.3   Performance of the Common Emitter Amplifier

What is of interest is the ability of the circuit to amplify AC signals. In order to follow this quantitatively, the transistor is replaced by its AC equivalent circuit. If the full complexity of the circuit is retained, the result is shown in figure 18.2. The way R_1 and R_C have been drawn requires comment. It has been assumed that the battery supplying power has zero impedance. Then from the point X of figure 18.1 there is a path up through R_1 and through the battery to earth. (Remember that for AC equivalent circuits DC voltage sources are omitted.) Likewise, from the collector there is a path through R_C then the battery to earth.

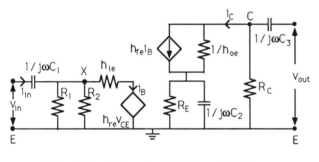

Fig. 18.2. Full equivalent circuit of the common emitter amplifier.

What changes are there if the power supply has resistance R_0?

Analysing figure 18.2 in full is possible but very laborious. It is better to make approximations that (a) $h_{re} = 0$ and (b) the capacitors have negligible impedance; at the end, the validity of these approximations can be checked and improved if necessary by back-substitution. Using these approximations the circuit simplifies dramatically to that of figure 18.3.

Input and output impedances
The interpretation of this circuit is simple. The input impedance r_{in} is that of h_{ie}, R_1 and R_2 in parallel. If $h_{ie} = 1\ k\Omega$ and R_1 and

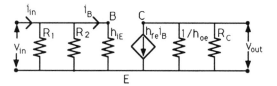

Fig. 18.3. Simplified (approximate)
equivalent circuit.

R_2 have the values chosen above (22 kΩ and 47 kΩ), the latter two carry little current and the input impedance is only 6.3% less than h_{ie}. Likewise i_B is 6.3% less than i_{in}.

The output circuit is equally simple: a current generator $h_{fe}i_B$ with an output impedance r_{out} given by R_C in parallel with $1/h_{oe}$. If the latter is 20 kΩ and $R_C = 2.7$ kΩ, the output impedance R_{out} is 12% less than R_C, i.e. 2.38 kΩ.

In summary, the bias resistors play only a small role in the performance of the common emitter amplifier providing their values are chosen so that R_1 and R_2 are large compared with h_{ie} and R_C is small compared with $1/h_{oe}$.

Current and voltage gain

If $h_{fe} = 100$, the output current $h_{fe}i_B$ is $100i_B \simeq 94i_{in}$. How much of this is delivered to the external load depends on the load impedance. For maximum current gain, the load should be small compared to R_C and the current gain of the amplifier is then 94. On the other hand, for maximum voltage gain, it should have a large impedance. The voltage gain is then $h_{fe}i_B r_{out}/h_{ie}i_B = 100 \times 2380/1000 = 238$.

If R_E and R_C are doubled, what is the effect on h_{iE} and the resulting change in the voltage gain?

Choice of capacitor values

Next, what is the effect of C_1 and C_3? The input voltage to the transistor (figure 18.2) is

$$v_X = \frac{v_{in} r_{in}}{r_{in} + 1/j\omega C_1}.$$

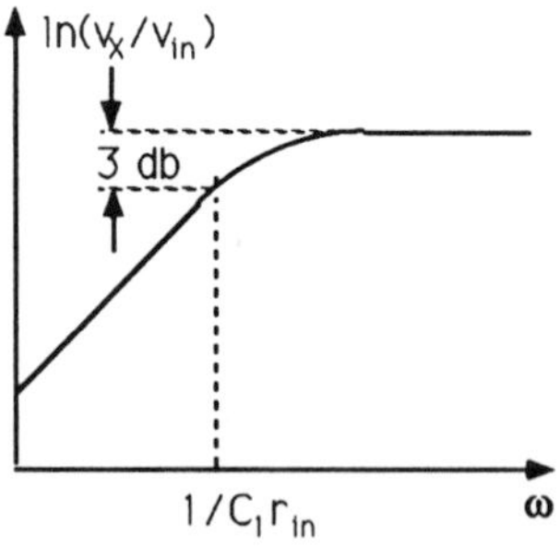

This falls off at low frequency, as shown in figure 18.4. The gain falls by 3 db (i.e. a factor $1/\sqrt{2}$) when in the denominator $\omega C_1 r_{in} = 1$, i.e. $\omega = 1/C_1 r_{in}$. Remember that $r_{in} \simeq h_{ie}$. For a cut-off at $\omega = 2\pi \times 50$ rad/s, say, a value $C_1 \simeq 3$ μF is required. Larger values give a lower cut-off frequency. Likewise C_3 and the output

Fig. 18.4. Low-frequency
cut-off due to C_1.

impedance r_{out} of the amplifier give a low-frequency cut-off at $\omega = 1/C_3 r_{out}$ and, since r_{out} is not very different from r_{in}, C_3 should be roughly similar to C_1.

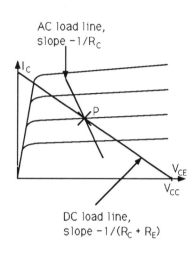

AC load line

The proper discussion of C_2 awaits the discussion of the emitter follower. Suppose, however, it is large enough to bypass R_E of figure 18.2. The DC load line in figure 18.5 shows the relation between I_C and V_{CE} for changes in the DC operating point. It has a slope of $-1/(R_C + R_E)$ because these two resistors appear in series with the transistor between supply voltage and earth. AC signals, however, bypass R_E. The transistor therefore responds to an AC input voltage by changes i_C and v_{CE} along the AC load line with slope $-1/R_C$.

It is recommended that you apply a small AC voltage to the circuit of figure 18.1 and follow the signal and the DC voltages through the circuit with an oscilloscope. If you increase the input signal you will see distortion when the output voltage is limited either by the power supply or by bottoming. What component change eases this problem? Vary the frequency and identify by trial and error which capacitor is responsible for the low-frequency cut-off. You can check the output impedance in the middle of the passband by putting load resistors across the output to halve the voltage. You can measure the input impedance likewise by putting resistors in series with the input to halve the output of the amplifier.

DC load line,
slope $-1/(R_C + R_E)$

Fig. 18.5. AC and DC load lines.

18.4   Feedback with the Common Emitter Amplifier

The input impedance h_{ie} of the transistor is very sensitive to I_E, and hence is sensitive to operating point, and so is the voltage gain. To remedy this, negative feedback is used to stabilise the gain, as described in Chapter 7. One very simple way of achieving this is to move R_1 of figure 18.1 to provide a feedback path from collector to base, as shown in figure 18.6. Because the amplifier inverts the input signal, this provides negative feedback. The value of R_1 is changed to provide the right biasing condition for the base and R_3 is chosen to provide the required amplification. The bias current of figure 18.1 now flows through R_C, making the DC current $I_C = 2.2$ mA and $V_C = 9.0$ V. To make $V_X = 4.2$ V, this requires $R_1 = 4.8/0.2 = 24$ kΩ; the nearest common value is 22 kΩ.

The voltage gain of the amplifier $-G/(1 + GB)$ becomes approximately $-1/B = -R_1/R_3$, providing this is much less than the voltage gain G of the transistor; it can be adjusted by varying R_3.

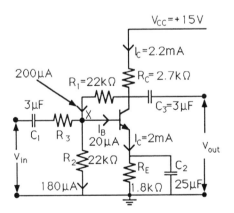

Fig. 18.6. Negative feedback with
the common emitter amplifier.

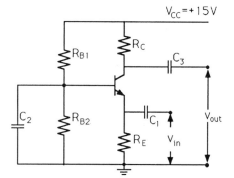

Fig. 18.7. Common base amplifier.

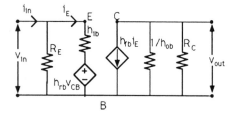

Fig. 18.8. Equivalent circuit of
the common base amplifier.

What are the input and output impedances?

18.5*   The Common Base Amplifier

The configuration of resistors for the common base amplifier is identical to that for the common emitter, and the input is simply put in a different place, across R_E of figure 18.7. A large capacitor C_2 shorts out AC signals across R_{B2}, so AC input appears between emitter and base. Likewise, the output, taken between collector and earth, appears between collector and base.

How large should C_2 be for a low cut-off frequency $\omega_0 = 100\pi$ rad/s?

If the usual simplifying assumptions are made of neglecting the impedances of the capacitors, the AC equivalent circuit is shown in figure 18.8. You are referred to Chapter 17 for the feedback effects between input and output arising from $h_{rb}v_{CB}$. These make it difficult to generalise about input and output impedances. However, the input impedance is h_{ib} in parallel with R_E, *providing* the effect of $h_{rb}v_{CB}$ can be neglected; since h_{ib} is only a few ohms, R_E can usually be neglected.

From collector to earth, R_C is connected via the battery, and if this has low impedance, R_{B1} appears between earth and base and is shorted out by C_2 as far as AC signals are concerned. As discussed in the previous chapter, the effect of $h_{rb}v_{CB}$ on output impedance is large unless the circuit is fed from a source with resistance $\gg r_\pi$. From equation (17.17), the current gain h_{fb} is ~ -1, but the voltage gain is equal to the output impedance divided by input impedance and is generally similar to that of the common emitter amplifier. The common base amplifier is used in situations where very low input impedance is required or particularly high output impedance; however, the feedback due to $h_{rb}v_{CB}$ is such that extreme values cannot be obtained for both simultaneously.

18.6   Emitter Follower

A circuit used extremely commonly is the emitter follower. The layout of the circuit, shown in figure 18.9, is similar to the common emitter amplifier, but (a) R_C is omitted and (b) R_E does not have a bypass capacitor. These distinctive features identify it in circuit diagrams. The DC bias conditions are the same as for the common emitter amplifier, except that $V_{CE} = V_{CC} - V_E$.

The AC operation is extremely simple. The base-emitter junction acts as a diode, so $V_E \simeq V_B - 0.6$ V. Consequently, the output follows the input, except for a DC drop of 0.6 V which has no

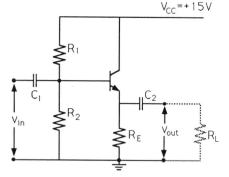

Fig. 18.9. The emitter follower.

effect on the AC behaviour. Hence for AC signals $v_E = v_B$, and the voltage gain is $\simeq 1$. There is, however, a current gain since $I_E = I_B + I_C = (\beta + 1)I_B$.

For AC signals, the battery behaves as a short circuit. Hence, as far as AC signals are concerned, the collector is at earth. Then the input signal is developed between base and collector and the output signal between emitter and collector. It is therefore the common collector amplifier.

It will emerge that the output impedance of the circuit is low, typically 20 Ω, and the input impedance r_{in} of the transistor $\simeq (\beta + 1)R_E$; i.e. output impedance is much less than input impedance. The emitter follower is therefore generally used to drive a low impedance load from a high impedance source. Typical values of R_E are ≤ 2 kΩ. The input impedance is high and R_1 and R_2 are generally made rather larger than r_{in}, so that most of the AC current flows into the transistor; typical values of R_1 and R_2 are ≥ 50 kΩ.

18.7   AC Equivalent Circuit of the Emitter Follower

An h equivalent circuit can be substituted for the transistor, but it is more instructive to use the hybrid-π. In figure 18.10, R_1 is connected from base to earth via the battery, otherwise the circuit is obvious. In order to simplify the analysis, the approximations will be made of neglecting the impedances of C_1 and C_2. Then the load resistor R_L appears in parallel with R_E and r_0. Suppose this parallel combination has impedance R'_E.

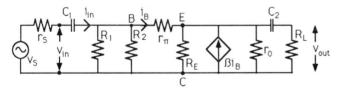

Fig. 18.10. AC equivalent circuit of the emitter follower.

Gains
With these simplifications, the current through R'_E is $i_E = (\beta+1)i_B$ and the voltage gain is

$$\frac{v_{out}}{v_{in}} = \frac{i_E R'_E}{i_B r_\pi + i_E R'_E} = \frac{R'_E}{R'_E + r_\pi/(\beta + 1)}. \qquad (18.1)$$

If $r_\pi/(\beta+1) \ll R_E'$, the voltage gain is ~ 1, as was found in qualitative fashion in the previous section.

Input impedance

In deriving the input impedance, suppose initially that R_1 and R_2 are large. Then $i_{in} \simeq i_B$ and the input impedance of the transistor is

$$r_{in} = \frac{v_{in}}{i_{in}} = r_\pi + (\beta+1)R_E' \simeq (\beta+1)R_E'. \qquad (18.2)$$

Since $\beta \simeq 100$, the input impedance can be quite large; for example, if $R_E' = 1\text{ k}\Omega$, the transistor itself presents an input impedance of 100 kΩ. In this case, it is not correct to ignore R_1 and R_2, which appear in parallel with r_{in}. Let us also remark in passing that when R_E' is bypassed, as in the common emitter amplifier, $r_{in} = r_\pi$.

Output impedance

The output impedance of the amplifier is obtained from the Thevenin equivalent:

$$r_{out} = v_{out}(\text{open circuit})/i_{out}(\text{short circuit}).$$

On open circuit, from equation (18.1)

$$v_{out} = v_S R_E'/[R_E' + (r_S + r_\pi)/(\beta+1)].$$

On short circuit

$$i_{out} = (\beta+1)v_S/(r_S + r_\pi).$$

Here r_S is the output impedance of the source feeding the emitter follower; since $r_\pi \simeq 1$ kΩ, r_S may not be negligible. Finally,

$$r_{out} = \frac{R_E'(r_\pi + r_S)/(\beta+1)}{R_E' + (r_\pi + r_S)/(\beta+1)}. \qquad (18.3)$$

If, for example, $R_E' = r_S = 1$ kΩ, and $\beta = 100$, then $R_{out} \simeq 20$ Ω.

The expression for r_{out} is just the resistance of R_E' in parallel with $(r_\pi + r_S)/(\beta+1)$ and it is quite revealing to see why. In figure 18.11(a), the equivalent circuit is redrawn, omitting R_1 and R_2. The current through r_S and r_π is a factor $(\beta+1)$ less than through R_E', so the voltage across them is reduced by the same factor. As far as impedance is concerned, the same result is given by figure 18.11(b). The impedance across AB (i.e. the Thevenin equivalent) is given by R_E' in parallel with $(r_S + r_\pi)/(\beta+1)$.

If for completeness we want to reinsert R_1 and R_2 into figure 18.11(a), as shown by the broken lines in figure 18.11(c), it

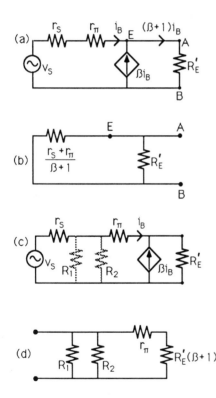

Fig. 18.11 (a) AC equivalent circuit omitting R_1 and R_2, (b) its equivalent for calculating r_{out}, (c) the effect of R_1 and R_2 on r_{out}, and (d) r_{in}.

is clear that they appear in parallel with r_S. figure 18.11(d) shows how to include the effect of R_1 and R_2 on input impedance.

The current gain is given by i_{out}/i_{in}. The maximum current gain is obtained by applying a short circuit across the output. In this case, $R'_E = 0$ and the input impedance becomes r_π, which is small compared with R_1 and R_2. Then the current gain is simply $(\beta + 1)$.

What is the maximum power gain?

Summary
The emitter follower has high input impedance, low output impedance, voltage gain $\simeq 1$ and current gain up to $(\beta + 1)$. It serves as an impedance transformer and is frequently used to drive a low-impedance load such as a loudspeaker. Since the voltage gain is 1, there is no Miller effect and the high-frequency performance is good.

18.8   Voltage Feedback in the Emitter Follower

From figure 18.9, the AC voltage applied between base and emitter of the transistor is $v_{in} - v_{out}$. In the language of Chapter 7, this arrangement has 100% negative feedback, $B = 1$. This is directly responsible for the fact that

$$\frac{v_{out}}{v_{in}} \simeq 1 = \frac{1}{B}$$

and also for the high input impedance and low output impedance.

The signature of this type of feedback is the unbypassed emitter resistance R_E, and one should be alert for it in scanning circuit diagrams. Sometimes it is used deliberately to stabilise the voltage gain of the common emitter amplifier. In figure 18.12(a), the resistor R_3 in series with the emitter is left unbypassed, and provides negative feedback to the input circuit. The equivalent circuit is shown in figure 18.12(b) and can be analysed with some labour. We shall not do this, but simply draw attention to the value required for C_2. Over the working frequency range of the amplifier, this should have an impedance small compared with $R_3 + r_\pi/(\beta+1)$, otherwise its dependence will distort the low-frequency response. If $I_E = 2$ mA and $\beta = 100$ for example, $r_\pi/(\beta + 1) \simeq 12.5\ \Omega$; if $R_3 = 0$ and the amplifier is to have a lower cut-off frequency of $\omega = 100\pi$ rad/s, a very large $C_2 \geq 250\ \mu$F is required. Even with $R_3 = 100\ \Omega$, $C_2 \simeq 30\ \mu$F is needed.

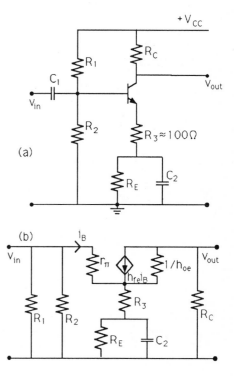

Fig. 18.12 (a) Common emitter amplifier with negative feedback from R_3 and (b) its equivalent circuit.

18.9   Short-circuit Protection

Resistors R_1 and R_2 in figure 18.9 clamp the base voltage V_B. If the output is accidentally short circuited, a voltage V_B appears across the base–emitter junction, and if this is much greater than 0.6 V, the current flow is likely to destroy the transistor. To prevent this happening, the current should be limited to a safe value by setting R_1 large. The base current I_B cannot be larger than V_{CC}/R_1 so $I_E \leq (\beta + 1)V_{CC}/R_1$, and I_E can be limited to a safe value.

An alternative trick to provide short-circuit protection is shown in figure 18.13. If the current through resistor r gets too large, it switches on the bypass transistor at the left and this protects the emitter follower.

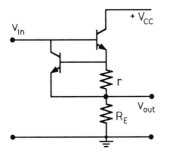

Fig. 18.13. Short-circuit protection.

18.10   The Darlington Pair

To get maximum input impedance in the emitter follower, one may dispense with R_2, as in figure 18.14(a); although this provides a less stable operating point, this is not crucial in the emitter follower. To get increased current gain and yet higher input impedance, figure 18.14(b) shows the use of two transistors in cascade. This is a very common arrangement and is known as the **Darlington pair**. Each transistor amplifies the current by a factor $(\beta + 1)$.

The AC equivalent circuit is shown in figure 18.14(c). There, it is safe to neglect r_{02} compared with R_L, so the second transistor has input impedance

$$r_{in2} = r_{\pi 2} + (\beta_2 + 1)[R_E R_L/(R_E + R_L)].$$

If this is small compared with r_{01}, the output current is $(\beta_1 + 1)(\beta_2 + 1)i_{B1}$; and the input impedance of the amplifier is the parallel combination of R_{B1} with $r_{\pi 1} + (\beta_1 + 1)r_{in2}$. For example, if $\beta = 100$, the input impedance is 1 MΩ in parallel with 1 MΩ. If r_{01} is not negligible, it syphons off some current, reducing the current gain and the input impedance.

The biasing arrangement is such that the voltage at the base of T1 is

$$V(B1) = V_{CC} - I_B R_{B1}$$
$$V(B2) = V(B1) - 0.6V$$
$$V(E2) = V(B2) - 0.6 = V_{CC} - 1.2 - I_B R_{B1}$$
$$= (\beta_1 + 1)(\beta_2 + 1)I_B R_E$$

from which it is easy to solve for I_B and $V(B1)$. Since the biasing conditions are not critical, it may be permissible to omit the decoupling capacitors.

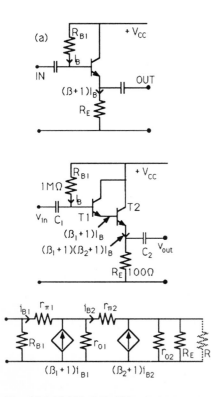

Fig. 18.14 (a) Modified bias to increase output impedance of the emitter follower, (b) the Darlington pair, (c) its equivalent circuit.

18.11 FETs

With an appropriate network of voltage supplies and bias resistors, the JFET can be inserted into all of the circuits discussed so far, making a common source amplifier (analogous to the common emitter), a common gate amplifier and a source follower (like the emitter follower). Its major practical difference from the bipolar transistor is its much higher input impedance. A second important difference is that the noise level of the FET is usually much lower than that of the bipolar transistor. Both features make the FET an obvious choice for the input stage of an operational amplifier. More generally, the first-stage amplifier (preamplifier) for weak signals usually uses an FET, because of the low noise.

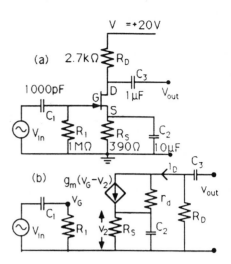

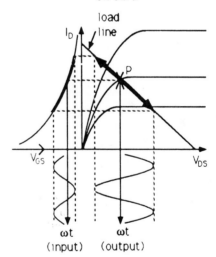

Fig. 18.15 (a) The common source amplifier and (b) its equivalent circuit

Biasing
Figure 18.15(a) shows a suitable arrangement for biasing the FET. The arrow on the transistor denotes the p-type gate (n-type channel). Suppose an operating point is required with $V_{GS} = -1$ V, $I_D = 2.5$ mA and $V_{DS} = 12$ V. The gate is at earth and the source at voltage $+R_S I_D$; hence we require $R_S = 400\ \Omega$ (or in practice 390 Ω). The slope $1/(R_D + R_S)$ of the load line in figure 18.16 is given by $(R_D + R_S) = (20-12)/2.5\times10^{-3} \simeq 3.2$ kΩ, so $R_D = 2.8$ kΩ (or 2.7 kΩ in practice). The purpose of R_1 is to prevent the gate from floating. Then C_1 and R_1 define the low-frequency cut-off.

If the input signal is very small, can you scale up R_S and R_D by a factor of say 30 and avail yourself of the full voltage gain available from R_D?

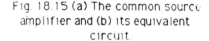

Fig. 18.16. Amplification in the JFET

Common source amplifier
The AC equivalent circuit of the resulting common source amplifier is shown in figure 18.15(b). Providing C_2 is large enough to bypass R_S, the output stage is simple. To see how large C_2 has to be, consider

$$i_D = g_m(v_G - i_D Z_2)$$

where Z_2 is the impedance of R_S and C_2 in parallel. So

$$i_D = \frac{g_m v_G}{1 + g_m Z_2} \tag{18.4}$$

and neglect of Z_2 requires $Z_2 \ll 1/g_m \simeq 300\ \Omega$. If $\omega = 100\pi$ rad/s, $C_2 \geq 10\ \mu$F. Then the output impedance r_{out} is given by R_D and r_d in parallel; if the load applied across the output is R_L, the voltage gain is $g_m R'_L$, where R'_L is the parallel combination of R_L with r_{out}. The input impedance is R_1.

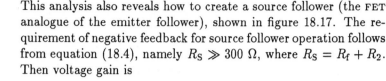

Source follower

This analysis also reveals how to create a source follower (the FET analogue of the emitter follower), shown in figure 18.17. The requirement of negative feedback for source follower operation follows from equation (18.4), namely $R_S \gg 300\ \Omega$, where $R_S = R_f + R_2$. Then voltage gain is

$$G_V = \frac{g_m R_S}{1 + g_m R_S} = \frac{1}{1 + 1/g_m R_S}.$$

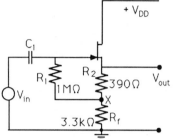

Fig. 18.17. The source follower.

The output impedance is

$$R_{out} = \frac{V_{out}(\text{open circuit})}{I_{out}(\text{short circuit})}$$

$$= \frac{R_S g_m v_G/(1 + g_m R_S)}{g_m v_G}$$

using equation (18.4), leading to

$$R_{out} = \frac{R_S}{1 + g_m R_S} = \frac{1}{g_m + 1/R_S}. \qquad (18.5)$$

This is equal to the parallel combination of R_S with a resistance $1/g_m$. If R_S is the larger of the two, $R_{out} \simeq 1/g_m \simeq 300\ \Omega$. This is larger than the output impedance of the emitter follower (typically 20 Ω), so the latter is normally preferred when a high output current is required. If the source follower is used to drive a very low resistance load ($R_S \leq 1/g_m$), the voltage gain $(1 + 1/g_m R_S)^{-1}$ drops significantly below 1.

It is recommended that you measure for yourself the characteristics of an FET and use it to construct a common source amplifier and source follower (by shorting R_D); check the DC levels in the circuit with an oscilloscope and then follow through the amplification of an AC voltage. You can obtain r_d either from the characteristic curves or by measuring the output impedance using a tuned circuit on resonance as load (very high AC impedance, low DC impedance).

18.12*   Bootstrapping

In figure 18.17, the resistor R_2 provides the DC bias between gate and source. The resistor R_1 is connected to the point X at the top of R_f. Why is it not connected to earth?

There are two reasons. The first is that splitting the source resistor into the two elements R_2 and R_f gives flexibility in setting the precise DC bias at the gate. The second and more important reason is that the input impedance of the amplifier is increased by the negative feedback across R_f. The point X follows the input by follower action, and the AC voltage across R_1 is consequently reduced from v_{in} to $v_{in}[1 - R_f/(R_2 + R_f)] = v_{in}R_2/(R_2 + R_f)$. Because of this, the current through R_1 is reduced and the input impedance is correspondingly increased by a factor $(R_2 + R_f)/R_2$. A more exact algebraic analysis is explored in exercise 17. This trick is known as **bootstrapping**.

An alternative scheme using the same principle is shown in figure 18.18(a). There R_2 is eliminated, and the DC bias is supplied by the potential divider R_A and R_B, which connect to the gate through R_1. If this potential divider were applied directly to the gate, it would contribute to the input impedance. As it is, the capacitor C_2 connects R_1 to X as far as AC signals are concerned, giving the bootstrap action which enhances the input impedance; the AC equivalent circuit is shown in figure 18.18(b), where R_A, R_B and r_d appear in parallel with R_f. If this parallel combination is denoted by R_f', $v_{out} = g_m v_{GS} R_f'$ and the input impedance is $R_1(1 + g_m R_f')$. If R_1 is increased to 10 MΩ, the input impedance of the amplifier can be as high as 10^9 Ω, but input capacitance then becomes significant at quite low frequency.

Similar bootstrapping techniques can be applied to the emitter follower to increase its input impedance.

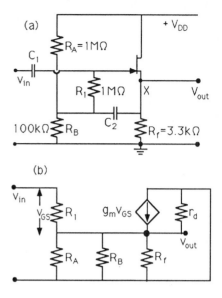

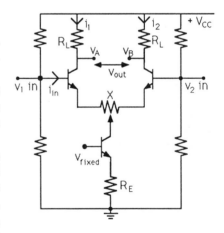

Fig. 18.18 (a) Bootstrapping and (b) the AC equivalent circuit.

18.13   The Difference Amplifier

Figure 18.19 shows an important variety of amplifier, called a difference amplifier or long-tailed pair. It produces an output proportional to the **difference** between two inputs v_1 and v_2.

Suppose the input impedance of the first amplifier is r_1. The AC current flowing into it is $i_{in} = (v_1 - v_X)/r_1$, and the output current is $i_1 = \beta(v_1 - v_X)/r_1$. Likewise, the second amplifier produces an output current $i_2 = \beta(v_2 - v_X)/r_2$.

The transistor at the bottom of the circuit is operated by a fixed voltage applied to its base; this fixed voltage may be derived from a pair of resistors between the power supply and earth. The performance of this transistor is insensitive to its collector voltage. Hence it acts as a constant current source. It holds $I_1 + I_2$ constant, and the changes in these currents, denoted by i, must be such that $i_1 = -i_2$. This is achieved by appropriate changes in v_X. Suppose, for example, that the transistors are selected so that $\beta_1 = \beta_2 = \beta$ and $r_1 = r_2 = r$. Then

Fig. 18.19. Difference amplifier.

$$i_1 = \beta(v_1 - v_X)/r$$
$$i_2 = \beta(v_2 - v_X)/r \tag{18.6}$$
$$i_1 + i_2 = \beta(v_1 + v_2 - 2v_X)/r = 0 \tag{18.7}$$

so

$$v_X = \tfrac{1}{2}(v_1 + v_2).$$

The AC output voltage of the amplifier is given by

$$v_A - v_B = -R_L(i_1 - i_2) = -R_L\beta(v_1 - v_2)/r \tag{18.8}$$

and is proportional to $(v_1 - v_2)$. It can be used as a DC amplifier by applying the input signal between the two inputs, ignoring the earth connection.

Common mode rejection

An important feature of the difference amplifier is its good common mode rejection. If the same voltage is applied to both inputs (common mode), it produces no output in the approximation used so far. In practice, a small difference in β or r between the two sides of the circuit will lead to a small term in $(v_A - v_B)$ depending on $(v_1 + v_2)$; it is easy to follow this through the algebra. In the old days, a great deal of trouble was necessary to select two transistors with parameters as nearly identical as possible. Nowadays, the whole circuit is laid down inside a single integrated circuit. Then automatically (a) the two transistors are very similar (in practice within $\frac{1}{2}\%$), since they are made on a single piece of silicon a fraction of a mm apart, and (b) the two transistors are at almost exactly the same temperature, eliminating the temperature sensitivity of β.

So far it has been assumed that the current source holds $I_1 + I_2$ absolutely constant. In reality it will have some high output impedance R_0 and the total current will vary slightly with V_X. The effect this has on a common mode signal will now be examined. The change in V_X is related to changes i in current by

$$v_X = (i_1 + i_2)R_0(1 + 1/\beta).$$

From equation (18.7),

$$i_1 + i_2 = \beta[v_1 + v_2 - 2R_0(1 + 1/\beta)(i_1 + i_2)]/r$$

so

$$i_1 + i_2 = \frac{\beta(v_1 + v_2)}{r + 2(\beta + 1)R_0}$$

and

$$v_X = \frac{\beta R_0(v_1 + v_2)}{r + 2(\beta + 1)R_0}.$$

If the two sides of the amplifier are perfectly symmetrical, this has the same effect on i_1 and i_2, so $v_A - v_B$ is still zero. However, the output is often taken from one side only, e.g. v_B. Then from (18.6)

$$v_B = -i_2 R_L$$

$$= -\beta\left(\tfrac{1}{2}(v_2 - v_1) + \tfrac{1}{2}(v_2 + v_1) - \frac{\beta R_0(v_1 + v_2)}{r + 2(\beta + 1)R_0}\right)\frac{R_L}{r}$$

$$= -\tfrac{1}{2}\beta\left((v_2 - v_1) + \frac{(v_2 + v_1)r}{r + 2(\beta + 1)R_0}\right)\frac{R_L}{r}.$$

The common mode rejection ratio (CMRR) is defined as the ratio of the two outputs produced by $(v_2 - v_1)$ and $(v_2 + v_1)$; this is

$$CMRR = 1 + 2(\beta + 1)R_0/r \simeq 2\beta R_0/r. \qquad (18.9)$$

The current source in figure 18.19 has its base fixed in voltage and signals appear at X and across R_E, so it functions as a common base amplifier. If $R_E \gg r_\pi$, $R_0 \simeq 1/h_{ob} \simeq 2 \times 10^6$ Ω. If $r = 1$ kΩ (common emitter) and $\beta = 100$, the common mode rejection ratio is $4 \times 10^5 = 112$ db. Operational amplifiers use difference amplifiers as the input stage; common mode rejection ratios are typically 90 db, but a ratio of 130 db is available (see table 7.1).

A minor point concerning the difference amplifier is that its gain depends on the input impedance r, which in turn varies as $1/I_E$. The gain of the amplifier may be varied by changing I_E by means of the voltage applied to the base of the current source.

18.14   The Current Mirror

An alternative constant current source used very widely in integrated circuits is the current mirror, shown in figure 18.20. It is used to supply a constant current I_2. If the two transistors in (a) are identical, the same emitter current will flow in each, because they share the same V_{BE}. The collector of T1 is tied to the base (0.6 V), so

$$I_\alpha = (V_{CC} - 0.6)/R_1 = I_1(1 + 2/\beta) = I_2(1 + 2/\beta).$$

So the value of R_1 controls I_2.

In an integrated circuit, a number of variants on this scheme are to be found. Firstly, it is easy to vary the ratio I_2/I_1 by means of the areas of the emitters of the two transistors. Secondly, it is desirable to keep resistors small, to economise on the area used to make them. In some applications I_2 is to be less than 100 μA and $V_{CC} = 10$ V, say; a value $R_1 \simeq 100$ kΩ would then be required in figure 18.20(a). This can be remedied by putting a small resistor R_2 in series with the emitter of T_2, as in figure 18.20(b). This reduces $V_{BE}(T2)$ and hence I_2, so the ratio of I_α to I_2 can be large, and hence both R_1 and R_2 can be small. The algebra of this trick will now be examined.

In figure 18.20(b) I_1 and I_2 are redefined so as to reduce the algebra slightly. Then, using

$$I_2 = I_{20} \exp(eV_{BE2}/kT)$$

the base voltage V_B is

$$V_B = V_{BE2} + R_2 I_2 \qquad (18.10)$$
$$= (kT/e)\ln(I_2/I_{20}) + R_2 I_2.$$

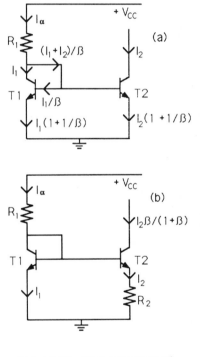

Fig. 18.20. Two versions of the current mirror.

IN   OUT

i_1 i_2

T1 i_2/β T2

$i_1 - \frac{i_2}{\beta}$ R   R $i_2(1+1/\beta)$

V_{fixed}

Fig. 18.21. Arrangement for
negative current feedback.

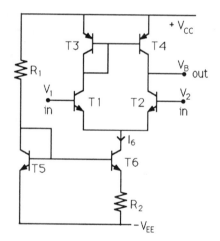

$+V_{CC}$

T3   T4

R_1 V_B out

V_1 V_2
in   T1   T2   in

I_6

T5   T6

R_2

$-V_{EE}$

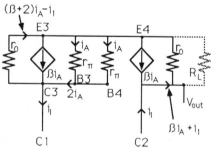

$(\beta+2)i_A - i_1$

E3   E4

i_A i_A

βi_A r_π βi_A R_L

B3 r_π
C3 $2i_A$ B4

i_1 V_{out}

C1   C2

$\beta i_A + i_1$

Fig. 18.22 (a) Current sources as
loads for the differential amplifier,
(b) equivalent circuit for T3 and T4.

Then

$$I_1 = I_{10} \exp\{(e/kT)\left[(kT/e)\ln(I_2/I_{20}) + R_2 I_2\right]\}$$
$$= (I_{10}/I_{20})I_2 \exp[(e/kT)R_2 I_2]. \qquad (18.11)$$

If $R_2 = 0$, the first term demonstrates that the ratio of the currents is defined by I_{10}/I_{20}, i.e. by the areas of the two emitters. Otherwise from equation (18.10)

$$I_2 = (V_{BE1} - V_{BE2})/R_2.$$

Suppose as an example that there is to be a factor of 20 difference in I_1 and I_2, and suppose $I_{10} = I_{20}$ so that the exponential of (18.11) dominates. Then

$$V_{BE1} - V_{BE2} \simeq \ln(20)/40 = 0.075 \text{ V}$$

and if $I_2 = 100 \ \mu A$, $R_2 = 750 \ \Omega$. If $V_{CC} = 10 \text{ V}$, $I_1 = 2 \text{ mA}$ and $R_1 = 4.7 \text{ k}\Omega$. The value of R_2 provides coarse control and R_1 provides fine control, because in equation (18.11) I_1 depends exponentially on R_2.

In Chapter 7, negative current feedback was discussed and it is now possible to see how this can be achieved. In figure 18.21 the current source at the bottom holds the sum of the currents constant through the two transistors, so

$$i_1 - i_2/\beta = -i_2(1 + 1/\beta)$$
$$i_1 = -i_2.$$

This achieves 100% negative current feedback. If a lower feedback fraction is required, the output of T2 may be split between two parallel resistors and the current feedback taken from one of them.

18.15   Constant Current Sources as Loads

In the differential amplifier of figure 18.19 the voltage gain is proportional to R_L. It helps to make R_L large, but this presents two problems. Firstly, DC levels limit the potential drop across R_L and hence limit its magnitude. Secondly, a large resistor occupies valuable space in an integrated circuit.

The alternative is to replace the resistors with constant current sources. One way of doing this is shown in figure 18.22, where a current mirror formed by T3 and T4 makes the loads for the differential amplifier formed by T1 and T2. A second current mirror, T5 and T6, drives the amplifier with a constant current I_6.

The mode of operation of the loads is most easily understood by replacing T3 and T4 by their hybrid-π equivalent circuits, shown in figure 18.22(b). Since their bases are tied together, the same AC current i_A flows in each. The differential amplifier provides equal and opposite AC currents i_1 at the collectors of T1 and T2. The currents through r_0 of T3 and the parallel combination of the load with r_0 of T4 are shown in the diagram, using Kirchhoff's current law at each collector. For T3, Kirchhoff's voltage law then gives

$$i_A r_\pi + [(\beta+2)i_A - i_1]r_0 = 0$$

or

$$i_A = \frac{i_1}{\beta + 2 + r_\pi/r_0}.$$

Then, for T4,

$$v_{\text{out}} = r_0'(i_1 + \beta i_A) = r_0' i_1 \left(1 + \frac{\beta}{\beta + 2 + r_\pi/r_0}\right) \simeq 2r_0' i_1$$

where r_0' is the parallel combination of r_0 and the load. It is clearly desirable that the load R_L should be of high impedance to achieve large v_{out}.

Figure 18.22 consists of six transistors and two resistors of order 1 kΩ. This is typical of one stage of an operational amplifier. However, the first stage of an opamp usually has each input transistor connected in an emitter follower configuration in order to achieve large input impedance. This is shown schematically in figure 18.23, following very similar principles to figure 18.22. The current mirror T5 and T6 provides constant current to the differential amplifier and T3 and T4 acts as active load. A later stage provides voltage amplification. A real operational amplifier is yet more intricate. The μA741C, for example, contains 22 transistors, ten resistors, 2 diodes and a small (30 pF) capacitor.

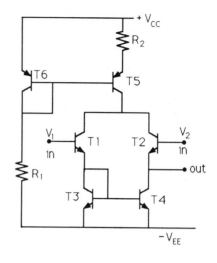

Fig. 18.23. A differential amplifier in the common emitter configuration with current mirror as active load.

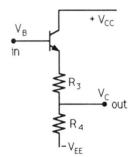

Fig. 18.24. An emitter follower as level shifter.

18.16 Level Shifting

It is inconvenient to include capacitors within an integrated circuit just to decouple stages. However, DC level shifts are required between stages. If figure 18.22 is taken as an example, the two input voltages will normally have a quiescent level at earth; so V_{out} will be well above earth potential, probably about 7.5 V if $V_{CC} = 15$ V. In order to connect to a following stage, the DC level has to be shifted down.

In principle this may be done with an emitter follower connected as in figure 18.24 to the negative supply voltage. The output V_C is taken off at a suitable voltage level. Suppose for example the quiescent level of V_B is 7.5 V and $-V_{EE} = -15$ V. If the quiescent level of V_C is to be at earth, $R_3/R_4 = 6.9/15$. The snag is that this involves an attenuation of the signal from V_B down the

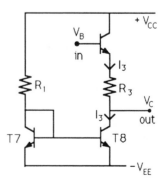

Fig. 18.25. Level shifting with a current mirror.

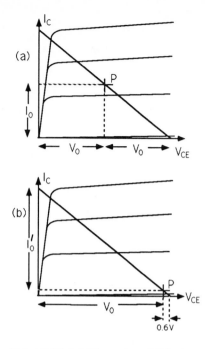

(a)

(b)

Fig. 18.26 (a) Class A and (b) class B operation of a power amplifier.

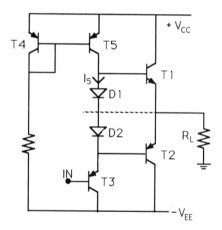

Fig. 18.27. A push-pull output stage.

potentiometer chain. A neat solution to this problem is shown in figure 18.25. The current mirror T7 and T8 provides a constant current I_3 through R_3. Then V_C follows V_B with constant voltage difference

$$V_B - V_C = 0.6 \text{ V} + I_3 R_3.$$

By suitable choice of I_3 and R_3, V_C can be set to any required quiescent level.

18.17   Class A, B and C Amplifiers

The operating point of the common emitter amplifier discussed at the beginning of this chapter was roughly in the middle of the characteristic curves, as in figure 18.26(a). This allows a voltage swing $V_0 \simeq \frac{1}{2} V_{CC}$ up to V_{CC} or down to zero. (As discussed earlier, the actual voltage swing is restricted slightly below this so as to allow the emitter resistor to set the DC operating point.)

The snag with this arrangement is that there is a large DC power dissipation $I_0 V_0$. This is wasteful. In an extreme case, if the amplifier is driving a discotheque loudspeaker, a peak audio signal of 100 W or more may be required but an equal DC power has to be dissipated continuously in the amplifier.

Class B operation

An alternative is the **push–pull** arrangement shown schematically in figure 18.27. Transistors T4 and T5 act as a current mirror providing a fixed current I_5 through the diodes and T3. The diodes provide a fixed 1.2 V between the bases of T1 and T2, hence guaranteeing that one of these two transistors must be on at any one time.

Suppose the input goes positive. Then T3 acts as an emitter follower and drives the base of T1 positive. This switches T1 on and transmits the input voltage to the load R_L. Conversely, if the input goes negative, T2 switches on and drives the output negative. By suitable choice of characteristics for T3 and $T5$, the quiescent output voltage can be chosen to be zero when the input is zero. This is indicated schematically by the broken line in figure 18.27, indicating zero DC voltage.

Each emitter follower T1 and T2 is now biased so that the centre of its operating range V_{CE} is about 0.6 V from zero (figure 18.26(b)). Each emitter follower can then operate on one sign of the input over the full voltage range (less 0.6 V) provided by the power supply. This is known as class B operation. The DC current is small and most of the power is dissipated in the load rather than in the transistor (78% in the optimum case, exercise 19, if the 0.6 V value of V_{BE} is ignored). A further refinement, shown in figure 18.28, is the addition of short circuit protection, by means of the resistors r and transistors T6 and T7.

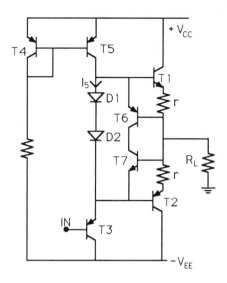

Crossover distortion
The diodes D1 and D2 are there to provide 1.2 V so that both transistors are just on when the input voltage is zero. If both emitter followers were instead biased strictly at zero by omitting D1 and D2, each would follow the input by 0.6 V and small inputs up to ±0.6 V would not be transmitted (figure 18.29). This leads to crossover distortion. Operation with the diodes removed (or with the amplifier biased into the cut-off region when quiescent) is called class C operation. If the input is a sine wave, the output contains large harmonic components. This is a useful trick for generating harmonics.

Fig. 18.28. Push–pull output stage with short-circuit protection.

18.18   Exercises

1.   What values do R_1, R_2, R_C and R_E take in the circuit of figure 18.30 if the operating point is to be at $V_{CE} = 8$ V, $V_{BE} = 0.6$ V, $I_C = 5$ mA, $I_B = 100$ μA and the base is to be at 5 V above earth? (b) You may assume that $h_{fe} = 50$, $1/h_{oe} = 12$ kΩ, $h_{ie} = 2$ kΩ and $h_{re} = 0$. Ignoring capacitances, but not the bias and load resistors, what are the input and output impedances of the amplifier and the maximum current and voltage gains available when feeding suitable loads? (c) What is the magnitude of C_1 which will reduce the voltage gain by 3 db at a frequency $f = 50$ Hz? *(d) What change in operating point is there if R_2 is 10% larger than it should be? (Ans: (a) $R_1 = 30$ kΩ, $R_2 = 12.5$ kΩ, $R_E = 860$ Ω, $R_C = 1.5$ kΩ; (b) $r_{in} = 1.6$ kΩ, $r_{out} = 1.35$ kΩ, maximum current and voltage gains 40 and 34, (c) $C_1 = 2$ μF, (d) V_B and V_E increased by 0.26 V, I_E and I_C raised by 6.0%, V_{CE} reduced by 0.7 V.)

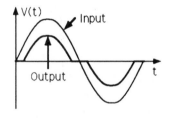

Fig. 18.29. Crossover distortion with class C operation.

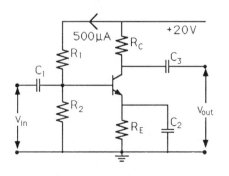

Fig. 18.30.

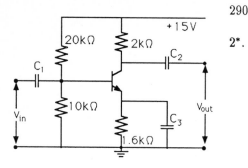

Fig. 18.31.

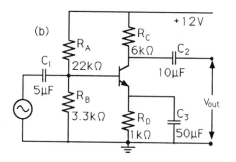

(a)

(b)

Fig. 18.32.

2*. A transistor with $h_{ie} = 1$ kΩ, $h_{re} = 5 \times 10^{-4}$, $h_{fe} = 50$ and $1/h_{oe} = 2 \times 10^4$ Ω is used in the common emitter amplifier of figure 18.31. Draw the h equivalent circuit assuming all the capacitors to have negligible impedance, and also initially neglecting $h_{re}v_{CE}$. Calculate the input and output impedances and the maximum available current and voltage gains of the amplifier. Now evaluate the magnitude of $h_{re}v_{CE}$. What is its effect on i_B? For the rest of this question neglect h_{re}. If the output of this amplifier feeds into an identical second stage amplifier, what are the overall current and voltage gains of the two stages? What values must all the capacitances have in this case for the first amplifier if they are individually to affect the gains by less than 3 db at $f = 50$ Hz? (Ans: $r_{in} = 870$ Ω, $r_{out} = 1.8$ kΩ, 43.5, 91, $h_{re}v_{CE} = -0.046v_{in}$; i_B increases by $\sim 4.6\%$; 1470, 2670; $C_1 = 3.6$ μF, $C_2 = 1.2$ μF, $C_3 = 160$ μF.)

3. Identify one part of figure 18.32(b) for which figure 18.32(a) is the AC equivalent circuit. In particular, describe the meaning of v and R_1 in this context and give the values of R_2 and C. (Two alternative answers: (i) input side of the transistor: $C = 5$ μF, $R_2 = 3.3$ kΩ in parallel with 22 kΩ, $R_1 = h_{ie}$, $v = h_{re}v_{CE}$; (ii) output side of the transistor: $C = 10$ μF, $R_2 = 6$ kΩ, $R_1 = 1/h_{oe}$, $v =$Thevenin equivalent $h_{fe}i_B/h_{oe}$.)

4. (Bedford 1974) Draw a low-frequency equivalent circuit for the amplifier of figure 18.33 and deduce an expression for the voltage gain of the circuit. If the parameters of the transistor are $\beta = 100$ and $r_B = 1.5$ kΩ show that the voltage gain of the circuit at 1 kHz is -5. The same circuit is used to amplify the output of a 1 kHz signal generator of 600 Ω output impedance. The output voltage, measured with an oscilloscope, is 300 mV. When the signal generator is connected to the amplifier, the output voltage from the amplifier, measured by the same oscilloscope, is -1.0 V. Show that this result is consistent with the amplifier gain quoted above, and indicate how the voltage gain could be increased. (Ans: increase 1.5 kΩ and 8.5 kΩ resistors or decrease 180 Ω resistor.)

5. (QMC 1974) In figure 18.34, what sort of feedback, for small signals, is provided by R_E? Draw the small-signal equivalent circuit of the amplifier and, assuming that h_{re} is zero and that $1/h_{oe}$ is infinite, determine an expression for the closed loop voltage gain. If the amplifier is required to have a closed loop voltage gain of -10 with $h_{ie} = 5$ kΩ and $h_{fe} = 250$, calculate a value for R_E. What is the voltage gain, with feedback, if a transistor with $h_{fe} = 500$ is used? (Ans: gain $= -2.03$ k$\Omega \times \beta/[h_{ie} + (\beta + 1)R_E]$; $R_E = 182$ Ω, gain $= -10.54$.)

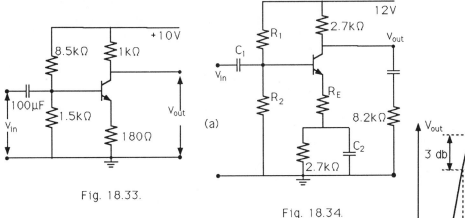

Fig. 18.33.

(a)

Fig. 18.34.

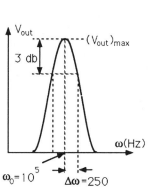

Fig. 18.36.

6**. (QMC 1967) For the circuit of figure 18.35, (a) sketch the frequency response, displaying relevant numerical parameters, (b) explain the response to an amplitude modulated signal $V_{in} = (10+\sin 5 \times 10^4 t) \sin(15 \times 10^4 t)$ mV, and (c) explain the response to square waves of angular frequency 5×10^3 rad/s. (Ans: figure 18.36, $(V_{out})_{max} = h_{fe} Z_{max} V_{in}/h_{ie}$ where $Z_{max} = 2 \times 10^5 \ \Omega$; signal $= 10 \sin 15 \times 10^4 t + \frac{1}{2}(\cos 10^5 t - \cos 2 \times 10^5 t)$: only the signal with $\omega = \omega_0$ has significant amplification; (b) only the harmonic at angular frequency ω_0 has significant gain, so the output is approximately a sine wave at $\omega = \omega_0$.)

7. (RHBNC 1987) The common emitter amplifier circuit in figure 18.37 has at least three errors in it; identify these errors and explain briefly how they may be corrected. (Ans: 0.1 μF capacitor should be in series with the input, transistor emitter in the wrong place, no collector resistor.)

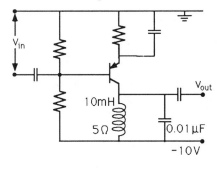

Fig. 18.35.

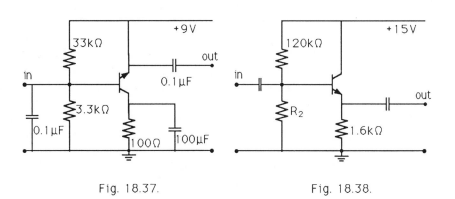

Fig. 18.37.   Fig. 18.38.

8. The transistor in figure 18.38 has $h_{ic} = 1000 \ \Omega$, $h_{fc} = 50$, $h_{oc} = 25 \times 10^{-6}$ mho and $h_{rc} = 0$. If the operating point is to be $I_E = 2.5$ mA and $I_B = 50 \ \mu$A, what should be the value

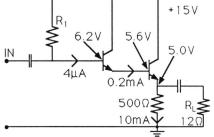

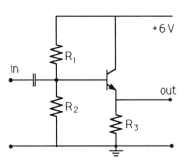

Fig. 18.39.

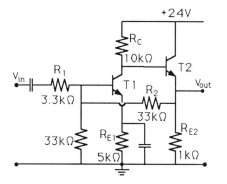

Fig. 18.40.

Fig. 18.41.

of R_2? What are the input and output impedances of this amplifier, and the maximum available current and voltage gains, (a) with $h_{oc} = 0$, (b) with h_{oc} given above? (Ans: $R_2 = 125$ kΩ; (a) 35.2 kΩ, 19.4 Ω, 51, 0.988, (b) 34.6 kΩ, 19.4 Ω, 51, 0.987.)

9. Using two transistors operating from a $+$ 15 V supply each with a current gain of 50, design a Darlington pair with the second transistor having an operating point $I_E = 10$ mA and an emitter resistor of 500 Ω connected to the second transistor. If the circuit drives a 12 Ω load through a coupling capacitor, what are the input impedance and the current and voltage gains? Assume transistor characteristics from the previous question. (Ans: figure 18.39, $R_1 = 2.3$ MΩ, $r_{in} = 77$ kΩ, 2360, 0.37.)

10. (QMC 1985) Describe the operation of the circuit shown in figure 18.40. Give an example of the use of this circuit. If the output is to drive a load of 1 kΩ, what would be suitable values of R_1, R_2 and R_3? If the transistor has a current gain of 50, what would the input impedance then be? Describe quantitatively the effect of C on the bandwidth. ($R_3 = 1$ kΩ, $R_1 = 40$ kΩ, $R_2 = 120$ kΩ, $r_{in} = 14$ kΩ; C acts with r_{in} as a high pass filter with cut-off $\omega = 1/Cr_{in}$.)

11**. Describe qualitatively, then quantitatively, the operation of the circuit shown in figure 18.41. You are given that (i) both transistors have a current gain of 50, (ii) T1 in the configuration shown has an input impedance of 700 Ω, and an output impedance of 20 kΩ, (iii) for T2 the value of r_B is 500 Ω and $r_0 = 20$ kΩ. You may also assume that v_{out} is applied to an identical circuit to that shown here.

Hints: (a) Throughout, assume component values have a tolerance of 5% and ignore quantities small compared with this tolerance, (b) What is the action of R_1 and R_2? (c) show that the static operating point is such that the base of T1 is at 5.5 V; you may assume a drop of 0.6 V between base and emitter in each transistor, (d) show that T2 has a large input impedance and a low output impedance, (e) ignoring temporarily the effect of R_2, what is the voltage gain across T1 and across T2? (f) Finally, what is the relation between v_{in} and v_{out}? What are the input and output impedances? (Ans: R_1 and R_2 form a feedback loop to stabilise the voltage gain at approximately $R_2/R_1 = -10$; (e) 73, 1; (f) $v_{out}/v_{in} = -8.8$; $r_{in} \simeq 4$ kΩ, $r_{out} \simeq 124$ Ω.)

12. (RHBNC 1989) The circuit of figure 18.42 shows a typical common source amplifier circuit. Explain the function of (a) the resistors R_G and R_S, (b) the capacitors C_G and C_D, and

(c) the capacitor C_S. The transfer and output characteristics of the transistor are shown in figure 18.43. Draw a load line for $V_{DD} = 30$ V, $R_D = 3.0$ kΩ and use it to explain how this amplifier operates when a small sinusoidal voltage is applied to its input. If the gate–source voltage is -1.0 V, calculate the value of the resistor R_S which is required to provide the gate–source bias voltage. With the circuit parameters given above, find (i) the DC voltage at the drain with respect to earth when no signal is applied to the input, (ii) the voltage gain of the amplifier when a sinusoidal input signal of amplitude 0.5 V is applied to its input. (Ans: $R_S = 250$ Ω, (i) 18 V, (ii) 18.)

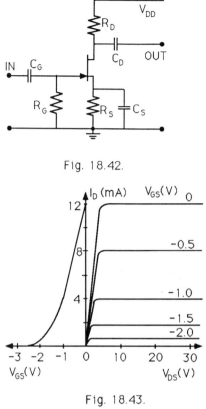

Fig. 18.42.

13. Using the characteristics of the BC264L FET shown in appendix D, draw a source follower circuit with a source resistor of 5 kΩ, a power supply of 25 V and a -1 V drop between source and gate. What is the DC gate voltage? What is the voltage gain and the output resistance? (Ans: figure 18.44, 11 V, 0.95, 285 Ω.)

Fig. 18.43.

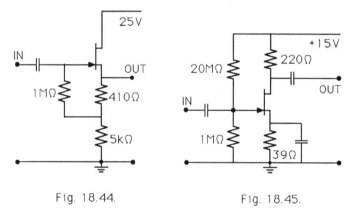

Fig. 18.44.   Fig. 18.45.

14. In the circuit of figure 18.45, the FET is the BC246 whose characteristics are given in appendix D. Verify that the operating point has $I_D = 45$ mA and find the voltage gain for AC signals. (Ans: -4.4.)

15. A common source amplifier has a load resistor of 22 kΩ and gives a voltage amplification of 42.5. When the load resistor is halved, the gain drops to 26. Find the output impedance of the FET and its g_m. (Ans: $r_d = 38$ kΩ, $g_m = 3.05$ mA V^{-1}.)

16. In the circuit of figure 18.46, the FET has $g_m = 3 \times 10^{-3}$ mho and $r_d = 30$ kΩ. The internal capacitance C_{DS} between drain and source is 5 pF and the circuit feeds into an oscilloscope with input impedance 1 MΩ in parallel with 2 pF. Sketch the magnitude of signal observed on the oscilloscope as a function of frequency. (Ans: midband gain = 22.5, low-ω

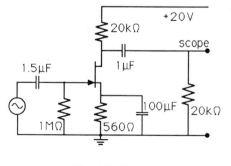

Fig. 18.46.

cut-off 30 rad/s, high-ω cut-off 2×10^7 rad/s.)

17. Show that for the source follower of figure 18.17,

$$i_{in} R_1 = R_2 i_D + v_{GS} \qquad\qquad v_{DS} + (R_2 + R_f) i_D \simeq 0$$

$$i_D = g_m v_{GS} + v_{DS}/r_d \qquad \text{and} \qquad v_{in} \simeq v_{GS} + (R_2 + R_f) i_D$$

where small i and v refer to small changes in currents and voltages. (The approximations involve neglecting the voltage drop across R_f due to i_{in}.) Hence show that the input impedance is $r_{in} = R_1[1 + g_m(R_2 + R_f) + (R_2 + R_f)/r_d]/[1 + g_m R_2 + (R_2 + R_f)/r_d]$. Evaluate this for the parameter values of figure 18.17 if $g_m = 1/300$ mho and $r_d = 20$ kΩ. (Ans: 5.4 MΩ.)

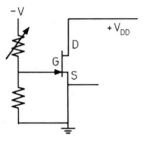

Fig. 18.47.

18. How may an FET be used as a variable resistor? Find the DC resistance of the BC264L FET from the characteristics of appendix D for small (≤ 2 V) voltages V_{DS} for (a) $V_{GS} = 0$, (b) $V_{GS} = -0.5$ V. (Ans: figure 18.47, 400 Ω, 720 Ω.)

19*. (QMC 1974) Figure 18.48(a) shows a three-stage amplifier. Describe briefly the small-signal properties (voltage gain, input and output resistance) of each stage and of the overall amplifier. (i) Explain why, for the DC conditions, the bias circuit for the FET is preferable to the simple form shown in figure 18.48(b), (ii) explain why capacitors C2 and C3 are used and draw a small-signal equivalent circuit for the first stage, (iii) C5 is used to 'bypass' the feedback of R9 for small signals; show that, for it to be effective, it must have a reactance much less than h_{ie}/h_{fe}. What is the input impedance of the circuit? (Ans: $R_2 R_3(1 + g_m R_5')/(R_2 + R_3)$, where R_5' is the parallel combination of R_5, r_d, R_1 and R_4.)

20. This exercise examines the efficiency of class A and class B power amplifiers. Suppose the amplifier of figure 18.27 is adjusted for ideal class A operation by inserting a resistor in series with the diodes, so that the operating point P is at voltage $\frac{1}{2}V_{CC}$, as in figure 18.26(a). Each transistor provides half the signal, and it is sufficient to consider one transistor. The largest undistorted AC output signal has amplitude $V_0 = \frac{1}{2}V_{CC}$ if (for simplicity) distortion in the region of saturation near $V_{CE} = 0$ is ignored. If the AC voltage developed by one transistor across the load resistor is $v_L = V_0 \cos \omega t$, show that the voltage V_C across the transistor is $V_0(1 - \cos \omega t)$ and the current through it is $I_0(1 + \cos \omega t)$. Show that the mean power dissipated over one cycle is $\frac{1}{2}I_0 V_0$ in the transistor and $\frac{1}{2}I_0 V_0$ in the load, and hence that the efficiency, $\epsilon =$ (power delivered to the load)/(total power dissipated), is 50%. If the signal falls to (a) $\frac{1}{2}V_0$, (b) zero, what happens to the power $\bar{P}$ dissipated in the transistor and to ϵ?

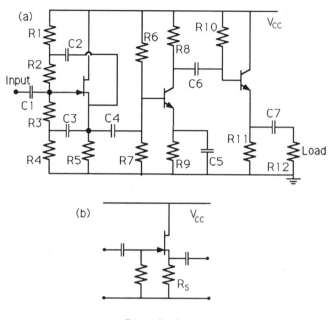

(b) V_{cc}

R_s

Fig. 18.48

Next suppose the bias resistor R_2 of figure 18.26(a) is adjusted for ideal class B operation, so that the operating point P is at voltage V_0, as in figure 18.26(b). Each transistor now provides the signal during one half-cycle, and it is sufficient to consider one transistor again. Show that if this transistor develops voltage $V_0 \cos \omega t$ across the load, $V_C = V_0(1 - \cos \omega t)$ and $I_C = I_0' \cos \omega t$. Show that the mean power dissipated over one half-cycle is $\frac{1}{2}I_0'V_0$ in the load and $(2/\pi)I_0'V_0$ in total, and hence that the efficiency is $\pi/4 = 0.785$. If the signal falls to (a) $\frac{1}{2}V_0$, (b) zero, what happens to the power dissipated in the transistor and to the efficiency? (Ans: (a) $\bar{P} = 7I_0'V_0/8$, $\epsilon = 1/8$, (b) $\bar{P} = I_0'V_0$, $\epsilon = 0$; (a) $\bar{P}$ reduced by a factor 4, $\epsilon \to \pi/8$, (b) $\bar{P} = 0$, $\epsilon \to 0$.)

21**. (IC 1988) Explain how the Miller effect occurs in simple transistor circuits. In which of the three configurations is it appreciable? In the simple common emitter amplifier circuit of figure 18.49(a), the transistor has an internal 1 pF capacitance between its collector and base and is fed from a signal source of output impedance 50 kΩ. The input signal is biased in such a way as to ensure that the quiescent collector voltage is 5 V. Ignoring any other capacitances and assuming the value of h_{fe} to be very large, estimate the gain at intermediate frequencies and estimate the high-frequency cut-off. What source characteristics would maximise the frequency

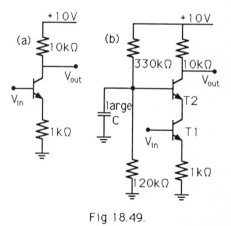

Fig 18.49.

response in a practical situation?

A practical solution to the Miller effect is the 'Cascode' circuit of figure 18.49(b): the common emitter circuit is connected to a second transistor in common base configuration. Given the same input signal conditions, and the fact that the collector of T2 is biased to 5 V, (a) calculate the voltage at the emitter of T2, (b) state the input impedance looking into the emitter of T2, (c) explain qualitatively why the gain of the circuit is the same as that of the single common emitter circuit used above, (d) show that the collector of T1 is at a fixed voltage independent of AC input signal, and (e) hence explain why this circuit does not suffer the Miller effect and estimate its frequency response, assuming both transistors to be the same as that used above. (Ans: -9.5, 1.8×10^6 rad/s, low output impedance; (a) 2.1 V, 50 Ω, (c) the same current flows in T_1 and T_2 and is governed by the 10 kΩ and 1 kΩ resistors, (e) cut-off at 2×10^7 rad/s.)

19

Oscillators

19.1   The Condition for Oscillation

Chapter 14 showed that a parallel LC combination with no resistance carries a current which executes undamped simple harmonic motion: $I \propto \cos\omega_0 t$, where $\omega_0 = 1/\sqrt{LC}$. However, if there is any resistance in the circuit, energy is dissipated in this resistance and the oscillation is damped out. In order to maintain a stable oscillation, power must be fed to the oscillator to replace the energy which is dissipated.

A simple way to achieve this is using positive feedback. Suppose the amplifier of figure 19.1 has gain $G(\omega)$ varying with frequency ω and suppose a fraction $B(\omega)$ of the output voltage is fed back positively to the input. The voltage amplification is

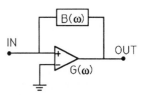

Fig. 19.1. Positive feedback with a non-inverting amplifier.

$$\frac{v_{\text{out}}}{v_{\text{in}}} = A(\omega) = \frac{G(\omega)}{1 - G(\omega)B(\omega)}. \qquad (19.1)$$

If $G(\omega)B(\omega) \to 1$, $A(\omega) \to \infty$ and the circuit is capable of producing an output without any external input. The principle of the sinusoidal oscillator is that this happens at a well defined frequency ω_0.

Chapter 7 showed that if the operational amplifier itself has input impedance r_{in}, the *circuit* has input impedance $(1 - G(\omega)B(\omega))r_{\text{in}}$. When $G(\omega)B(\omega) \to 1$, the input impedance of the circuit goes to zero, and the oscillation is undamped. If $(1 - G(\omega)B(\omega))$ is negative, the amplitude actually grows with time until the amplifier begins to saturate; then G drops to the point where $\langle GB \rangle = 1$ averaged over the cycle.

The crudest example of such a circuit is shown in figure 19.2. The LC parallel circuit used as load has a large impedance only in a narrow band of frequencies around $\omega = 1/\sqrt{LC}$ and the amplifier has large gain only over a correspondingly narrow range of frequencies. If the sign of the mutual inductance is chosen so that the feedback to the base is positive and if the feedback fraction B is large enough to satisfy the condition $GB = 1$, the circuit will

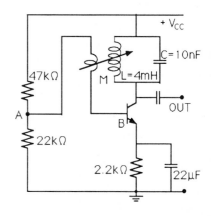

Fig. 19.2. A crude form of oscillator

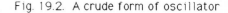

oscillate with a frequency at or close to the peak of the resonance curve.

It is recommended that you try making this oscillator yourself. A few turns of wire wrapped loosely round the inductor provides the necessary feedback between A and B. You may need to flip the turns over to get the right polarity for positive feedback. As the number of turns increases, the oscillator saturates. This saturation may be controlled by a variable resistor between A and B. If you remove C, stray capacitance across the coil leads to oscillation at a higher frequency. Try putting your hand close to the coil. Its stray capacitance alters the oscillation frequency noticeably.

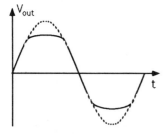

Fig. 19.3. Saturation of the amplifier (full curve) and consequent distortion of the pure sine-wave output (broken).

Saturation

The phase of the current in a tuned circuit varies rapidly with frequency (figure 14.2). For oscillation to occur, $\langle G(\omega)B(\omega)\rangle$ must be real and equal to one. What determines the oscillation *frequency* is the condition that $\langle G(\omega)B(\omega)\rangle$ is real; the condition $|GB| = 1$ determines the *amplitude* of oscillation. The ideal is that G does not vary over the cycle of oscillation and B is just large enough to ensure $GB = 1$; the output is then a pure sine wave. If, however, the feedback is larger, $G(\omega)$ will fluctuate over the cycle, due to saturation or cut-off of the amplifier at the maximum of the swing. In this case, the condition determining the amplitude is that the mean value $\langle G(\omega)B(\omega)\rangle$ over the cycle is one; the output waveform may be deformed from a pure sine wave, as in the full curve of figure 19.3.

If the output does deform, will the first harmonic contribute or will the second harmonic come in first?

19.2   Frequency Stability

For many purposes, the frequency of the oscillator should be stable against changes in operating conditions such as load and temperature. This is obviously the case, for example, with the frequency of a radio or TV station. When the output of the oscillator is applied to another circuit, the input impedance of that circuit appears in the AC equivalent circuit in parallel with the tuned circuit of the oscillator. (Likewise, the impedance between points AB of figure 19.2 is 'reflected' by the transformer into the tuned circuit, see Chapter 22). The equivalent circuit of the transistor load is then as shown in figure 19.4. If either resistance in the circuit changes, the oscillation frequency will change slightly. Even worse, if either has a reactance which changes (e.g. a change in stray capacitance), the oscillation frequency is affected.

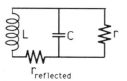

Fig. 19.4. Equivalent circuit of the load.

To alleviate these problems, it is essential that $\text{Re}(G(\omega)B(\omega))$ varies rapidly with ω, so that the oscillation frequency is well defined. This was the reason for choosing a tuned circuit as load in figure 19.2. Secondly, it is usual to connect the oscillator to the outside world through an emitter follower (or similar high-input-impedance circuit). Thirdly, the transistor of the oscillator is often used in common base configuration rather than common emitter in order to achieve high output impedance and low input impedance (hence low resistance reflected into the tuned circuit).

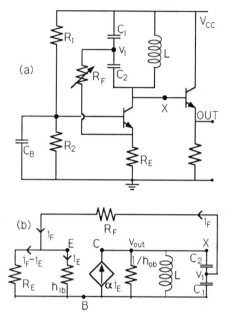

Fig. 19.5 (a) Colpitts oscillator and (b) its equivalent circuit.

Colpitts oscillator

The Colpitts oscillator is shown in figure 19.5(a) as an example with these features. Positive feedback is achieved by tapping off some of the output signal from the capacitive arm of the tuned circuit and feeding it back across R_E.

Why does this achieve positive feedback?

The operation of the circuit may be understood with the help of the AC equivalent circuit of figure 19.5(b). We follow it through first qualitatively, then algebraically. The base is effectively grounded for AC signals by the large capacitor C_B. Also, the battery acts as a short for AC signals across R_1+C_B, so the input signal is developed between emitter and base and the output between base and collector. Suppose there is a small AC signal v_{in} at the emitter; then $i_E = v_{in}/h_{ib}$. On resonance, the tuned circuit has large impedance, so the voltage amplification is large ($\sim 10^4$). Large currents flow in the tuned circuit (see Chapter 14) and if these currents are large compared with the feedback current i_F, $v_1 \simeq v_{out}/(1 + C_1/C_2)$. If R_F were small, the current i_F would be large (unless C_1/C_2 is very large). The purpose of R_F is to regulate the feedback current to the small value required to maintain oscillation. By pursuing this argument algebraically, an appropriate value of R_F can be found.

Firstly, neglecting $h_{rb}v_{CB}$ for simplicity

$$\begin{aligned} v_{in} &= h_{ib}i_E \\ &= R_E(i_F - i_E) \\ i_F &= (1 + h_{ib}/R_E)i_E. \end{aligned} \qquad (19.2)$$

Also

$$i_F = \frac{v_1 - v_{in}}{R_F} = \frac{v_{out}}{R_F(1 + C_1/C_2)} - \frac{h_{ib}i_E}{R_F}. \qquad (19.3)$$

Equating these two expressions,

$$\frac{v_{out}}{R_F(1 + C_1/C_2)} = i_E\left(1 + \frac{h_{ib}}{R_E} + \frac{h_{ib}}{R_F}\right). \qquad (19.4)$$

The current from the generator in the output circuit flows through the three arms of the output circuit and

$$\alpha i_E = v_{out}(h_{ob} + j\omega C + 1/j\omega L) \qquad \text{where} \qquad C = C_1 C_2/(C_1 + C_2). \tag{19.5}$$

Substituting from (19.4) for v_{out},

$$\alpha i_E = i_E R_F(1 + C_1/C_2)\left(1 + \frac{h_{ib}}{R_E} + \frac{h_{ib}}{R_F}\right)[h_{ob} + j\omega(C - 1/\omega^2 L)]. \tag{19.6}$$

For the imaginary part of the right-hand side to be zero, $\omega^2 = 1/CL$; this confirms the expected result that the oscillation frequency is at the peak of the resonance curve. From the real parts,

$$\alpha = R_F(1 + C_1/C_2)\left(1 + \frac{h_{ib}}{R_E} + \frac{h_{ib}}{R_F}\right)h_{ob}.$$

In practice the terms involving h_{ib} are small and $\alpha \simeq 1$, so the circuit just oscillates if

$$R_F(1 + C_1/C_2) \simeq 1/h_{ob} \simeq 10^6 \ \Omega.$$

What is the effect of resistance in the coil L? See Chapter 14 for how to include its effect on the tuned circuit approximately. In reality, the resistance of the tuned circuit at resonance may be $\leq 1/h_{ob}$, lowering the right-hand side significantly.

In practice R_F is adjusted to be rather smaller than this, so that there is always enough feedback for the circuit to oscillate; limiting occurs by the transistor turning off for part of the cycle. With the transistor off, current continues to flow round the tuned circuit; the transistor, while it is on, tops up losses in $1/h_{ob}$ and R_F. If the tuned circuit has a high Q value, harmonics of the resonant frequency will have low gain and the distortion from a pure sine wave is very small; typically harmonics are less than or equal to 0.1% of the fundamental.

If a high output power is required, it is desirable to bias the transistor differently, so that it operates instead as a class C amplifier and is off most of the cycle; it comes on only at the peak of the oscillation in order to make up the power losses. Tricks to accomplish this are discussed in more advanced texts.

Quartz crystal

The oscillation frequency is given by $\omega_0 = 1/\sqrt{LC}$. However, C may be quite temperature sensitive; a typical variation is 2×10^{-4} per ° C. A means of improving frequency stability is to replace the feedback resistor R_F by a quartz crystal, as in figure 19.6. Quartz

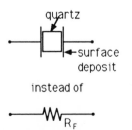

Fig. 19.6. A quartz crystal as substitute for R_F.

has piezoelectric properties such that if it is squeezed, an electrical potential is developed across it. Conversely, an AC voltage across it makes the crystal vibrate mechanically. The crystal transmits a signal readily at its mechanical resonant frequency, which depends on crystal dimensions. Its resonance curve is very narrow, since losses in the crystal are low. A Q of 10^5 is typical, compared with ~ 250 for an LC circuit. The LC circuit now plays the role of a crude tuning load, but the precise frequency of the oscillator is governed by the narrower mechanical resonance of the quartz crystal. The crude oscillator has a stability of typically 2 parts in 10^4. With careful temperature stabilisation, the quartz crystal oscillator achieves a stability of 1 part in 10^8. Digital watches use a quartz crystal oscillator operating at 2^{15} Hz.

19.3   The Wien Bridge Oscillator

An alternative to using a tuned circuit is to use a feedback loop which depends sensitively on ω. An example is the Wien bridge oscillator of figure 19.7. The voltages at negative and positive inputs of the amplifier are

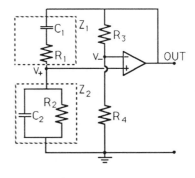

Fig. 19.7. The Wien bridge oscillator.

$$v_- = v_{out}R_4/(R_3 + R_4) = v_{out}/(1 + R_3/R_4)$$
$$v_+ = v_{out}Z_2/(Z_1 + Z_2) = v_{out}/(1 + Z_1/Z_2)$$

where

$$Z_2 = \frac{R_2/j\omega C_2}{R_2 + 1/j\omega C_2} = \frac{R_2}{1 + j\omega C_2 R_2}$$
$$Z_1 = R_1 + 1/j\omega C_1 = (1 + j\omega C_1 R_1)/j\omega C_1.$$

If $R_1 = R_2 = R$ and $C_1 = C_2 = C$,

$$v_+ = \frac{v_{out}}{1 + (1 + j\omega CR)^2/j\omega CR} = \frac{v_{out}j\omega CR}{1 + 3j\omega CR - \omega^2 C^2 R^2}. \quad (19.7)$$

This is real if $\omega = 1/CR$ and this condition defines the oscillation frequency. (The opamp needs to be chosen so that the slew rate is not a limitation at the operating frequency.) With this value of ω, $v_+ = v_{out}/3$. Oscillation requires $v_+ \geq v_-$, i.e. that the negative feedback is less than the positive feedback; this requires $R_3 \geq 2R_4$. In practice, R_3 is made slightly greater than $2R_4$. A trick is often used to stabilise the magnitude of the output waveform by making R_3 (or part of it) a temperature-sensitive resistor. If its resistance drops with increasing current (a thermistor), the output signal will stabilise in magnitude at a particular output current with R_3 warmed up by the feedback current until it is exactly equal to $2R_4$.

The frequency stability may again be improved with a quartz crystal in one of the feedback loops. An oscillator like this, having no inductance in it, is convenient for packaging in an integrated circuit. The resistors and capacitors in the feedback loop may be unequal (see exercise 1), but the definition of the oscillation frequency then becomes rather less sharp.

What happens if R_3 is significantly greater than $2R_4$?

19.4   The Nyquist Diagram

In discussing oscillators, it is important to focus attention on the phase of the feedback signal. For this reason, it is convenient to plot the loop gain $G(\omega)B(\omega)$ as a curve on the complex GB plane, i.e. the plane where Re GB and Im GB are axes. Such a plot is known as a **Nyquist diagram**.

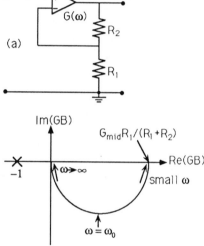

Fig. 19.8 (a) A non-inverting feedback oscillator and (b) its Nyquist diagram.

Single-stage amplifier
Let us consider a few examples. First we consider the simple non-inverting amplifier with **negative** feedback shown in figure 19.8(a). For this,

$$G(\omega)B(\omega) = \frac{R_1}{R_1 + R_2}G(\omega)$$

and

$$\frac{v_{\text{out}}}{v_{\text{in}}} = A(\omega) = \frac{G}{1 + GB}.$$

Because the feedback is now negative, the point of instability where the system might oscillate is $GB = -1$.

At low frequency $G(\omega)$ is constant at the mid-band gain G_{mid}, but at high frequency capacitance causes its gain to fall, as discussed in Chapter 4. For a single-stage amplifier, the frequency dependence is

$$G(\omega) = \frac{G_{\text{mid}}}{1 + j\omega/\omega_0} \tag{19.8}$$

where ω_0 is the cut-off frequency. The Nyquist diagram is shown in figure 19.8(b). Equation (19.8) may be rewritten in terms of modulus and phase as

$$G(\omega) = \frac{G_{\text{mid}}e^{j\phi(\omega)}}{[1 + (\omega/\omega_0)^2]^{1/2}} \qquad \text{where} \qquad \tan\phi(\omega) = -\omega/\omega_0.$$

As $\omega \to 0$, $G(\omega) \to G_{\text{mid}}$ and $\phi(\omega) \to 0$. For small ω, the square root in the denominator is close to 1 and $G(\omega)$ develops a small negative imaginary part: $\phi(\omega) \simeq -\omega/\omega_0$. The denominator begins to matter when $\omega \simeq \omega_0$; at this frequency $|G(\omega)| = G_{\text{mid}}/\sqrt{2}$ and $\phi(\omega) = -45°$. At high frequency, $G(\omega) \simeq -j(\omega_0/\omega)G_{\text{mid}}$, and the curve approaches zero tangentially to the negative Im GB axis.

For this system, $G(\omega)B(\omega)$ goes nowhere near the point -1, so it will not oscillate. If we wanted it to oscillate, we would have to make the feedback fraction $B(\omega)$ depend on ω in such a way as to take GB to the point -1. This implies a phase lag of more than $90°$ in B.

Sketch the Nyquist diagram for the oscillator of figure 19.2.

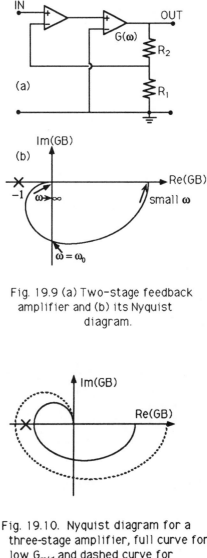

Fig. 19.9 (a) Two-stage feedback amplifier and (b) its Nyquist diagram.

Two stages

Next consider the two-stage amplifier of figure 19.9(a), supposing that each amplifier has a gain which varies with frequency according to equation (19.8). The resulting Nyquist diagram is sketched in figure 19.9(b); the overall amplification is

$$BG = BG_1(\omega)G_2(\omega) = \frac{BG_{mid}^2}{(1 + j\omega/\omega_0)^2}$$

$$= \frac{BG_{mid}^2 e^{j\phi'}}{[(1 - \omega^2/\omega_0^2)^2 + 4\omega^2/\omega_0^2]^{1/2}}$$

where $\tan\phi' = -2\omega/(1 - \omega^2/\omega_0^2)$ and $B = R_1/(R_1 + R_2)$. This time, $\phi' \to 180°$ as $\omega \to \infty$, because there is a phase lag of $90°$ in each amplifier at high frequency. The product GB still does not reach the point of instability at -1, but it is getting closer. At frequencies above $\omega = \omega_0$, the overall phase lag in the two amplifiers has become greater than $90°$, and what was intended as negative feedback has become positive feedback.

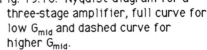

Three stages

The extrapolation to further stages of amplification is clear. With feedback over three stages of amplification, the Nyquist diagram is as shown in figure 19.10. In this case, the amplification over three stages is

$$GB = \frac{BG_{mid}^3}{(1 + j\omega/\omega_0)^3}$$

Fig. 19.10. Nyquist diagram for a three-stage amplifier, full curve for low G_{mid} and dashed curve for higher G_{mid}.

and the product GB may go close to -1 (full curve in the figure) or encircle it (broken curve). This is one way of making an oscillator. Figure 19.11 shows a circuit where three RC networks are joined in sequence to produce a phase lag of $60°$ each, making $180°$ in total. Exercise 2 shows that this circuit oscillates at $\omega = \sqrt{6}/CR$ with $B = 1/29$.

There is a clear warning in figure 19.10. If several steps of amplification are cascaded, care is needed that the feedback to the input from the third or subsequent stage is not enough to cause oscillation. Most people who have tried to build a powerful amplifier have encountered the galling experience that it oscillates. The

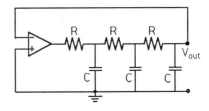

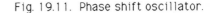

Fig. 19.11. Phase shift oscillator.

unintended feedback can be via stray capacitance, via the power supply or via earth loops. Another familiar example is a public address system; most of us will have witnessed occasions when the amplification is turned up and the system suddenly breaks into a high pitched scream.

In this case the time delay between the sound going out from the loudspeakers and returning to the microphone plays a crucial role. Can you see how?

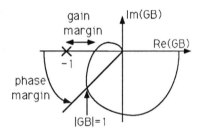

Fig. 19.12. Conditional stability.

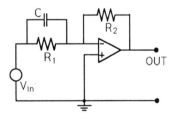

Fig. 19.13. Gain and phase margin.

Stability criterion

Nyquist showed quite generally that the amplifier oscillates if the curve for GB encircles the point -1 with ω increasing in a clockwise sense. This is so for the broken curve of figure 19.10; at the frequency where the curve crosses the negative real axis, the circuit oscillates and the amplifier saturates or cuts off so that $|G(\omega)B(\omega)|$ averages 1 over a cycle. Exercise 3 shows that a three-stage amplifier will oscillate if $G_{\mathrm{mid}}^3 B > 8$.

With a clever choice of $B(\omega)$, we might make $G(\omega)B(\omega)$ follow the curve shown in figure 19.12, skirting below the point of instability. Technically, this system will not oscillate. However, if the gain is increased or decreased, oscillation will occur, so it is not good practice.

Whenever GB goes close to -1, the overall gain of the system $G/(1 + GB)$ becomes very large. This introduces a nasty blip into the frequency dependence of the amplifier (see exercise 8 for a numerical example). Also, if the circuit is disturbed, transient oscillations occur with small damping. It is generally wise to avoid the point of instability in either or both of two ways. Firstly, it can be arranged that $|GB| \ll 1$ when the curve crosses the negative real axis. The amount by which $|GB|$ falls below 1 is known as the **gain margin** (figure 19.13) and a useful rule of thumb is that a gain margin of 8 to 10 db is satisfactory (i.e. a factor of three). Alternatively, it can be arranged that when $|GB| \geq 1$ the overall phase lag ϕ is well below $180°$. This gives a **phase margin**. The rule of thumb is that a phase margin of 30 to $40°$ is desirable.

Phase advance compensation

The network shown in figure 19.14 produces a transfer function

$$V_{\mathrm{out}}/V_{\mathrm{in}} = -(R_2/R_1)(1 + \mathrm{j}\omega C R_1).$$

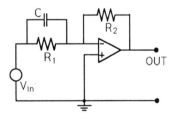

Fig. 19.14. Network for phase advance compensation.

At low frequencies the gain is constant, but at high frequencies there is a phase *advance* of the output, eventually reaching $90°$; it is $45°$ at $\omega_1 = 1/CR_1$. Essentially what happens is that the capacitor bypasses R_1 above this frequency. This may be used to increase

the phase margin at high frequencies and convert an otherwise unstable system to a stable one. The value of ω_1 is chosen to leave
the low frequency behaviour unchanged but to provide **phase advance compensation** at frequencies where the gain margin would
otherwise be inadequate. The algebra is discussed in exercise 11.
The resistor R_1 inhibits the circuit from displaying the noisy behaviour of the differentiator circuit of figure 8.7, where $R_1 = 0$.

19.5   Uncompensated Operational Amplifiers

You may have wondered why the μA741C operational amplifier is
designed with a gain which falls linearly with frequency over most
of its working range, as shown in figure 7.3. Why not keep the
gain high to high frequencies, then cut it off quickly? The answer
to the first question is that it reduces the risk of high-frequency
oscillation, while the latter suggestion increases that risk.

Suppose an amplifier is made with a constant gain G_1 at intermediate frequencies. Eventually, capacitance somewhere will force
the gain to fall away beyond a cut-off frequency ω_1. Initially the
fall-off is at -6 db/octave. Then, when a second capacitance becomes important at frequency ω_2, it falls off at -12 db/octave (see
exercise 4.8). The general behaviour is shown in figure 19.15.

Algebraically,

$$G(\omega) = \frac{G_1}{(1 + j\omega/\omega_1)(1 + j\omega/\omega_2)\dots}. \qquad (19.9)$$

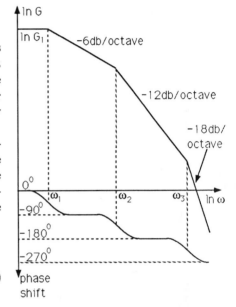

Fig. 19.15. Bode diagram of a
general amplifier.

Associated with every -6 db/octave is a phase lag of 90°. It turns
out to be a general consequence of causality and Fourier's theorem that this connection between gain and phase shift cannot be
avoided; to do so would require a circuit which responded before
the signal arrived. So if you want the gain to fall rapidly with frequency, the penalty is a large phase lag and this is likely to cause
oscillation.

Operational amplifiers are designed to go into feedback loops.
If they are not to oscillate, it is essential that $|GB|$ should fall to 1
before the phase lag reaches 180°. In practice it is desirable, in order to achieve reasonable phase margin, that the critical frequency
ω_c where the phase shift reaches say 150° is above the frequency
where $|GB|$ falls to 1. Compensated operational amplifiers like the
μA741C are deliberately given a roll-off of -6 db/octave such that
the gain falls to 1 at frequency ω_c in the worst case with $B = 1$.

Pole zero cancellation
If it can be guaranteed that $B < 1$, this design is unduly conser-
vative: the roll-off could be moved to a higher frequency such that

GB rather than G reaches 1 at frequency ω_c. 'Uncompensated' amplifiers are available giving higher gain at high frequencies. For example, the 748 is the uncompensated version of the 741. But it is then the responsibility of the user to arrange that the feedback loop is such that the circuit does not oscillate. The manufacturer helps by providing pin connections where compensation networks can be applied within the amplifier to stabilise it. Suppose, for example, you are desperate to extend the operation of a circuit to a frequency above ω_2, where the natural phase lag of the amplifier approaches 180°. It is possible to apply the compensation network shown in figure 19.16, having transfer function

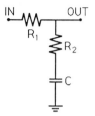

IN —WW— OUT
R_1
R_2
C

Fig. 19.16. Pole zero network.

$$T = \frac{V_{out}}{V_{in}} = \frac{R_2 + 1/j\omega C}{R_1 + R_2 + 1/j\omega C} = \frac{1 + j\omega C R_2}{1 + j\omega (R_1 + R_2)C}. \quad (19.10)$$

If it is arranged that $CR_2 = 1/\omega_2$, the numerator of (19.10) can be used to cancel the pole term $(1 + j\omega/\omega_2)$ in the denominator of (19.9). This is called **pole zero cancellation**. There are two related penalties. Firstly, a new pole is substituted at higher frequency ω_2' due to the denominator of (19.10). Secondly, the gain is kept high to higher frequencies. Unless steps are taken to reduce G_0, the circuit will probably oscillate near frequency ω_2'. That is, the circuit can be made to operate at frequency ω_2, but not with unlimited gain.

An alternative to using a pole zero network is to use phase advance compensation (figure 19.14) in the feedback loop.

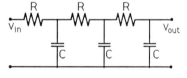

R R R
V_{in} —WW—⊤—WW—⊤—WW—⊤— V_{out}
C C C

Fig. 19.17.

19.6   Exercises

1.   For the Wien bridge oscillator of figure 19.7, show that if C_1 and C_2 are different and R_1 and R_2 are different, the oscillation frequency is $\omega = (C_1 C_2 R_1 R_2)^{-1/2}$, and that the condition on R_3 and R_4 for oscillation is that $R_3/R_4 \geq (C_1 R_1 + C_2 R_2)/C_1 R_2$. If the phase lag of the feedback voltage is $\phi(\omega)$, find $d\phi/d\omega$ at resonance for the case where $C_1 = C_2 = C$ and $R_1 = R_2 = R$. (Ans: $-2/(3\omega)$.)

2.   Show that, for the phase shifting network of figure 19.17, $1/B = v_{in}/v_{out} = (1 - 5\omega^2 C^2 R^2) + j(6\omega CR - \omega^3 C^3 R^3)$ and hence that for the oscillator of figure 19.11 the frequency of oscillation is $\omega = \sqrt{6}/CR$ and G_{mid} must be greater than or equal to 29 for oscillation.

3.   (Kings 1974) A wideband amplifier has three identical stages each having a mid-band gain of -10. The high-frequency amplification of each stage is limited by a time constant of $10^{-6}/2\pi$ s. The output of the amplifier is connected to the negative input through a resistive attenuator and a large capacitance. Calculate the voltage attenuation such that the

amplifier will just oscillate and determine the frequency of oscillation. Sketch the Nyquist plot for the system. (Ans: 125, $f = 10^6\sqrt{3}$ Hz.)

4. (Westfield 1974) The differential amplifier in the circuit of figure 19.18 has a very high, frequency-independent gain, a high input resistance and a low output resistance. Derive an expression for the transfer function $T(\omega) = v_0/v_{in}$ of the circuit. Show that the circuit can oscillate when its output Y is connected to the input X. Determine the minimum value of R_2 required for oscillation and calculate the oscillation frequency if $R = 5$ kΩ and $C = 1000$ pF. (Ans: $T = (R_1 + R_2)/[R_1(3 + j\omega CR + 1/j\omega CR)]$; $R_2 = 10$ kΩ, $\omega = 2 \times 10^5$ rad/s.)

5. In figure 19.19, A is a wideband amplifier, producing no phase shift, and having a gain $A = 2$. Will the system oscillate? Give reasons for your conclusions. (Ans: Yes, at $\omega = 1/CR$; oscillation requires $A > \frac{3}{2}$.)

6. (IC 1988) By constructing the Nyquist plot of the system with unity negative feedback and gain $G(\omega) = K/[j\omega(j\omega+a)]$, where K and a are positive constants, justify the statement that this system is stable.

7. (QMC 1978) Draw a circuit diagram for a free-running oscillator which generates a sine wave with an angular frequency of 2×10^3 Hz. Outline briefly how the circuit works. Comment on the features of your circuit which affect the purity of the output sine wave and its frequency stability.

8. (IC 1988) An amplifier with large gain G is connected in a negative feedback system. The Nyquist plot is shown in figure 19.20. Is it stable? The overall amplification is $A = G(\omega)/[1 + G(\omega)B(\omega)] = (1/B)[GB/(1 + GB)]$. Sketch the magnitude of A against ω. If $|GB/(1 + GB)|$ is to be less than 1 for all ω, what constraint must be obeyed by the locus in the Nyquist plot? (This is *not* satisfied by figure 19.20.) (Ans: figure 19.21; Re$(GB) > -\frac{1}{2}$.)

9. (IC 1989) Sketch the Bode diagrams for the following loop gains GB_1, using straight-line approximations for the magnitude plots: (i) K/s, (ii) $K/(Ts + 1)$, (iii) $K/[s(T_1s + 1)]$, where $s = j\omega$. Indicate the phase margin in each case. Sketch the Bode plot of the phase lead compensation network represented by $B_2 = (T_2s + 1)/(T_3s + 1)$ where $T_2 > T_3$; if $0 < 1/T_1 < 1/T_2 < 1/T_3$, show how such a network can increase the phase margin of the system with BG given by (iii). (Ans: figure 19.22(a)–(e).)

10*. (Bedford 1974) Draw the Bode diagram (both gain and phase

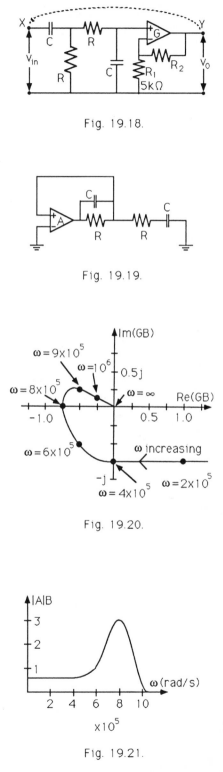

Fig. 19.18.

Fig. 19.19.

Fig. 19.20.

Fig. 19.21.

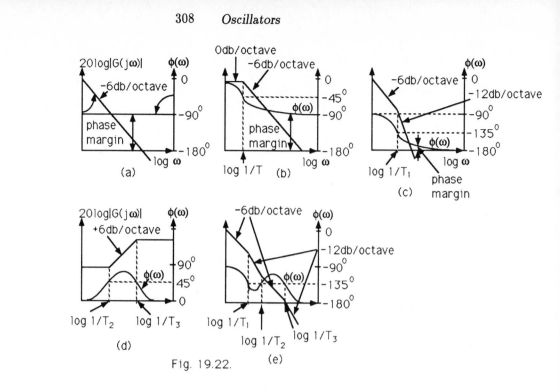

Fig. 19.22.

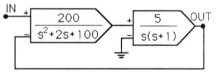

Fig. 19.23.

shift) of the unity feedback system of figure 19.23, where $s = j\omega$, showing clearly the asymptotic approximations used. Estimate the gain and phase margins for this system. (Ans: 7.2 db, 12.8°.)

11. The amplifier in figure 19.24(a) has a gain $G(\omega) = G_0/(1 + j\omega\tau_0)$ where $\tau_0 = 0.01$ s. Show that the amplification of the circuit is $A = V_{out}/V_{in} \simeq G_0/[1 + j(\omega\tau_0 - 1/\omega\tau_2)]$, where $\tau_2 = C_1 R_F/G_0$. Plot $A(\omega)$ on the complex plane and find the angular frequency at which $|A|$ is a maximum if $G_0 = 10^5$. Find the amplification of the circuit in figure 19.24(b) and plot the Nyquist diagram. (Ans: figure 19.25(a), 10^6 rad/s, $A = G_0/[1 + x + j\omega(\tau_0 - x\tau_1)]$ where $\tau_1 = C_1 R_1$ and $x = G_0 R_1/[R_2(1 + \omega^2\tau_1^2)]$; figure 19.25(b).)

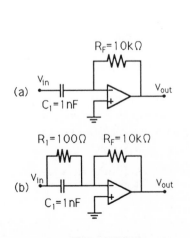

Fig. 19.24.

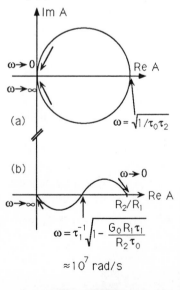

Fig. 19.25.

20

Control Systems and Synchronisation

20.1   Introduction

Electronics is used widely for automatic control. This is a complex subject and the present chapter does no more than introduce basic ideas. The intention is to provide an entry point to the extensive literature on questions such as precision of control and stability.

Attention will be focused initially on an example familiar enough that many properties are intuitively obvious. Suppose that a mass M is to be suspended a height h_R above the ground, as shown in figure 20.1. The value of h_R may vary with time; that is, the mass may have to follow some programmed path with time. The mass is partially supported by a spring providing an upward force $\lambda(L - h)$, where L is the natural length of the spring and h is the actual height of the mass. The rest of the upward force F is provided by a rope attached to a crane or winch. This force F is varied by means of a control system or **servomechanism**. As the mass moves vertically, there is some friction which is indicated schematically by a dashpot providing a retarding force $k\,\mathrm{d}h/\mathrm{d}t$ proportional to velocity.

The equation of motion is

$$M\frac{\mathrm{d}^2h}{\mathrm{d}t^2} = F - Mg + \lambda(L - h) - k\frac{\mathrm{d}h}{\mathrm{d}t}$$

or

$$M\frac{\mathrm{d}^2h}{\mathrm{d}t^2} + k\frac{\mathrm{d}h}{\mathrm{d}t} + \lambda h = F + \lambda L - Mg. \tag{20.1}$$

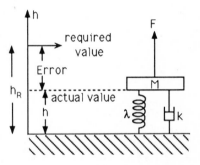

Fig. 20.1. Illustrative example.

Electromechanical analogy
Do not be intimidated by this equation. For the most part, we shall not be concerned with solving it fully, but with limiting cases. It describes damped simple harmonic motion if F is a constant. In

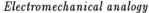

passing, notice that this equation is very similar to that for an LCR series circuit driven by a potential V, equation (14.1):

$$L\frac{d^2I}{dt^2} + R\frac{dI}{dt} + \frac{I}{C} = \frac{dV}{dt}.$$

Mass is analagous to inductance, k to resistance, λ to $1/C$ and h to I. It implies that it is possible to model the behaviour of a mechanical system by an electrical one obeying the same differential equation. This is known as the **electromechanical analogy**. It can be useful to simulate a mechanical control system by means of an electrical circuit whose parameters can be varied straightforwardly, observing the performance on an oscilloscope.

Consider first the equilibrium when $F = 0$ in equation (20.1). The system settles to rest at height h_0 where

$$\lambda h_0 = \lambda L - Mg. \qquad (20.2)$$

The original equation (20.1) may be simplified by measuring heights from this equilibrium position by setting

$$z = h - h_0 \qquad dz/dt = dh/dt \qquad \text{and} \qquad d^2z/dt^2 = d^2h/dt^2.$$

Then

$$M\frac{d^2z}{dt^2} + k\frac{dz}{dt} + \lambda z = F + \lambda L - Mg - \lambda h_0 = F. \qquad (20.3)$$

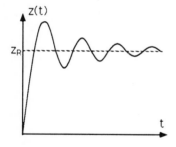
$z(t)$

Fig. 20.2. Damped oscillations about the required height.

To keep the problem simple initially, suppose the required height is constant, so that the required value of z is $z_R = h_R - h_0$. If a force F of the right magnitude is switched on, the mass will execute damped oscillations about z_R and eventually settle there (figure 20.2). It is desirable that these oscillations should be critically damped, so that the system settles as quickly as possible.

To describe the oscillations, let us substitute into equation (20.3)

$$z = z_R + Ae^{st}$$

with the result

$$(Ms^2 + ks + \lambda)Ae^{st} + \lambda z_R = F. \qquad (20.4)$$

The solution is

$$F = \lambda z_R \qquad (20.5)$$

$$z = z_R + A_1 e^{s_1 t} + A_2 e^{s_2 t} \qquad (20.6)$$

where s_1 and s_2 are the two solutions of

$$Ms^2 + ks + \lambda = 0 \qquad (20.7)$$

namely

$$s_{1,2} = [-k \pm j(4M\lambda - k^2)^{1/2}]/2M. \qquad (20.8)$$

This should be familiar from Chapter 14. For critical damping, the square root must be zero.

20.2   Feedback

This solution requires making $F = \lambda z_R$. Setting F by dead reckoning is likely to be unreliable. The result is sensitive to any miscalculation or any wind that blows. The obvious answer is to measure the discrepancy or error between z_R and z and adjust the force F to make this zero. This is where the control system comes in. The value of F, and hence z, is to be controlled by a feedback loop.

Suppose the force is

$$F = F_0 H(z_R - z) \qquad (20.9)$$

where F_0 is a constant of proportionality. In the simplest case, H is just 1 and

$$F = F_0(z_R - z). \qquad (20.9a)$$

In this case the force is proportional to the error. However, it will be necessary to explore the possibility of making H depend on differentiation or integration of z, for example

$$H(z_R - z) = \left(1 + \alpha\frac{d}{dt}\right)(z_R - z)$$

$$= (z_R - z) + \alpha\frac{d}{dt}(z_R - z). \qquad (20.10)$$

The shorthand notation

$$Qz = \left(M\frac{d^2}{dt^2} + k\frac{d}{dt} + \lambda\right)z$$

will be used to rewrite equations (20.3) and (20.9) as

$$Qz = F_0 H(z_R - z). \qquad (20.11)$$

Figure 20.3 displays the feedback in a way similar to Chapter 7. At the left is the required value z_R (analogous to an input voltage). The force F applied to the mass is $F_0 H(z_R - z)$. This force is derived from a feedback loop with z_R fed in positively and z negatively.

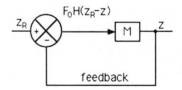

Fig. 20.3. Schematic of the feedback loop.

20.3   Zero Error

The question to be addressed next is how successful the control system is. Does it indeed drive the mass to the required height? The answer is not quite.

Suppose H is given by equation (20.10) and suppose z_R is constant for simplicity. Then equation (20.11) written out in full reads

$$M\frac{d^2z}{dt^2} + (k + \alpha F_0)\frac{dz}{dt} + (\lambda + F_0)z = F_0 z_R. \qquad (20.12)$$

This is an equation of precisely the same form as (20.3) but with different coefficients. The use of feedback has transformed the parameters of the differential equation and it is necessary to examine how $z(t)$ changes.

First consider the equilibrium after transients have died away. The terms involving dz/dt and d^2z/dt^2 are then zero and

$$z = \frac{F_0 z_R}{\lambda + F_0}. \qquad (20.13)$$

If $F_0 \gg \lambda$, $z \simeq z_R$ as required. If not, there is an **offset error**: the control system fails to meet its objective. It is easy to see how this arises physically. The force F is proportional to $(z_R - z)$ so disappears when $z_R = z$, and the mass is pulled down by the spring until the error is big enough for F to balance the spring. The solution is clearly to eliminate the spring (if the feedback control is dependable).

Next consider the transients in the approach to equilibrium. With the additional terms in (20.12) due to $F_0 z$ and $\alpha F_0 \, dz/dt$, the transients are given by

$$z = z_R + A_1' e^{s_1' t} + A_2' e^{s_2' t} \qquad (20.6a)$$

where s_1' and s_2' are the two solutions of the new quadratic

$$Ms^2 + (k + \alpha F_0)s + (\lambda + F_0) = 0 \qquad (20.7a)$$

i.e.

$$s_{1,2} = [-(k + \alpha F_0) \pm j[4M(\lambda + F_0) - (k + \alpha F_0)^2]^{1/2}]/2M. \quad (20.8a)$$

Suppose firstly that $\alpha = 0$. The effect of F_0 is to increase the frequency of oscillation, which depends on the square root. This is because of the additional restoring force on the system: the force F acts like an extra spring. For critical damping this is actually a disadvantage, since we require

$$k^2 = 4M(\lambda + F_0)$$

i.e. increased damping to combat the increase due to F_0.

Damping via friction is wasteful: the force F is dissipating energy against friction. Furthermore, there is the familiar problem that the system may jam somewhere. However, if $\alpha \neq 0$, the requirement for critical damping is instead

$$(k + \alpha F_0)^2 = 4M(\lambda + F_0) \qquad (20.14)$$

and if k and λ are small,

$$\alpha \simeq (4M/F_0)^{1/2}. \qquad (20.14a)$$

This is why H was chosen to contain the term $\alpha\, \mathrm{d}(z_R - z)/\mathrm{d}t$. By making the force proportional to velocity, the damping has been improved. This technique is called **velocity feedback**. Electronically, it is achieved by differentiation familiar from Chapter 3. The subtleties possible with feedback are beginning to appear.

20.4   Velocity Error

Thus far we have achieved a control system which is critically damped and settles to $z = z_R$ if the spring constant $\lambda = 0$. This is a good start. Next it is necessary to examine how the system responds if z_R is not constant. We shall find that the system devised so far now lags behind the required value.

Suppose that the target height h_R moves with constant velocity β so that

$$z_R = \beta t. \qquad (20.15)$$

How well does z follow?   When transients have died away, $\mathrm{d}^2 z/\mathrm{d}t^2 = 0$ and z follows z_R with the same velocity β; from (20.12), if $\lambda = 0$

$$(k + \alpha F_0)\beta = F_0(z_R - z)$$

or

$$z_R - z = \beta(\alpha + k/F_0). \qquad (20.16)$$

This situation is shown in figure 20.4; z follows z_R with an error proportional to β, i.e. to the velocity. This is called **velocity error**. It is less serious than zero error, but it does imply that the system cannot keep up with its intended position.

With a little ingenuity in the feedback loop, this error can be eliminated. But the moral is that each differential contributing to the target height $z_R(t)$ introduces some new error and requires further complexity in the feedback system.

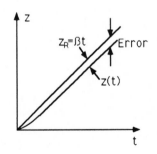

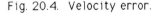

Fig. 20.4. Velocity error.

The three-term controller

The elimination of velocity error is not too difficult. The discrepancy in figure 20.4 between z_R and z settles to a constant value, so if $z_R(t) - z(t)$ is integrated and a component is applied to F

depending on this integral, z will be forced to converge on the required value $z_R(t)$. Suppose $\lambda = 0$ (no spring) and

$$H(z_R - z) = \left(1 + \alpha \frac{d}{dt} + \gamma \int dt\right)(z_R - z). \qquad (20.17)$$

Then from equation (20.11)

$$M\frac{d^2 z}{dt^2} + (k + \alpha F_0)\frac{dz}{dt} = F_0(z_R - z) + \alpha F_0 \frac{dz_R}{dt} + \gamma F_0 \int (z_R - z)dt$$

or

$$M\frac{d^3 z}{dt^3} + (k + \alpha F_0)\frac{d^2 z}{dt^2} = F_0 \frac{d(z_R - z)}{dt} + \alpha F_0 \frac{d^2 z_R}{dt^2} + \gamma F_0(z_R - z).$$
$$(20.18)$$

After transients have died away, $d^3 z/dt^3$ and $d^2 z/dt^2$ are zero; also z moves with the same velocity as z_R, so the first and second terms on the right hand side are zero. Finally, $z_R - z = 0$, as required.

The form of feedback expressed by equation (20.17) is very popular and is known as a **three-term controller**. The term $\alpha\,d/dt$ is derived by differentiation with a CR circuit and the integral by an integrator. Beware, however, offset currents in the integrator, which lead to some velocity error.

20.5   Stability

So far attention has been concentrated on the steady-state solutions after transients have decayed. There must be a nagging doubt that these transients may be unpleasant, and may indeed not decay at all, but grow into an oscillation. For a control system, this would be a disaster. The previous chapter described Nyquist's criterion for the stability of a feedback amplifier. If the algebra can be rearranged to look like the system considered there, the solution can be carried over to the servo-system.

For this it is necessary to return to figure 20.3 and equation (20.11):

$$(Q + F_0 H)z = F_0 H z_R. \qquad (20.19)$$

This is analogous to the equation for negative feedback over an amplifier:

$$v_{out} = G(v_{in} - B v_{out})$$

or

$$v_{out}(1 + GB) = G v_{in}. \qquad (20.20)$$

Here v_{out} is analogous to z and v_{in} to z_R, and from figure 20.3, B is clearly 1, since the whole output signal is fed back. To make equation (20.19) read like (20.20) it is necessary to divide through by Q and set

$$G = F_0 H/Q \qquad (20.21)$$

giving

$$(1+G)z = Gz_R. \qquad (20.22)$$

Then the mathematics of the previous chapter will carry over directly, since in both cases we are dealing with the solution of the same differential equation.

Recapitulating, the Nyquist criterion for stability is that the curve for $G(\omega)$ when plotted on the complex plane (figure 20.5) should not enclose the point -1 inside a clockwise loop. What does this mean for the servosystem? There

$$Hz = \left(1 + \alpha \frac{\mathrm{d}}{\mathrm{d}t} + \gamma \int \mathrm{d}t\right) z \qquad (20.23)$$

$$Qz = \left(M \frac{\mathrm{d}^2}{\mathrm{d}t^2} + k \frac{\mathrm{d}}{\mathrm{d}t} + \gamma\right) z. \qquad (20.24)$$

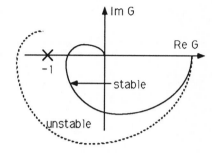

Fig. 20.5. The Nyquist plot again.

So for an oscillatory system where $z \propto \mathrm{e}^{\mathrm{j}\omega t}$,

$$G(\omega) = \frac{F_0(1 + \mathrm{j}\omega\alpha + \gamma/\mathrm{j}\omega)}{-M\omega^2 + \mathrm{j}\omega k + \lambda}. \qquad (20.25)$$

To examine the behaviour of $G(\omega)$ on the complex plane, suppose first that $\alpha = 0$ and $\gamma = 0$ for simplicity. Then

$$G(\omega) \to \frac{F_0}{-M\omega^2 + \mathrm{j}\omega k + \lambda}. \qquad (20.26)$$

The curve for this function is given by the full curve of figure 20.6. It is precisely analogous to figure 19.9(b). Although this curve does not encircle the point -1, it goes close if F_0 is large; the phase margin is then poor, and the system will exhibit large oscillatory transients at the frequency where the curve is closest to the point -1. So large F_0 may be undesirable.

Next suppose $\gamma \neq 0$. Its effect is indicated by the broken curve at the left of figure 20.6. For large ω (near the origin) it has very little effect on the numerator of equation (20.25). The effect of the term $\gamma/\mathrm{j}\omega$ is to add a phase lag to the full curve. As $\omega \to 0$, the term $1/\mathrm{j}\omega$ diverges, so the phase lag becomes 90° and the broken curve approaches the negative imaginary axis asymptotically. As $\omega \to 0$, $|G| \to \infty$, so from equation (20.22):

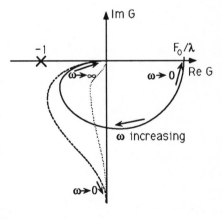

Fig 20.6. $G(\omega)$ for equation (20.25).
Full line, $\alpha = \gamma = 0$; broken line $\alpha = 0$; dotted line, α and γ both non-zero.

$$z \to \frac{G}{1+G} z_R \to z_R.$$

This term eliminates the zero error for steady conditions ($\omega = 0$). This is called **phase lag compensation**. We emphasise that its dominant effect is on the low-frequency response of the servosystem.

Finally, what is the effect of the term $j\omega\alpha$ in the numerator of equation (20.25)? At low frequency it has little effect. As $\omega \to \infty$, its effect is that

$$G(\omega) \to \frac{F_0\alpha}{j\omega M}.$$

It advances the phase of $G(\omega)$ by 90° as $\omega \to \infty$ and makes the broken curve approach 0 asymptotically along the negative imaginary axis. It pulls the curve decisively away from the point -1 unless α is very small, i.e. it restores the phase margin. This corresponds to the fact that it contributes to critical damping in equation (20.14). It is just the **phase lead compensation** of Chapter 19.

Summarising the virtues of the three-term controller, (a) it has no zero error or velocity error, because of the integrating term in the numerator of H, and (b) it is stable by virtue of the velocity feedback $\alpha\,d/dt$ and if α is correctly adjusted it can be critically damped. The effect of further terms in $H(\omega)$ on (a) stability, and (b) the steady-state solution ($\omega \to 0$) can be followed easily using the methods developed above. It is amusing to observe parallels in the economy and taxation. Any form of wealth tax or capital gains tax is analogous to the integrator; a parallel to velocity feedback is the use of high interest rates to damp inflation.

One further general remark is that any form of delay introduces phase lag and hence tends to lead to instability. For further details on this, Laplace transforms, non-linear effects, sampling, the effect of noise, and the design of optimised control systems, you need to refer to specialist texts. This is a large and complex subject, with much ongoing research.

20.6   Phase-locked Loops—Synchronisation

The previous sections have examined how to control the *position* of a system. Another problem which arises commonly is control over its *frequency* and even its *phase*. This is required when two signals are to be synchronised in frequency. An arrangement which performs this feat is called a **phase-locked loop**.

Phase-locked systems are quite common in everyday life. A very familiar example is the Moon, which always keeps one face towards the Earth (figure 20.7). This implies that, as it follows its orbit around the Earth, it spins about its axis with an angular velocity ω_S equal in magnitude and direction with ω_L, its orbital angular velocity. The Earth and Moon did not start that way. The two angular velocities have been brought into synchronism by tides on the Moon, which have dissipated other components of ω_S by friction in the rocks.

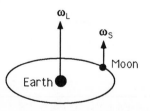

Fig. 20.7. Phase locking of the Moon's rotation with its orbital angular velocity.

Fashion, whether in women's clothes or the business world, is a second example of synchronisation. The flight of birds in a flock is yet another. A juggler is phase locked.

In the electrical world, a familiar example is the tuning of a radio set or TV. As you turn the tuning knob, the receiver locks into synchronism with the incoming signal and remains locked over an appreciable range of the knob. A second example is the synchronisation of the speed of a tape recorder or record player with a local oscillator.

The basic scheme is outlined in figure 20.8. The incoming signal at the left has phase $\omega_1 t + \phi_1$; the output signal at the right has phase $\omega_2 t + \phi_2$. The output is fed back to the input, where some form of phase detector produces an error signal depending on the difference between input and output phases. We shall find that the simplest variety of detector produces beats between the two frequencies; one component is at the difference frequency $(\omega_1 - \omega_2)$ and the other at the sum frequency $(\omega_1 + \omega_2)$. A low-pass filter eliminates this second term. The remaining difference signal drives a voltage-controlled oscillator (VCO) whose frequency ω_2 varies linearly with applied voltage. Over a certain range of frequencies, ω_2 can be brought into synchronism with ω_1. When this happens, the error signal goes to zero or, at worst, to a constant value.

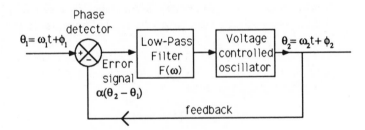

Fig. 20.8. A phase-locked loop.

A common form of phase detector is the four quadrant multiplier, which forms the product of its two inputs. Suppose the two input signals are $V_1 = A_1 \sin(\omega_1 t + \phi_1)$ and $V_2 = A_2 \cos(\omega_2 t + \phi_2)$. The output of the multiplier is the error signal

$$
\begin{aligned}
V_E &= K A_1 A_2 \sin(\omega_1 t + \phi_1) \cos(\omega_2 t + \phi_2) \\
&= \tfrac{1}{2} K A_1 A_2 \{ \sin[(\omega_1 - \omega_2)t + \phi_1 - \phi_2] \\
&\qquad + \sin[(\omega_1 + \omega_2)t + \phi_1 + \phi_2] \} \qquad (20.27) \\
&= A \left[\sin(\theta_1 - \theta_2) + \sin(\theta_1 + \theta_2) \right]
\end{aligned}
$$

where $A = \tfrac{1}{2} K A_1 A_2$ and K is a constant and $\theta = \omega t + \phi$.

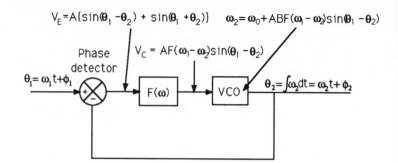

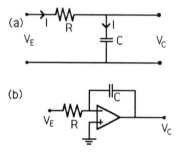

Fig. 20.9. Details of the phase-locked loop.

Figure 20.9 gives the signals at various points of the loop. The filter gives an output depending on the low-frequency component of the error signal V_E, so the control input to the vco is

$$V_C = AF(\omega_1 - \omega_2) \sin(\theta_1 - \theta_2) \qquad (20.28)$$

where $F(\omega)$ is the frequency response of the filter. The oscillator produces an output $A_2 \sin(\omega_2 t + \phi_2)$ at frequency

$$\omega_2 = \omega_0 + BV_E = \omega_0 + ABF(\omega_1 - \omega_2) \sin(\theta_1 - \theta_2) \qquad (20.29)$$

where B is a constant; ω_0 is an adjustable frequency corresponding to zero input voltage to the vco. Normally ω_0 will be equal to the input frequency ω_1 or close to it.

Broadly speaking, there are two classes of filter. One is the simple passive filter of figure 20.10(a) and the other is the integrator of figure 20.10(b). For these two

$$F_a(\omega) = 1/(1 + j\omega\tau) \qquad (20.30)$$

$$F_b(\omega) = 1/j\omega\tau \qquad (20.30a)$$

where $\tau = CR$. We shall follow the behaviour of the former and mention the differences produced by (b).

The approach will be to find a differential equation for $V_C(t)$, the voltage applied to the vco, and examine its properties. To do this, it is first necessary to relate the frequency response of the filter $F_a(\omega)$ to a differential equation giving the time dependence of the signals through the filter. For output V_C from the first type of filter, the current I is given by

$$V_C = (1/C) \int I \, dt$$

$$I = C \frac{dV_C}{dt}$$

$$V_E = IR + V_C = CR \frac{dV_C}{dt} + V_C = A \sin(\theta_1 - \theta_2) \qquad (20.31)$$

Fig. 20.10. Two low-pass filters.

if the high-frequency term in equation (20.27) is ignored. In order to relate θ_2 to

$$\omega_2 = d\theta_2/dt = \omega_0 + BV_C \tag{20.32}$$

we differentiate (20.31) with respect to t, with the result

$$CR\frac{d^2V_C}{dt^2} + \frac{dV_C}{dt} = A\cos(\theta_1 - \theta_2)\frac{d(\theta_1 - \theta_2)}{dt}$$
$$= A(\omega_1 - \omega_0 - BV_C)\cos(\theta_1 - \theta_2)$$

$$CR\frac{d^2V_C}{dt^2} + \frac{dV_C}{dt} + AB\cos(\theta_1 - \theta_2)V_C = A(\omega_1 - \omega_0)\cos(\theta_1 - \theta_2). \tag{20.33}$$

For a given input signal $\theta_1(t)$, this equation and (20.32) make a pair of differential equations for V_C coupled through θ_2. They cannot be solved analytically, so it is necessary to study numerically the dynamics by which a signal becomes phase locked, i.e. V_C settles to a steady value. Here the results are summarised with comments on the static situation after a signal has become locked.

Steady state
First consider the steady state with the loop locked and input and output frequencies the same, $\omega_1 = \omega_2$. The value of $\theta_1 - \theta_2$ is constant and V_C settles at

$$V_C = (\omega_1 - \omega_0)/B. \tag{20.34}$$

This voltage is what is required to bias the VCO to provide an output frequency ω_2 synchronised with ω_1. Notice that this result is independent of A, and hence independent of the strength of the input signal. So if this strength fluctuates in magnitude or even disappears momentarily, V_C remains steady. This is important in locking to weak signals buried in a high noise level, one of the very important uses of the phase-locked loop. The oscillator is capable of 'freewheeling' for a time at the correct frequency. Noise is usually uniformly distributed over all frequencies and on average does not affect the VCO. However, if the input signal disappears for too long, random fluctuations in the noise disturb V_C and the loop drifts out of lock.

Next, what is the value of the discrepancy $\theta_1 - \theta_2$ between input and output? In the steady state, $dV_E/dt = 0$ and $dV_C/dt = 0$ so from equation (20.31) $V_E = V_C$; then from (20.34) and (20.31)

$$BA\sin(\theta_1 - \theta_2) = \omega_1 - \omega_0. \tag{20.35}$$

If the input frequency ω_1 happens to be equal to ω_0, the unbiased frequency of the VCO, then $\theta_1 = \theta_2$ and the input signal $A_1\sin\omega_1 t$ and the output $A_2\cos\omega_2 t$ are locked 90° apart in phase. If ω_1

and ω_0 differ, $\theta_1 \neq \theta_2$. The only major difference between the two filters of figure 20.10 is that for (a) $F(\omega) \to 1$ as $\omega \to 0$ while for (b) $F(\omega) \to \infty$ as $\omega \to 0$. This implies that for the latter filter the feedback becomes very strong, equation (20.30a), when the loop is locked and $\omega_1 - \omega_2 \to 0$; it guarantees that $\theta_1 = \theta_2$, regardless of $\omega_1 - \omega_0$, when the loop is locked. This is like phase lag compensation, figure 20.6, for a servosystem.

Let us return to equation (20.33) and consider the approach of V_C to its equilibrium value. For simplicity, suppose that $\theta_1 - \theta_2$ is small, so that the cosine terms are 1. It is desirable that the lock should operate as quickly as possible. This implies that the oscillations of V_C should be critically damped. This is achieved by adjusting the product $CR = \tau$ according to the available values of A and B so that

$$1 - 4AB\tau = 0. \tag{20.36}$$

However, this condition depends on the magnitude of A, and hence on the strength of the input signal; it is not possible to maintain ideal conditions if this strength varies. From equation (20.33), the natural frequency of oscillations of V_C is $\omega_n = (AB/\tau)^{1/2}$, and if the loop is critically damped equation (20.36) gives $\omega_n = 2AB$. Then V_C locks to its final value in a time roughly $1/\omega_n$. Large loop gain AB is desirable for rapid locking.

Hold range

Suppose next that the loop is locked and the input frequency ω_1 is changed very slowly. The VCO will follow over the range of frequencies for which equation (20.35) can be satisfied; $\sin(\theta_1 - \theta_2)$ cannot exceed ± 1, so the frequency range is in principle

$$\omega_1 - \omega_0 = \pm BA. \tag{20.37}$$

This is called the **hold range**, $\Delta\omega_H$. (In practice, if A and B are large, the available range may be limited instead by the operating range of the VCO.) If ω_1 is adjusted beyond the hold range, the output drops out of synchronism with the input. Note that the hold range is proportional to A and decreases for weak input signals.

If instead the input θ_1 changes in jerks, there are transients in V_C which may drive the system instantaneously beyond the hold range; hence it is possible that the input and output will fall out of synchronism when $(\omega_1 - \omega_0)$ is less than the hold range. To study this, the time dependence of V_C and ω_2, hence the time dependence of the error signal, have to be studied numerically. Results depend on how rapidly θ_1 changes, but if it does so instantaneously (the worst case) it is possible to compute a value $\Delta\omega_{PO}$ beyond which the circuit loses synchronism. This is called the **pull-out range**. It is of course less than the hold range.

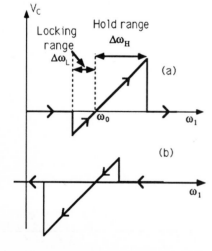

Fig. 20.11. The behaviour of V_C as the input signal is swept in frequency (a) upwards through ω_0 (b) downwards.

Next consider the process by which the system becomes locked in the first instance. Suppose that the input and output are out of synchronism and ω_1 is slowly swept through ω_0. Over a certain range of frequencies called the locking range, $\Delta\omega_L$, the loop will lock within one cycle of ω_n, i.e. very quickly. If ω_1 is swept slowly through ω_0 from below, the sequence of events is shown in figure 20.11(a). Within a narrow locking range, the input signal is 'captured' and ω_2 then tracks it over the wider hold range, $\Delta\omega_H$. If ω_1 is swept downwards in frequency, the sequence of events is as shown in figure 20.11(b).

It is difficult to find a precise formula for the locking range, but we will outline what the factors are. The value of the output frequency ω_2 at any instant is given by equation (20.29):

$$\omega_2 = \omega_0 + ABF(\omega_1 - \omega_2)\sin(\theta_1 - \theta_2).$$

If the signals are unlocked, ω_2 sweeps rapidly back and forth with $\sin(\theta_1 - \theta_2)$ over the range

$$\omega_2 = \omega_0 \pm ABF(\omega_1 - \omega_2)$$

as shown in figure 20.12(a). If this fails to reach the input frequency ω_1, there is no possibility of locking. When ω_1 and ω_2 are well apart, $F(\omega_1 - \omega_2)$ is small, so ω_2 varies over a rather narrow range. As ω_1 and ω_2 come closer together at the top of the cycle in figure 20.12, $F(\omega_1 - \omega_2)$ increases, and this provides positive feedback driving ω_2 still closer to ω_1. As this happens, ω_2 and ω_1 spend longer and longer in close proximity. The upshot is that one can define (i) a **locking range** $\Delta\omega_L$ where synchronisation occurs very rapidly, within the critical damping time of V_C, and (ii) a wider **pull-in range** $\Delta\omega_{PI}$ over which the synchronisation occurs gradually over many cycles of figure 20.12. The details depend on the damping within the filter circuit, and are quite different for different filters. Qualitatively, results are as shown in figure 20.13.

The vco does not have to be a sine-wave generator; indeed, it usually is not. If it is a square-wave oscillator, its output can be Fourier analysed into fundamental and harmonics. The algebra given above applies with the appropriate Fourier coefficient giving the output amplitude A_2. It is possible, indeed easy, to lock to harmonics or subharmonics of the input signal. This may be used for multiplying or dividing frequencies. It is used, for example, to generate equally spaced carrier frequencies from a single master oscillator; these can provide the carriers for frequency division multiplexing, discussed in Chapter 15.

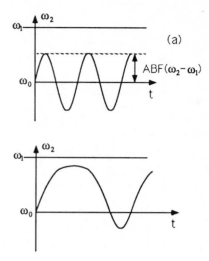

Fig. 20.12. $\omega_2(t)$ (a) when unlocked, (b) approaching lock.

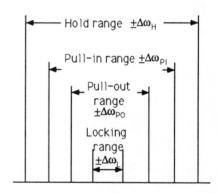

Fig. 20.13. Ranges of different locking processes.

Digital systems

If both input and output signals are digital, an exclusive OR gate may be used as phase detector (figure 20.14(a)). Figures 20.14(b)–(d) display output signals for three phase relations between the

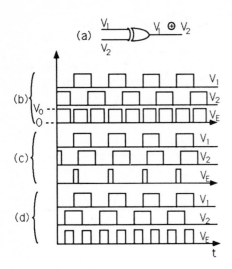

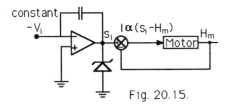

Fig. 20.14 (a) The XOR gate as phase detector, (b) – (d) waveforms for three phase differences $\theta_1 - \theta_2$.

two inputs to the detector. If these outputs are fed to the low-pass filter, the average voltage level controls the VCO. It varies over the range 0 to V_0, the height of the output pulses from the XOR gate. A minor annoyance is that when the signals are synchronised, as in figure 20.14(d), the mean output level is $\frac{1}{2}V_0$ and has to be shifted down to zero. Other more complex digital systems eliminate this problem, and are also capable of providing unlimited pull-in range: whatever the difference in input and output frequencies $\omega_1 - \omega_2$, the VCO moves into synchronism, provided this is within its operating range. For details of such systems, you need to consult specialised texts (see, for example, Roland E Best, 1984, *Phase-Locked Loops* (New York: McGraw-Hill)).

Ways of making the voltage-controlled oscillator will be discussed in the next chapter. The VCO itself may be used to create a frequency modulated signal: the modulating signal is the voltage applied to the oscillator, whose output frequency responds to the modulation. Conversely, a phase-locked loop acts as an FM demodulator: the signal V_C at the output of the low-pass filter is the low-frequency error signal, and hence comprises the demodulated signal.

Many brands of VCO and phase-locked loop are available commercially as integrated circuits. One can find them for any frequency range from 1 Hz to 1 GHz, though no one device covers such a large frequency range.

20.7   Exercises

1.   (QMC 1967) A large electromagnet has a non-linear relation between field and driving current. You are given (a) a probe which produces an output voltage which is linear with the applied magnetic field, and (b) a DC current supply whose large output current is proportional to its input voltage. Describe a system capable of producing a slow linear increase of field with time (from zero up to a maximum field). Point out the features essential to the success of your system. (Ans: figure 20.15.)

2.   (QEC 1974) If the open loop transfer function of a positional control servo system is $f_1(p) = K_v/p(1 + p\tau_m)$, where $p = d/dt$, K_v is the velocity error constant and τ_m is the motor time constant, derive expressions for the damping factor γ and the natural frequency ω_0 of the closed loop system in terms of K_v and τ_m.

If velocity feedback is employed giving a feedback transfer function $(1 + K'p)$, obtain new expressions for γ and ω_0. For $\tau_m = 0.25$ s determine K_v for the 'best' step response when no velocity feedback is present. Calculate the value

Fig. 20.15.

of K' required to halve the risetime but still maintain the 'best' step response. (Ans: $\omega_0^2 = K_v/\tau_m$, $\gamma = 1/2\tau_m$; $\gamma = (1 + K_v K')/2\tau_m$; $K_v = 1$ s^{-1}; $K' = 0.25$ s.)

3.   (IC 1988) Determine the transfer function of the control system shown in figure 20.16. For what values of K is this system stable? (Ans: $T(s) = K/(s^2 + 4s + K)$; $K > 0$.)

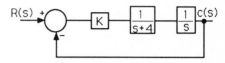

Fig. 20.16.

4*.   (Bedford 1974) In a certain torque amplifier, the angular movements of the output shaft are caused to follow those of the input shaft with the aid of a motor whose torque is controlled by the difference in displacements between the two shafts. A voltage V, numerically equal to this difference, is fed through an integrating network consisting of a large resistor R_1 in series with a capacitor C and a small resistance R_2; the voltage V is applied across the complete network and the output to the motor is taken from across the capacitor C and resistance R_2. The motor exerts a torque μ times the voltage applied to it, the motor and load together have a moment of inertia J, and there is a frictional resistance B times the angular velocity of the load. Show analytically that the system is stable if $J < BR_2C$ and determine the frequency of oscillation of the marginally stable system. (Ans: $1/(2\pi CR_2)$.)

21

Digital Circuits

21.1 The Schmitt Trigger

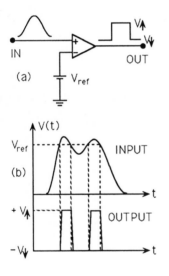

Fig. 21.1 (a) A comparator,
(b) input and output waveforms.

At the input to a digital system, it is frequently necessary to trigger from a slowly rising or falling input pulse. The requirement is to generate a standard square pulse when the input rises above some well defined threshold level. The crudest way of achieving this is to use an operational amplifier as a **comparator**, as in figure 21.1. In the absence of an input, the reference voltage drives the output to negative saturation $-V_\downarrow$. When the input signal goes above V_{ref}, the output swings to positive saturation, $+V_\uparrow$.

The problem with this simple arrangement is that, if the input dithers around V_{ref} (for example because of ringing), the output may switch back and forth several times (figure 21.1(b)). The Schmitt trigger circuit improves on this by triggering when the input exceeds a threshold level, but resetting only when the input falls below a distinctly lower level, as in figure 21.2(a). The relation between V_{in} and V_{out} now follows the hysteresis loop shown in figure 21.2(b). The circuit symbol for the Schmitt trigger circuit adopts this loop as a logo (figure 21.2(c)).

Trigger voltages.
Like many digital circuits, the Schmitt trigger circuit uses positive feedback to achieve decisive switching. One form of the circuit is shown in figure 21.3. To analyse it, let us begin by working out the voltage at the positive input Y when the amplifier is saturated positive or negative. In the latter case,

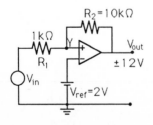

Fig. 21.3. The Schmitt
trigger circuit.

$$V_Y = -V_\downarrow + \frac{R_2}{R_1 + R_2}(V_{\mathrm{in}} + V_\downarrow)$$

$$= \frac{R_2 V_{\mathrm{in}} - R_1 V_\downarrow}{R_1 + R_2}. \tag{21.1}$$

If the output is to be stable and negative, $V_Y < V_{\mathrm{ref}}$. For the values in the figure this requires $V_{\mathrm{in}} < 3.4$ V.

If a positive input larger than this is applied, the amplifier will switch to positive saturation. The positive feedback applied to Y through R_2 reinforces the switching action. The value of V_Y increases to

$$V_Y' = \frac{R_2 V_{\text{in}} + R_1 V_\uparrow}{R_1 + R_2} \tag{21.2}$$

This remains above V_{ref} providing $V_{\text{in}} \geq 1.0$ V. The amplifier will not switch back to its original state until V_{in} falls below this value. When this happens, V_Y falls back to the value given by equation (21.1) and positive feedback again reinforces the switching action. The upper and lower trigger levels may be adjusted by varying any two of V_{ref}, R_1 and R_2. If V_{ref} is negative, the trigger circuit responds to negative input pulses.

This circuit is intended to deal with slowly changing inputs. If the input is a fast pulse, the output of the amplifier cannot change faster than the slew rate, and may not reach $V_\uparrow$ before the input pulse disappears. For this reason, operational amplifiers are mostly unsuitable for high-speed circuits (faster than a few μs). Circuits using individual transistors may be designed using the same principles and are illustrated in the exercises at the end of the chapter. It is necessary to work through the DC levels in the circuit to follow the switching action.

Speed-up capacitors

A second snag is that there is always some capacitance C_{in} between the feedback terminals of the operational amplifier (shown dotted in figure 21.4). The necessity to charge this capacitance slows down the risetime of the input to the amplifier. This may be remedied by compensation, as discussed in Chapter 8. It is done using the **speed-up capacitor** C_1. The feedback network is then as shown in figure 21.5. With exact compensation, the RC networks form a potential divider between V_{in} and earth, via Y. However, it is often desirable to make C_1 somewhat larger, so as to put the full change of V_{in} across C_{in} and charge it at the maximum rate. This is called **overdrive** and will be discussed later in this chapter. However, $C_1 R_1$ must not be so large that the charging time of C_1 is greater than the interval between pulses.

21.2   The Monostable

The duration of the output pulse from the Schmitt trigger circuit is governed by the length of the input pulse. For many purposes it is convenient to devise a circuit which gives an output pulse with a preset duration independent of the input. Such a circuit is called a **monostable** or **univibrator**.

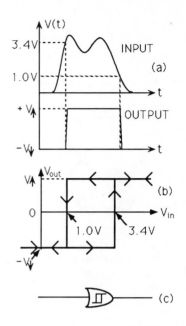

Fig. 21.2 (a) Input and output waveforms of the Schmitt trigger circuit, (b) the relation between V_{in} and V_{out}, (c) the circuit symbol.

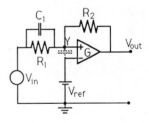

Fig. 21.4. Use of a speed-up capacitor.

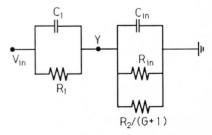

Fig 21.5. Feedback network.

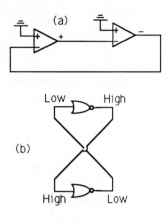

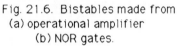

Fig. 21.6. Bistables made from
(a) operational amplifier
(b) NOR gates.

In order to introduce this circuit, consider first the idealised arrangement shown in figure 21.6. This is a **bistable**. It has two stable states. Suppose at some instant the output of the first amplifier is positive. The output of the second is negative. This is fed back to the first amplifier, providing positive feedback to enhance the original state, which is therefore stable. However, the two amplifiers are symmetrical and there is clearly an alternative stable state with the output of amplifier 1 negative and that of amplifier 2 positive. The circuit may be switched from one state to the other by suitable set or reset pulses applied to the positive terminals of the amplifiers. An alternative arrangement based on NOR gates is shown in figure 21.6(b). This is the set–reset flip-flop familiar from Chapter 13.

To turn the circuit into a monostable requires that one of the two states should be unstable and the second stable. This is achieved with the circuit of figure 21.7. (A very similar circuit can be constructed from NOR gates or NAND gates.) The bias voltage $-V$ keeps the output of amplifier 2 normally positive and that of amplifier 1 normally negative. However, a large enough positive input pulse to the non-inverting input of amplifier 1 drives its output positive. The output is transmitted through C, which cannot change its voltage instantly, and amplifier 2 changes state to a negative output; its output feeds back to amplifier 1, enhancing the switching action there.

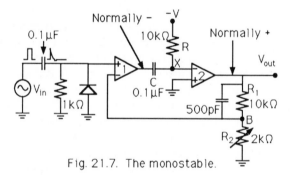

Fig. 21.7. The monostable.

Fig. 21.8. Charging of C.

Pulse length

Clearly this creates an unstable state. What then happens is that C charges through R and when the voltage at X passes zero, amplifier 2 switches back to its normal stable state and takes amplifier 1 with it. In order to find the length of the output pulse, it is necessary to study the charging of C (figure 21.8).

Suppose the output of amplifier 1 switches from $-V_\downarrow$ to $+V_\uparrow$. When the trigger pulse arrives and amplifier 1 switches, the capacitor drives X positive by an amount $V_\uparrow + V_\downarrow$. Then the subsequent voltage V_X is given by

$$V_X = -V + (V_\uparrow + V_\downarrow)e^{-t/CR}.$$

V_X reaches zero when

$$V = (V_\uparrow + V_\downarrow)e^{-t_0/CR}$$

$$t_0 = CR\ln\left(\frac{V_\uparrow + V_\downarrow}{V}\right). \tag{21.3}$$

The sequence of waveforms is shown in figure 21.9. The length of the output pulse may be varied via V, R or C.

When the first amplifier switches back to its quiescent state, the capacitor C has to charge again to its normal level. Until this is complete, the circuit is not fully primed to receive another trigger pulse, though it will respond to a second trigger. However, because the capacitor is abnormally charged, the length of the output pulse will be altered. The time required for the capacitor to recharge is called its **settling time**. Various refinements are possible to return the capacitor to its steady state more quickly, for example, by putting a diode across C to allow rapid charging in one direction.

Suppose the output pulse is differentiated by putting it through a CR circuit. A diode may then be used as in figure 21.10 to transmit the negative spike at the end of the square wave. This can be used in a subsequent circuit to generate a controlled **time delay**.

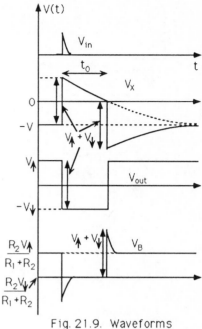

Fig. 21.9. Waveforms in the monostable.

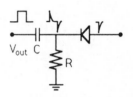

Fig. 21.10. Generating a controlled time delay.

21.3   The Astable

The astable circuit is one with no stable state. It flips back and forth continuously between two states. The circuit is shown schematically in figure 21.11(a) and the output waveforms are shown in figure 21.11(b). The ratio τ_1/τ_2 is called the **mark-to-space ratio** for obvious reasons. Unless diodes are used to recharge capacitors C_1 and C_2 quickly in one direction, there is no simple expression for τ_1 and τ_2, but they can be adjusted experimentally by varying C_1 or R_1 and C_2 or R_2. The astable or **multivibrator** is a convenient pulse generator.

21.4   Voltage-controlled Oscillators

In the previous chapter, a voltage-controlled oscillator (vco) was a vital element in a phase-locked loop. A crude way of making a vco is to use the voltage level to control the trigger level of an astable, as shown in figure 21.12, using equation (21.3). However, because

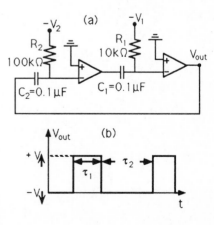

Fig. 21.11 (a) The astable, (b) its waveform.

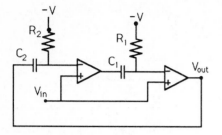

Fig. 21.12. Voltage-controlled astable.

the capacitors C_1 and C_2 do not charge linearly with time, the frequency will not be quite linear with V_{in}. A constant charging current is needed.

Figure 21.13(a) outlines one method of achieving this. The circuit to the left of the point X mixes V_{in} with a reference voltage guaranteeing that V_X is positive:

$$V_X = \alpha V_{ref} + \beta V_{in}.$$

α and β can be adjusted by means of resistors R_1 to R_4. It is V_X which controls the frequency of the oscillator. It controls the charging of an integrator which feeds a Schmitt trigger circuit. The integrator charges one way until point Y passes the threshold of the Schmitt trigger circuit. Then it discharges until point Y passes the reset level of the Schmitt trigger. The result is a triangular waveform at Y of figure 21.13(b) and a square wave at the output. The subtlety lies in arranging that the capacitor charges and discharges at the same rate, making the output mark-to-space ratio 1:1. This is achieved using the network of resistors labelled R and $2R$.

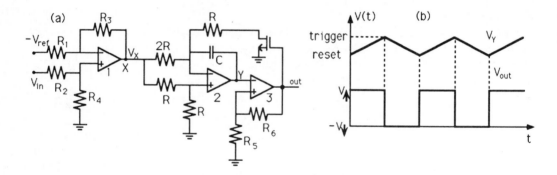

Fig. 21.13 (a) Voltage-controlled oscillator, (b) waveforms.

The voltage at the positive input terminal of amplifier 2 is $\frac{1}{2}V_X$, by the potentiometer action of the two resistors R. The voltage at the negative terminal is likewise $\frac{1}{2}V_X$. Suppose V_{out} is negative, so that the FET is off. The charging current through resistor $2R$ is $\frac{1}{2}V_X/2R = V_X/4R$. Point Y moves negative until the Schmitt trigger resets positive, switching on the FET. Thereafter, there is a current $\frac{1}{2}V_X/R = V_X/2R$ through the FET and away from the negative terminal of amplifier 2. The net current through the capacitor is $-V_X/4R$, and point Y moves positive until the Schmitt trigger fires and the cycle resets.

Sophisticated VCOs are readily available commercially as integrated circuits, with a linearity down to 0.002% and frequencies up to 200 MHz.

21.5 The 555 Timer

A very popular integrated circuit is the 555, which will function as either monostable or astable. Its internal logic is shown in figure 21.14(a). It consists of a set–reset flip-flop which is set when the trigger input falls below $V_{CC}/3$ and is reset when the terminal labelled 'threshold' rises above $2V_{CC}/3$. These voltage levels may be varied if required by means of a control voltage at terminal 5.

Monostable

Its use as a monostable is shown in figure 21.14(b). When it is quiescent, the input must be high, the output is low and $\overline{Q}$ is high; the latter turns on transistor T1 and keeps the capacitor discharged, so that the input to pin 6 is also low. There are then negative signals applied to both S and R. If an input pulse drives the input below $V_{CC}/3$, comparator 1 generates a set signal and $\overline{Q}$ goes low, switching off T1. The capacitor then charges through resistor C towards V_{CC}. When it reaches $2V_{CC}/3$, comparator 2 generates a reset level, resetting Q to 0 and $\overline{Q}$ to 1, hence discharging C again. The waveforms are shown in figure 21.14(c). The length of the output pulse is $CR\ln 3$. The input pulse must be narrower than the output or the waveforms will repeat. The flip-flop may be reset at any time by a low signal at terminal 4.

Astable

For operation as an astable, resistors R_1 and R_2 are connected as shown in figure 21.15. Let us follow the operation from the instant when Q goes high, hence $\overline{Q} = 0$ and T1 is off. Because the flip-flop has just been set, pins 2 and 6 are at voltage $V_{CC}/3$. From this point, C charges towards V_{CC} through R_1 and R_2 with time constant $C(R_1 + R_2)$. When it reaches $2V_{CC}/3$, comparator 2 resets the flip-flop, $\overline{Q}$ goes high and T1 switches on, pulling pin 7 down close to zero. Then C discharges through R_2 until it reaches voltage $V_{CC}/3$ again, and the cycle restarts.

This timer can be used for pulse lengths up to about 1 s. For longer intervals, it may be used together with a counter which extends the time interval using external logic. More elaborate chips such as the 74HC4060 have this counter built in and can be used to generate pulses with a duration of minutes.

21.6 Separation of Signal from Noise

In many experimental situations, there are noise pulses present as well as signal. If the experiment is well designed, noise is smaller than signal, as in figure 21.16(a). The genuine pulses are of fairly

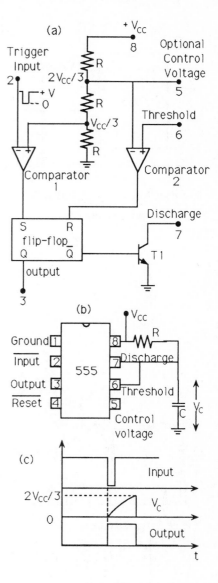

Fig. 21.14 (a) Logic of the 555 timer, (b) connections for monostable operation, (c) waveforms.

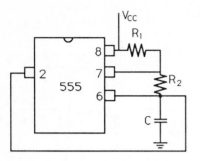

Fig. 21.15. Circuit for astable operation.

uniform pulse height. The smaller ones are noise arising from thermal generation of small signals in the detector or from stray pickup.

Figure 21.16(b) shows the corresponding pulse height spectrum. By setting the trigger level at the broken line, all the genuine signals are kept and the noise is eliminated. Such a circuit is a version of the monostable called a **discriminator**.

Efficiency

If the discriminator threshold is set too high in order to reject noise, some of the genuine pulses may be lost and the system becomes inefficient. Efficiency ϵ may be defined as the ratio

$$\epsilon = \frac{\text{number of monostable pulses}}{\text{number of true signals}}.$$

Sometimes two true signals arrive so close in time that the monostable has not fully recovered from an earlier pulse. This also gives rise to inefficiency and to minimise it the pulse length should be kept down to the minimum required for subsequent operations, for example, driving a scaler successfully. The time during which the monostable is unresponsive is known as its **deadtime**. Notice that if the threshold is set so low that noise is counted, genuine signals may be lost because the monostable has just fired on a noise pulse. This is known as **pile-up** of pulses at the discriminator.

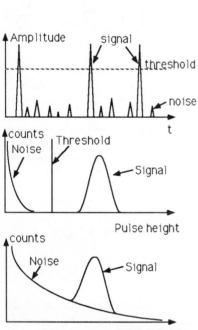

Fig. 21.16 (a) Signal and noise pulses and their spectrum when noise is (b) <<signal, (c) comparable.

Plateau curve

In order to find the right setting of the threshold, counting rate is measured against threshold setting. The result will be schematically as in figure 21.17. This is the integral of figure 21.16(b). Ideally, it shows a flat region or **plateau** where all true signals are accepted but no noise.

Suppose the noise and signal are not cleanly separated in a single detector, as in figure 21.16(c). What is to be done? If the same signal can be viewed with a second detector, there is the possibility of arranging **coincidences** between the two detectors. Such an arrangement may be possible, for example in a nuclear physics experiment (figure 21.18(a)), where two γ-rays from a single process (e.g. $e^+e^- \rightarrow \gamma\gamma$) are detected in two separate detectors. One may even arrange to view the same source with two independent detectors, as in figure 21.18(b).

In this circumstance, the discriminator of each detector may be set low enough to detect all or almost all of the signal, at the expense of counting some noise pulses. Then the signal separation is improved by demanding that the signal is simultaneously present

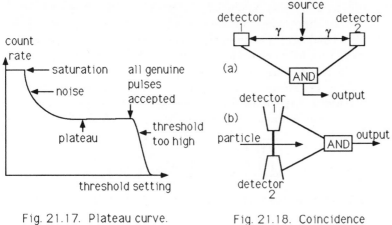

Fig. 21.17. Plateau curve.

Fig. 21.18. Coincidence arrangements.

in both detectors, using an AND gate (Chapter 11) or coincidence circuit, as it is often called in this context.

Time walk
In a real detector, there is always some **time jitter**, which arises as shown in figure 21.19. Suppose one detector gets a big pulse and the other one a small one. If both detectors have the same threshold, the big pulse gives an earlier output than the small one. The difference is called time walk.

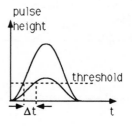

Fig. 21.19. Time walk.

Resolution time
In order to aid separation of noise pulses from signals, it is desirable to keep pulse lengths short, but still long enough that time walk is insufficient to destroy the coincidence. The total length of time over which two pulses will give a coincidence output is known as the **coincidence resolution time**, τ. This time can be measured by deliberately changing the timing of one detector A with respect to the other B and plotting the rate of coincidences $A.B$. This is called a **delay curve** (figure 21.20). Eventually the timing is set in the middle of the curve, where $t_2 - t_1 = 0$.

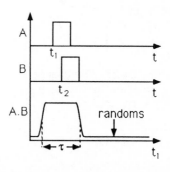

Randoms
Off the correct timing, the $A.B$ rate does not drop to zero. This is because of **random coincidences** between a true signal in A and a noise pulse in B or vice versa (or even between two noise pulses). It is instructive to evaluate the signal/noise ratio in these coincidences. Suppose A carries S signals per second and N_1 noise pulses, while B carries S signals and N_2 noise pulses. Then $A.B$

Fig. 21.20. A delay curve between pulses A and B.

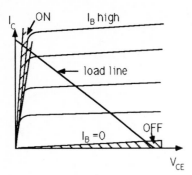

Fig. 21.21. Quiescent states, shown cross–hatched, before and after switching.

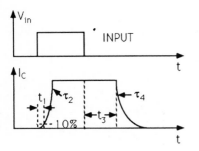

Fig. 21.22. Response of I_C to a square voltage pulse at the base.

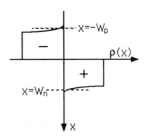

Fig. 21.23. The charge distribution across a pn junction.

will count all true S coincidences (providing the detectors are efficient). How many random coincidences are there? For any one true signal on A, the probability that there is a noise pulse present on B within the coincidence resolution time is τN_2. Since there are S true signals/s on A, the rate of these random coincidences is $S\tau N_2$ per second. Likewise, the rate of random coincidences between a true signal on B and noise on A is $S\tau N_1$ per second, and the number of random noise coincidences is $\tau N_1 N_2$. Finally,

$$\boxed{\frac{\text{signal}}{\text{noise}} = \frac{S}{\tau[N_1 N_2 + S(N_1 + N_2)]}}. \tag{21.4}$$

It is obviously desirable to minimise τ, N_1 and N_2.

21.7*   Switching Speed

In digital circuits a bipolar transistor switches between the cross-hatched regions of figure 21.21. In both regions, the power developed in the transistor itself is small: in the off state $I_C \simeq 0$, while in the on state $V_{CE} \leq 0.4$ V. However, it does not switch on or off instantly, for a number of reasons. If an ideal square pulse is applied at the base, the response of the collector current I_C is schematically as shown in figure 21.22. To understand this diagram, it is necessary to consider storage of charge within the transistor.

Time delay

Figure 21.23 recalls the charge distribution across the base–emitter junction. This charge distribution acts just like a capacitor C_{jE} between base and emitter. There is a similar charge distribution at the base–collector junction acting as C_{jC}. As the transistor moves from reverse bias to forward bias, the depletion layers contract and the mobile carriers (i.e. stored charge) increase outside the depletion layer; formulae are given in appendix C. The initial rise t_1 on figure 21.22 is given by (i) the time constant for changing the charge stored in C_{jE} and C_{jC}, and (ii) the time required for drift of charge across the base region to the collector. Normally the latter is a small effect, but it becomes important in extremely high frequency devices.

Risetime

The subsequent rise of I_C is governed by the bandwidth of the transistor. Suppose its gain is

$$G = \frac{G_{\text{mid}}}{1 + j\omega/\omega_0}$$

where G_{mid} is the mid-band gain and ω_0 the bandwidth (Chapter 8). The output voltage begins to rise along a trajectory shown in figure 21.24 with time constant $1/\omega_0$ towards a current $h_{fe}I_B$. This limit is, however, far beyond the saturation value. This disparity is called **overdrive**. The result is that τ_2 in figure 21.22 is of order $0.01/\omega_0$ and therefore of order 1 ns.

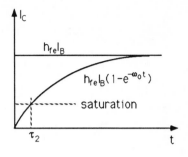
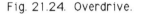

Fig. 21.24. Overdrive.

Storage time

While overdrive is advantageous in switching the transistor on, it works the opposite way during switching off. While the transistor is saturated, the emitter is injecting carriers into the base region and a large charge is stored there. As discussed in appendix C, the stored charge is very large for forward bias and is proportional to I_C. The time required to sweep this charge away, t_3, can be tens or even hundreds of ns. During this time, a reverse current flows to the base as the charge is removed.

Falltime

In the final step of figure 21.22, V_{in} falls to zero or a negative value. Again there are two effects. The obvious one is that I_C decays with a time constant τ_4 governed by the series resistance R_L and capacitance C_{BC} (plus any capacitative load). Secondly, because V_C is rising, the capacitance C_{BC} may induce a voltage at the base sufficient to keep it above 0.6 V and hence keep base current flowing. If so, τ_4 is increased.

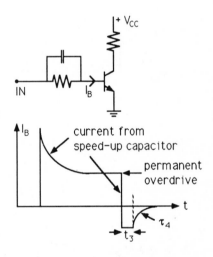

Fig. 21.25. Temporary overdrive due to speed-up capacitors.

Remedies

There are several remedies. Firstly, speed-up capacitors discussed earlier provide **temporary** overdrive combatting t_1, t_3 and the second component of τ_4, as shown in figure 21.25. The capacitor acts as a source of charge. This remedy carries two penalties: (a) the capacitor has a settling time, and (b) the capacitor makes the circuit act like the differentiator of Chapter 8 and makes it more sensitive to noise.

Secondly, it is desirable to prevent the collector current from reaching saturation. This can be done, as in figure 21.26, by means of a special diode which conducts when the voltage across it exceeds 0.25 V. This diverts base current to the collector. It eliminates t_3, which is often the most serious limitation to switching speed. An ordinary diode will not do, since it stores charge when forward biased just like the base region of the transistor. The Schottky diode is made from a junction between aluminium and n-type material. The metal behaves like a p-type material and when the junction is forward biased electrons flow from the n-type material into the aluminium, where they are majority carriers. Because of

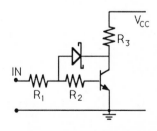

Fig. 21.26. Clamping the collector.

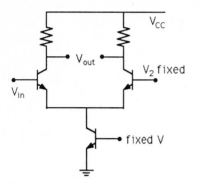

Fig. 21.27. Current steering logic.

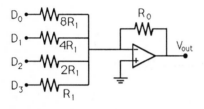

Fig. 21.28. Digital to analogue conversion.

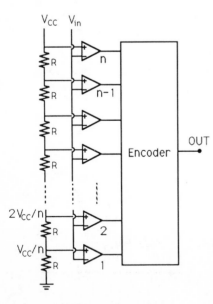

Fig. 21.29. A flash ADC.

the high conductivity of the metal, charge storage is eliminated and the switching time is very small (ns). The circuit symbol for the Schottky diode is given in figure 21.26. Schottky diodes are used in TTL gates to prevent saturation.

The third technique is to prevent the transistor from saturating by using **current steering logic** (figure 21.27). A difference amplifier is fed with constant current insufficient to saturate the transistors; current is steered between them. This eliminates t_3. This scheme is the fastest and is used in ECL gates.

21.8   Digital to Analogue and Analogue to Digital Conversion (DAC and ADC)

With the ubiquitous use of computers today, digitisation of analogue signals (voltage or current) is an everyday problem. Likewise, control of equipment by a computer requires the reverse conversion of a digital code to analogue form. A wide variety of converters is available commercially. Here the principles will be outlined briefly. Selection of a particular DAC or ADC is a question of cost, accuracy, speed and the number of bits (dynamic range).

DAC

A straightforward way of converting from binary to a DC voltage is to use an analogue adder (Chapter 7), as in figure 21.28. Suppose a four-digit binary number is held so that a one is represented by 5 V and a zero by 0 V. If the most significant digit (denoted D_3 in the figure) is fed to the operational amplifier through resistance R_1, it produces an input current $D_3 \times 5/R_1$, where $D_3 = 0$ or 1. Likewise the other three digits contribute input currents depending on D_2, D_1 and D_0 and the output voltage is

$$V_{out} = -5(R_0/R_1)\left(D_3 + \frac{D_2}{2} + \frac{D_1}{4} + \frac{D_0}{8}\right).$$

The circuit of figure 21.28 has the disadvantage of requiring several precision resistors over a wide range of resistance. An alternative resistor network achieving the same result with just two resistor values R and $2R$ is discussed in exercise 9.

Flash ADC

There is a wide variety of techniques for the converse process of analogue to digital conversion. Four examples are described here. The fastest is the flash ADC shown in figure 21.29. The chain of n resistors provides n reference voltages to n comparators. If the input voltage is smaller than the reference voltage, the comparator

gives a positive output. The encoder identifies the lowest comparator which is set. With this technique, digitisation times of tens of
nanoseconds are available, at a price.

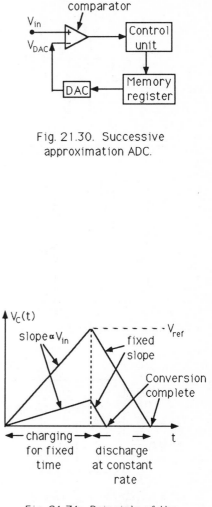

Fig. 21.30. Successive
approximation ADC.

Successive approximation

A second method, called successive approximation, is shown in
figure 21.30. Suppose the converter is set to digitise signals within
the range 0 to 5 V; and suppose for example that the voltage
to be digitised is 3.13 V. The result of the conversion is to be
represented by a binary number where the most significant digit
stands for 2.5 V, the second for 1.25 V and so on. The control
unit begins a conversion by loading the binary number $1000\ldots000$
into the memory register and the DAC converts this to a 2.5 V
analogue signal. The comparator compares the input signal to be
digitised with the output of the DAC and produces an output which
is positive, since $V_{DAC} < V_{in}$. In this case, the control unit adds a
1 to the next binary digit and repeats the comparison. It now finds
$V_{DAC} = 3.75$ V, which is greater than V_{in}, so it subtracts a one from
the next digit and $V_{DAC} \rightarrow 3.125$ V. The procedure continues to
the least significant digit. ADCs working on this principle complete
a digitisation in typically 1 μs.

Slope converter

A snag with this technique is that the voltage step between one
digit and the next depends on the resistor values in the DAC, so
variations may arise from one digit to the next. For pulse height
analysis this is undesirable; there, a vital requirement is that each
channel should be accurately of the same width. Consider the following hypothetical scheme. An integrator is charged by a current
proportional to the input signal and a counter starts at the same
instant as the integrator. The counter is stopped by a comparator
when the voltage of the integrator passes the input voltage. The
larger the input current, the smaller the counter reading. This
technique has its own snags. The conversion depends on the absolute rate of the counter and the absolute slope of the integrator;
the capacitor in it is likely to be temperature sensitive. There is
also an offset or 'pedestal' problem if the integrator does not start
exactly from zero.

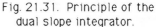

Fig. 21.31. Principle of the
dual slope integrator.

These problems are eliminated in the **dual-slope integrator**
(figure 21.31). The integrator charges for a fixed *time* at a rate
proportional to the input voltage. Then it is discharged at a fixed
rate and a counter measures the time interval before the integrator
reaches zero. Since the same counter defines the charging process
and measures the discharge, the result is independent of the rate of
the counter. The same is true for the magnitude of the capacitor.
Shifts in the zero level have no effect since they are identical for

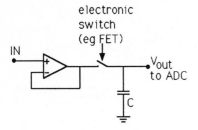

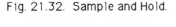

Fig. 21.32. Sample and Hold.

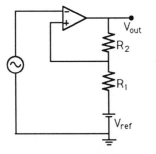

Fig. 21.33

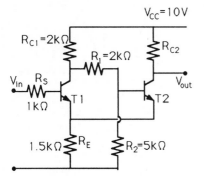

Fig. 21.34.

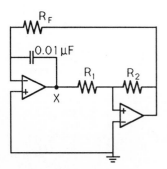

Fig. 21.35.

charging and discharging. Accuracy depends only on the reference voltage, which can be precise.

TDC
The single-slope integrator is also used for time to digital conversion (TDC). It may be necessary to digitise a very short time interval (a few ns) too short to use a counter. During this time, a constant current feeds an integrator between start and stop pulses; then the voltage level is digitised by any of the normal ADC techniques.

Sample and hold
In all types of ADC, there is the problem that the input signal may be fluctuating over the length of time required to digitise it. Figure 21.32 shows schematically an arrangement known as **sample and hold**. When the switch is closed, the buffer amplifier charges the capacitor rapidly to the same voltage as the analogue input. The switch is opened at the start of the digitisation procedure and the capacitor holds the sampled signal. Sample and hold circuits are available as integrated circuits, to which the capacitor C is attached externally. It should obviously be a low loss capacitor.

A practical ADC may also contain elaborations such as filters to remove or average out unwanted backgrounds like noise and mains hum.

21.9   Exercises

1.   In the Schmitt trigger circuit of figure 21.33, $V_{out} = \pm 10$ V. Find values of R_1/R_2 and V_{ref} for the trigger to fire at 3 V and reset when the input drops to 1 V. (Ans: $R_1/R_2 = 9$, $V_{ref} = 2.2$ V.)

2.   (QMC 1974) In the Schmitt trigger circuit of figure 21.34, the gain h_{fe} of the transistors is 50. Find (i) the value V_1 of V_{in} when T1 comes ON as V_{in} is increased, (ii) the value of R_{C2} to give the maximum swing without saturating T2, (iii) the value of V_{in} when T1 goes off as V_{in} decreases. (Ans: 5.5 V, 1.4 kΩ, 3.5 V.)

3.   (RHBNC 1988) The combination of an integrator with a Schmitt trigger can be used to generate a triangular waveform using the circuit shown in figure 21.35. Assuming that the outputs of the operational amplifiers saturate at ± 13 V, calculate the values of the components for a triangular wave of frequency 500 Hz and amplitude 2 V peak-to-peak at the point X if the integration capacitor is 0.01 μF. (Ans: $R_2/R_1 = 13$, $R_f = 650$ kΩ.)

4. (Westfield 1974) Describe the action of a univibrator oscillator circuit capable of producing square waves. Find the values of components and voltage supply needed to produce 10 V pulses of width 50 μs.

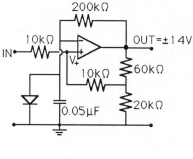

Fig. 21.36

5*. (QMC 1978) What function does the circuit of figure 21.36 perform? Outline how the circuit works and describe one application; assume that V_{in} is a generator of low output impedance. What magnitude of V_{in} is required to initiate the operation of the circuit if $V_{out} = \pm 14$ V and the diode switches at $+0.6$ V? Hints: (a) With $V_{in} = 0$, show that $V_+ = V_{out}/14$, (b) consider the quiescent conditions with V_{out} positive, then negative, (c) use the superposition theorem to find V_+ for a given V_{in}. (Ans: monostable; (b) with V_{out} positive, $V_- \simeq 0.6$ V (stable) and with V_{out} negative, $V_- \simeq -V_{out}$ (unstable); (c) $\Delta V_+ = 5\Delta V_{in}/7$; fires from a negative input pulse larger than -0.56 V; then $V_- = 0.6 - 14.6 \times (1 - e^{-t/\tau})$ where $\tau = 10^{-2}$ s. Switches back when $V_- = -1$ V, or $t = 1.2$ ms.)

6*. (Kings 1974) The free-running multivibrator of figure 21.37 uses transistors which conduct when $V_{BE} = 0.6$ V. The circuit is required to produce collector voltage waveforms having an amplitude of nearly 5 V, a repetition frequency of 2.5 kHz and a mark-to-space ratio of 3:1. (a) Sketch, with dimensions, the collector and base waveforms of one transistor. Calculate the required values of C_1, C_2, R_{B1} and R_{B2}, and (b) suggest a method of reducing the risetimes of the collector waveforms. (Ans: (a) figure 21.38, $C_1 R_{B1} = 156$ μs, $C_2 R_{B2} = 468$ μs; choose R_{B1} and R_{B2} of order 100 kΩ so that C_1 and C_2 charge rapidly through R_{C1} and R_{C2}. Speed-up capacitors across R_{C1} and R_{C2}, and/or limiting diodes from base to collector.)

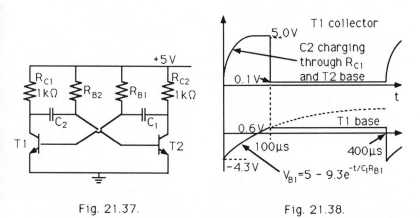

Fig. 21.37.   Fig. 21.38.

7. What function does the circuit of figure 21.39 perform?

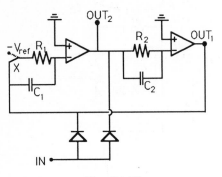

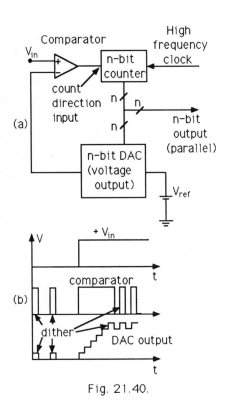

Fig. 21.39

(a)

(b)

Fig. 21.40.

Describe in outline the operation of the circuit. If the connection at the switch X is broken and the resistor R_1 is connected instead to the negative reference voltage, how is the operation of the circuit altered? (Ans: Bistable; when switched to $-V_{ref}$, monostable.)

8.   (IC 1989) The circuit shown in figure 21.40(a) is a type of analogue to digital Converter (ADC) synthesised from a DAC, a comparator and a digital counter. The counter is fed with a continuous fast digital clock signal of frequency $f_{ck} \gg f_{max}$, the maximum frequency at which V_{in} is sampled; it either increments or decrements the binary count by one unit at every rising edge of the clock, depending on the state of its direction input. A high level indicates 'count up' and a low level 'count down'. The DAC gives a voltage output between 0 V and $+V_{ref}$. The input voltage is held at zero for a long period and then changed abruptly from 0 V to $+V_{in}$ at time $t = 0$; make a detailed sketch showing the input voltage, the DAC output voltage and the comparator output in the correct relative phases. Explain how the system responds to a slowly varying input signal. Show that if the DAC has n-bit resolution and the input is a sine wave of maximum possible amplitude, its frequency must not exceed $f_{max} \leq f_{ck}/2^n \pi$ if accurate conversion is to be achieved. (Ans: figure 21.40(b); follows if signal changes more slowly than slew rate of DAC.)

9*.   An alternative resistor network for a digital to analogue converter is shown in figure 21.41. Verify that the voltage at the input to the amplifier is $V_4/2 + V_3/4 + V_2/8 + V_1/16 + V_0/32$.

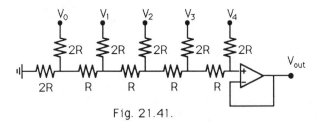

Fig. 21.41.

22

Transformers and Three-phase Supplies

22.1   Introduction

Electrical power is distributed nationally as alternating current. At the generator the voltage is typically 20 kV. The National Grid, however, operates at 132 and 270 kV in the UK and at similar values in other countries. This is to reduce power losses (I^2R) in the cables; it keeps cable diameters to reasonable dimensions.

In order to keep the voltage loss down the cable to a reasonable ($\leq 5\%$) level, estimate roughly the cable diameter required.

The voltage is stepped up from the generator to the Grid by a **transformer**. At the user end of the Grid, a second transformer steps the voltage down again, usually via intermediate 33 and 11 kV stages.

The principle of the transformer is illustrated in figure 22.1. A **primary** coil of n_1 turns is wound on to an iron former, and a **secondary** coil of n_2 turns feeds the output. It will emerge that

$$\frac{\text{secondary voltage}}{\text{primary voltage}} = \frac{n_2}{n_1} \quad \text{and} \quad \frac{\text{secondary current}}{\text{primary current}} \simeq \frac{n_1}{n_2}.$$

The transformer is the electrical analogue of the gearbox. In the first approximation the power is the same in both circuits, though there is a small loss, typically 1–4%, due to eddy currents in the iron core of the transformer.

Transformers are also used to step the mains voltage up or down. In a television set, a transformer generates a high voltage of typically 10 kV for the electron gun of the tube. In electronic power supplies, the voltage is stepped down to typically 5 or 12 V.

Relation between L_1, L_2 and M
In figure 22.1, the two coils have self-inductances L_1 and L_2 and mutual inductance M, which will be included explicitly in circuit

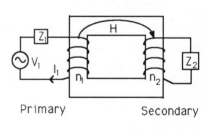

Primary   Secondary

Fig. 22.1. The transformer.

diagrams. Any further impedances in the primary and secondary circuits are lumped into Z_1 and Z_2. These include the losses due to hysteresis in the iron core. Let us first discuss the relation between L_1, L_2 and M, making the assumption that all the magnetic field H circulates through both coils. This is the condition for a **perfect transformer**. Subsequently, it is easy to allow for field which bypasses one or other coil.

If there are n_1 turns in the primary and n_2 in the secondary

$$H = \alpha_1 n_1 I_1 + \alpha_2 n_2 I_2$$

where α_1 and α_2 are geometrical constants; we shall not need to know them unless L and M are to be calculated from first principles. The back-EMF in circuit 1 is

$$V_1 = \beta_1 n_1 \frac{dH}{dt} = \beta_1 n_1 \left(\alpha_1 n_1 \frac{dI_1}{dt} + \alpha_2 n_2 \frac{dI_2}{dt} \right)$$

where β_1 is a further geometrical constant. This relation may be written

$$V_1 = L_1 \frac{dI_1}{dt} + M_{12} \frac{dI_2}{dt} \tag{22.1}$$

with

$$L_1 = \alpha_1 \beta_1 n_1^2 \tag{22.2}$$

$$M_{12} = \alpha_2 \beta_1 n_1 n_2. \tag{22.3}$$

Likewise, in the second circuit there is a back-EMF

$$V_2 = \beta_2 n_2 \frac{dH}{dt} = \beta_2 n_2 \left(\alpha_1 n_1 \frac{dI_1}{dt} + \alpha_2 n_2 \frac{dI_2}{dt} \right)$$

$$= M_{21} \frac{dI_1}{dt} + L_2 \frac{dI_2}{dt} \tag{22.4}$$

where

$$L_2 = \alpha_2 \beta_2 n_2^2 \tag{22.5}$$

$$M_{21} = \alpha_1 \beta_2 n_1 n_2. \tag{22.6}$$

From these relations

$$\boxed{M_{12} M_{21} = L_1 L_2} \tag{22.7}$$

for a perfect transformer, and it will be shown below from energy considerations that $M_{12} = M_{21} = M$. If the second coil is wound in the opposite sense, M changes sign.

For an **imperfect transformer**, some field produced by circuit 1 does not go through circuit 2, and vice versa. The mutual inductance is due only to that part of the field which goes through both, while the self-inductances are due to the whole field through each circuit. Thus for an imperfect transformer

$$M^2 = k^2 L_1 L_2 \qquad \text{where} \qquad k^2 < 1.$$

In this case L_1 is still $\propto n_1^2$, $L_2 \propto n_2^2$ and $M \propto n_1 n_2$.

22.2   Energy Stored in a Transformer

Suppose current I_1 is first established in the primary and then I_2 is brought up from zero, keeping I_1 constant. The work done against the back-EMFs is

$$E = \int L_1 \frac{\mathrm{d}I_1}{\mathrm{d}t} I_1 \, \mathrm{d}t + \int L_2 \frac{\mathrm{d}I_2}{\mathrm{d}t} I_2 \, \mathrm{d}t + \int M_{12} \frac{\mathrm{d}I_2}{\mathrm{d}t} I_1 \, \mathrm{d}t$$
$$= \tfrac{1}{2} L_1 I_1^2 + \tfrac{1}{2} L_2 I_2^2 + M_{12} I_1 I_2.$$

The first term corresponds to the power due to current I_1 and the back-EMF $L_1 \, \mathrm{d}I_1/\mathrm{d}t$ in coil 1; likewise for the second term in coil 2. The third term comes from the power in coil 1 due to current I_1 and the back-EMF $M_{12} \, \mathrm{d}I_2/\mathrm{d}t$ due to increasing current I_2.

If the process is carried out in reverse order, first establishing current I_2 in the secondary then bringing up current I_1 from zero in the primary, the indices swop over for M:

$$E = \tfrac{1}{2} L_1 I_1^2 + \tfrac{1}{2} L_2 I_2^2 + M_{21} I_1 I_2.$$

These two energies must be identical, otherwise a perpetual motion machine could be devised. This requires $M_{12} = M_{21}$. The same result may be obtained from electromagnetic theory by evaluating the geometrical constants α and β appearing in equations (22.2)–(22.6). Finally

$$E = \tfrac{1}{2} L_1 I_1^2 + \tfrac{1}{2} L_2 I_2^2 + M I_1 I_2. \qquad (22.8)$$

22.3   Circuit Equations and Equivalent Circuits

It is a tricky business keeping track of the signs of voltages in a transformer (and often it does not matter). We consider briefly how to follow these signs. In figure 22.2, the field H between the coils is indicated schematically. The back-EMF $L_1 \, \mathrm{d}I_1/\mathrm{d}t$ is such as to oppose changes in current I_1. The direction of I_2 and the sign of $M \, \mathrm{d}I_1/\mathrm{d}t$ depends on the sense of the windings. If they are as in figure 22.1, the polarities are as shown in figure 22.2. Because both coils are wound in the same sense with respect to

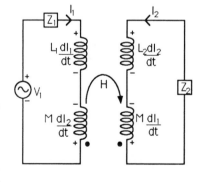

Fig. 22.2. Circuit diagram of Fig. 22.1.

H, the polarities of $M\,\mathrm{d}I_1/\mathrm{d}t$ and $L_1\,\mathrm{d}I_1/\mathrm{d}t$ with respect to H are the same. The positive 'terminal' of $M\,\mathrm{d}I_1/\mathrm{d}t$ is indicated by the dot in the second circuit. The self-inductance of the second coil $L_2\,\mathrm{d}I_2/\mathrm{d}t$ has the opposite polarity, so as to oppose the change in I_2. Finally, the current I_2 induces a voltage $M\,\mathrm{d}I_2/\mathrm{d}t$ in the primary oriented with respect to the field like $L_2\,\mathrm{d}I_2/\mathrm{d}t$. The sign of this induced EMF is indicated by the dot in the primary circuit.

If the secondary winding has the opposite sense, $M\,\mathrm{d}I_1/\mathrm{d}t$ changes polarity and so do $L_2\,\mathrm{d}I_2/\mathrm{d}t$ and I_2. (The dot in the secondary would then move to the top of the secondary coil.) The polarity of $M\,\mathrm{d}I_2/\mathrm{d}t$ likewise changes sign with I_2.

Circuit equations

After these details about sign, the circuit analysis can proceed straightforwardly. For circuits 1 and 2,

$$V_1 = Z_1 I_1 + L_1\frac{\mathrm{d}I_1}{\mathrm{d}t} - M\frac{\mathrm{d}I_2}{\mathrm{d}t} \tag{22.9}$$

$$M\frac{\mathrm{d}I_1}{\mathrm{d}t} = L_2\frac{\mathrm{d}I_2}{\mathrm{d}t} + Z_2 I_2. \tag{22.10}$$

These are the fundamental equations giving the response of the two circuits to AC voltages or steps in voltage. They will be manipulated into the form of an equivalent circuit.

Secondary equivalent circuit

Suppose an alternating voltage $V_1\mathrm{e}^{\mathrm{j}\omega t}$ is applied to the primary. There will be alternating voltages and currents of the same frequency ω in the secondary, and equation (22.10) gives

$$I_2 = \frac{\mathrm{j}\omega M}{Z_2 + \mathrm{j}\omega L_2} I_1. \tag{22.11}$$

Substituting for I_1 in equation (22.9),

$$V_1 = \frac{(Z_1 + \mathrm{j}\omega L_1)(Z_2 + \mathrm{j}\omega L_2)}{\mathrm{j}\omega M} I_2 - \mathrm{j}\omega M I_2$$

or

$$\frac{V_1}{I_2} = \frac{Z_1 Z_2}{\mathrm{j}\omega M} + \frac{L_1 Z_2}{M} + \frac{Z_1 L_2}{M} + \mathrm{j}\omega\frac{L_1 L_2}{M} - \mathrm{j}\omega M. \tag{22.12}$$

For a perfect transformer, the last two terms cancel and

$$\frac{M}{L_1}\frac{V_1}{I_2} = Z_2 + \frac{Z_1 L_2}{L_1} + \frac{Z_1 Z_2}{\mathrm{j}\omega L_1}.$$

Remember that $L_2 \propto n_2^2$, $L_1 \propto n_1^2$ and $M \propto n_1 n_2$, so if the ratio of turns in the secondary and primary is

$$r = n_2/n_1$$

then

$$r\frac{V_1}{I_2} = Z_2 + r^2 Z_1 + \frac{Z_1 Z_2}{j\omega L_1}.$$

This equation may be interpreted in terms of the equivalent circuit shown in figure 22.3. The driving voltage is rV_1, i.e. V_1 **stepped up** by a factor r. Note carefully that V_1 is the voltage driving the whole primary circuit, not just the voltage across the primary of the transformer. The terms $r^2 Z_1$ and $L_1/Z_1 Z_2$ allow for the voltage drop in the primary across Z_1. If Z_2 is small, equation (22.11) gives

$$\boxed{I_2 \simeq I_1/r.}$$

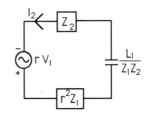

Fig. 22.3. Equivalent circuit of the secondary of a perfect transformer.

That is, the current in the secondary is **stepped down** from that in the primary. In this approximation, $V_1 I_1 = V_2 I_2$ and the transformer simply converts power from one voltage to another.

Apart from the capacitor in figure 22.3 (often negligible), the circuit is the Thevenin equivalent of a generator with output impedance $r^2 Z_1$. Thus the transformer changes the matching between the primary voltage source and the load. Small transformers are sometimes used specifically to achieve matching of a high-impedance source to a low-impedance load (e.g. in driving a loudspeaker, where the standard values of load resistance are 4, 8 and 12 Ω).

What happens to the equations if the secondary coil is wound in the opposite sense to figure 22.1?

Primary

An equivalent circuit for the primary may be obtained by eliminating I_2 from equation (22.9) using (22.11):

$$V_1 = (Z_1 + j\omega L_1)I_1 + \frac{\omega^2 M^2 I_1}{Z_2 + j\omega L_2}$$

or

$$\frac{V_1}{I_1} = Z_1 + \frac{j\omega L_1(Z_2 + j\omega L_2) + \omega^2 M^2}{Z_2 + j\omega L_2}$$

$$= Z_1 + \frac{j\omega L_1 Z_2 + \omega^2(M^2 - L_1 L_2)}{Z_2 + j\omega L_2}. \qquad (22.13)$$

Again the term $(M^2 - L_1 L_2)$ disappears for a perfect transformer and

$$\frac{V_1}{I_1} = Z_1 + \frac{j\omega L_1 Z_2}{Z_2 + j\omega r^2 L_1} = Z_1 + \frac{j\omega L_1 Z_2/r^2}{Z_2/r^2 + j\omega L_1}.$$

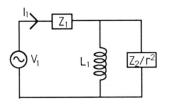

Fig. 22.4. Equivalent circuit
of the primary of a perfect
transformer.

This has the equivalent circuit shown in figure 22.4. The load Z_2/r^2 appears across the inductor and is said to be 'reflected' from the secondary. Likewise, in figure 22.3, the load $Z_1 r^2$ is said to be 'reflected' from the primary into the secondary.

Can you understand the appearance of Z_2/r^2 in the primary in terms of energy dissipation in Z_2?

If the secondary is shorted ($Z_2 = 0$), what is the magnetic field in the transformer?

Summary
If a perfect transformer has load Z_2 applied to the secondary where $Z_2 \ll j\omega L_2$,

$$\boxed{\frac{\text{secondary voltage}}{\text{primary voltage}} = r} \tag{22.14}$$

$$\boxed{\frac{\text{secondary current}}{\text{primary current}} \simeq 1/r} \tag{22.15}$$

and the load seen by the primary $\simeq Z_2/r^2$. The load seen by the secondary is $r^2 Z_1$. When the transformer is not perfect, it is possible (exercise 7) to find more complicated equivalent circuits involving $(1-k)$, but unless you are working day to day with transformers it is easier to solve problems direct from equations (22.9) and (22.10).

Floating AC voltages
There is no direct connection between primary and secondary, so there can be a difference in DC voltage between them. If the primary is earthed, the secondary is said to be floating. Sometimes this is undesirable (e.g. in commercial power supplies) and they are earthed to a common point. It does, however, represent a simple means of eliminating (or introducing) DC biases.

22.4   Three-phase Systems

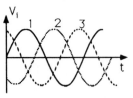

Fig. 22.5. Voltages v. t
in a three-phase supply.

The AC generators considered so far produce a voltage $V_0 \cos \omega t$ between two terminals. This is what is provided by an ordinary wall-plug between the 'live' and 'neutral'. It suffers the disadvantage that the voltage passes through zero twice per cycle, and so does the power. For some purposes, notably driving an electric motor, it is preferable if the power is more nearly constant. This can be achieved using three supplies differing in phase by 120°, as in figure 22.5. Because of the phase difference of 120°, no two voltages pass through zero simultaneously.

At the generator, this is achieved (see figure 22.6) by winding three fixed coils on the **stator** (the stationary part) and providing a rotating magnetic field by means of a **rotor** in the centre. With suitable geometrical design, the magnetic field in each stator coil varies approximately sinusoidally, and (after some smoothing considered in Chapter 9) induces a sinusoidal voltage in each stator coil. The rotor is supplied with direct current via slip-rings; the current required in the rotor is small compared with the currents generated in the stator coils. In a power station the rotor is driven by a turbine; in a local generator it is driven from the mains. The design of a power generator is complicated, because of the large mechanical forces on the conductors, the large energy stored in the rotor and the large voltages and currents.

The three coils are connected together as shown in figure 22.7 with one common terminal, called the **neutral**. This plays two roles. Firstly, it economises on cables by reducing the number of output lines from six to four. Secondly, it avoids unpredictable and unwelcome floating voltages between the three generator coils.

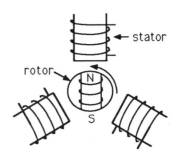

Fig. 22.6. Schematic layout of a three-phase AC generator.

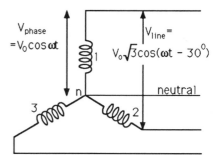

Fig. 22.7. Three-phase supply.

Line voltage
The voltages generated in the three coils vary with time as shown in figure 22.5. On a phasor diagram, their magnitudes and relative phases are as in figure 22.8. The voltage between lines 1 and 2 is $V_{12} = V_1 - V_2$, and if V_2 and V_1 are of equal magnitude V_0, V_{12} lags 30° behind voltage V_1;

$$|V_{12}| = 2V_0 \cos 30° = V_0\sqrt{3}. \tag{22.16}$$

This is known as the **line voltage**. That between live and neutral terminals is known as the **phase voltage**. Household supplies are taken from single phases, live to neutral, i.e. RMS 110 V in the USA, 220 V in Europe and 240 V in the UK; line voltages are a factor $\sqrt{3}$ larger.

22.5   Balanced Loads

Suppose three loads are connected between live and neutral, as shown in figure 22.9. This is known as a **star** or **Y** configuration (the latter from its shape when drawn the other way up). The currents flowing in the three loads are

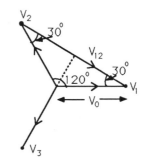

Fig. 22.8. Vector diagram of phase voltages.

$$I_1 = V_1/Z_1$$
$$I_2 = V_2/Z_2 \tag{22.17}$$
$$I_3 = V_3/Z_3.$$

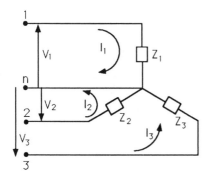

Fig. 22.9. Star connection of loads.

If the three loads are the same and equal to Z, the resulting current in the neutral is

$$I_n = I_1 + I_2 + I_3 = \frac{V_1}{Z}(1 + e^{j120°} + e^{j240°}) = 0.$$

This is called a **balanced load**. A three-phase motor is normally configured this way. Household supplies are arranged on alternate phases door to door, so as to balance the neutral current approximately.

Suppose your house is supplied by phase 1 and is metered between this phase and neutral. Suppose your neighbour is likewise supplied from phase 2. If you connect a heater between your phase and your neighbour's, can you defraud (a) the power company, (b) your neighbour?

Power
The total power supplied to the load is

$$P = V_1^2/Z_1 + V_2^2/Z_2 + V_3^2/Z_3. \tag{22.18}$$

In each term, Z is to be interpreted as a complex number:

$$Z = R + jX = |Z|e^{j\phi}$$

and individual terms in equation (22.18) take the form

$$P = V_i^2 \cos\phi_i/|Z_i|. \tag{22.19}$$

For a resistive load, $Z = R$ and $\cos\phi = 1$. For a purely inductive or capacitive load, $Z = jX$ and $\cos\phi = 0$, so no power is dissipated; in this case, current and voltage are 90° out of phase. When an electrical motor is running with no load on it, the phase of the rotation adjusts so that voltage and current are 90° apart and no power is drawn. When a load is applied, the motor is retarded so that a phase angle develops between current and voltage and power is absorbed from the mains. The greater the load, the larger the phase retardation.

If the loads are balanced and the neutral voltage is strictly zero, the instantaneous power P is

$$P = (V_0^2/Z)[\cos^2\omega t + \cos^2(\omega t + 120°) + \cos^2(\omega t + 240°)]$$
$$= (V_0^2/2Z)[3 - \cos 2\omega t - \cos(2\omega t + 240°) - \cos(2\omega t + 480°)]$$
$$= 3V_0^2/2Z. \tag{22.20}$$

The three cosines add up to zero and P is independent of t. This result, that constant power can be supplied, is the essential reason for choosing to deliver power nationally via a three-phase system.

Would a four- or five-phase supply give a constant power? How about two phases? If the mains voltage contains harmonics, what is their effect on the power?

22.6 Unbalanced Loads

If the loads are not balanced, currents I_1, I_2 and I_3 are different (in magnitude or phase or both) and have to be calculated individually from equations (22.17). Examples are given in the exercises at the end of the chapter.

Suppose the neutral line is disconnected. In this case, the voltage V_n at the centre of the star will move away from zero until the voltages $I_1 Z_1$, $I_2 Z_2$ and $I_3 Z_3$ across the three loads satisfy the vector relation illustrated in figure 22.10.

$$V_1 - V_n = I_1 Z_1$$
$$V_2 - V_n = I_2 Z_2$$
$$V_3 - V_n = I_3 Z_2$$
$$I_1 + I_2 + I_3 = 0.$$

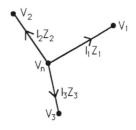

Fig. 22.10. Unbalanced loads and neutral voltage floating.

This is not an arrangement the average user would choose. However, between the power station and the local substation, the neutral is normally eliminated completely. If the net loads across the individual phases are unbalanced, the result is that the neutral will float up or down. To minimise this, the neutrals of generator and sub-station are earthed, and any unbalance of current at the centre of the star has a return path through the earth, which is quite a good though not perfect conductor. This being so, the neutral presented to users is at earth potential, except for the small voltage caused by the impedance of the neutral back to the substation and the current in this line due to unbalanced loads. Normally, the neutral is within one or two volts of earth potential.

22.7 The Delta Configuration

An alternative way of connecting the loads is shown in figure 22.11. This is called the **delta** or **pi** configuration (the latter from its appearance when drawn upside down with the connection between Z_A and Z_C splayed out). Here the neutral is not connected, and loads are applied directly between live terminals. In this case,

$$I_A = V_{12}/Z_A$$
$$I_B = V_{23}/Z_B$$
$$I_C = V_{31}/Z_C.$$

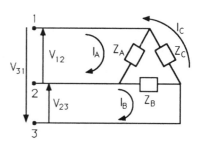

Fig. 22.11. Delta configuration of loads.

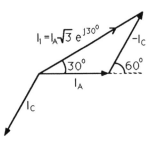

Fig. 22.12. Vector addition of I_1 and I_3.

These are called phase currents. The net current through terminal 1, called the line current, is

$$I_1 = I_A - I_C = (V_{12}/Z_A) - (V_{31}/Z_C).$$

If the loads are balanced and $V_{31} = V_{12}\exp(j240°)$, then

$$I_1 = (V_{12}/Z)(1 - e^{j240°}).$$

The resultant current is given by the vector diagram of figure 22.12 and is of magnitude $I_A\sqrt{3}$. Thus the line current is $\sqrt{3}$ times the currents through the individual loads. It leads I_A by 30°. If the loads are resistive, it is in phase with the phase voltage, as expected. If the loads are unbalanced, it is necessary to calculate the individual currents I_A, I_B and I_C and add them vectorially.

The power delivered to the loads is

$$P = (V_{12}^2/Z_A) + (V_{23}^2/Z_B) + (V_{31}^2/Z_C)$$

and for a balanced load

$$P = 3|V_{\text{line}}|^2/Z. \tag{22.21}$$

Comparing with equation (22.20), the power delivered to the loads is a factor of three larger for this configuration than for the star configuration because $V_{\text{line}} = V_0\sqrt{3}$, equation (22.16).

22.8   Worked Example

Suppose a three-phase system with 240 V RMS on each phase is applied to the star network of figure 22.13(a). Let us take $t = 0$ so that

$$V_1 = 240\sqrt{2}\cos\omega t \quad \text{or} \quad 240\angle 0° \text{ for short}$$
$$V_2 = 240\sqrt{2}\cos(\omega t + 120°) \quad \text{or} \quad 240\angle 120° \text{ for short}$$
$$V_3 = 240\sqrt{2}\cos(\omega t + 240°) \quad \text{or} \quad 240\angle 240° \text{ for short.}$$

The phase currents are $I_1 = 6\angle -40°$ A, $I_2 = 6\angle 80°$ A and $I_3 = 6\angle 200°$ A. These phase currents lag 40° behind phase voltages. The mean power in each arm is $240 \times 6\cos 40°$ W and in all three arms together is 3309 W.

We might want to know the relation between line current and line voltages. From figure 22.13(b), the voltage of 1 with respect to 2 is $V_{12} = V_0\sqrt{3}\angle -30°$. So current I_1 lags 10° in phase behind V_{12}.

Next, suppose the delta load of figure 22.13(c) is connected. Current $I_A = 4\sqrt{3}$ A and leads V_{12} by 10°. The total current

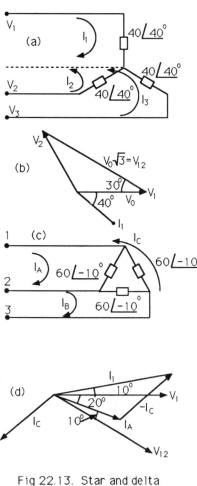

Fig 22.13. Star and delta balanced loads.

through terminal 1 is $I_1 = I_A - I_C$ and is given by the vector diagram (d); I_1 leads the phase voltage by 10° and is of magnitude $I_A\sqrt{3} = 12$ A. The mean power dissipated in each arm is $240\sqrt{3} \times 4\sqrt{3}\cos 10°$ W and the total power dissipated is 2836×3 W $= 8509$ W.

If both networks are connected simultaneously, the total power disspated is $3309 + 8509 = 11818$ W. The total current I_1 is $6\angle-40° + 12\angle10°$ A, where the angles are with respect to the phase voltage. The result is $16.51\angle-6.2°$ A. As a check, total power consumption $= 3 \times 240 \times 16.51\cos 6.2°$ W, which agrees with the previous result.

22.9   The Star–delta Impedance Transformation

A useful trick is to replace a star network of impedances by an equivalent delta configuration, or vice versa. The algebra is straightforward if approached in the right way.

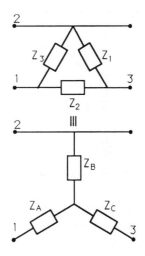

Fig. 22.14. Star and delta configurations.

Delta to star
Consider the impedance Z_{12} between terminals 1 and 2 of figure 22.14. In (a) it is given by Z_3 in parallel with $Z_1 + Z_2$ so

$$Z_{12} = Z_A + Z_B = \frac{Z_3(Z_1 + Z_2)}{Z_1 + Z_2 + Z_3}.$$

Also

$$Z_{23} = Z_B + Z_C = \frac{Z_1(Z_2 + Z_3)}{Z_1 + Z_2 + Z_3}$$

$$Z_{31} = Z_C + Z_A = \frac{Z_2(Z_1 + Z_3)}{Z_1 + Z_2 + Z_3}.$$

It is easy to solve for Z_A by multiplying the second equation by -1 and adding the three equations:

$$\boxed{Z_A = \frac{Z_2 Z_3}{Z_1 + Z_2 + Z_3}.} \tag{22.22}$$

Then

$$\boxed{Z_B = \frac{Z_3 Z_1}{Z_1 + Z_2 + Z_3}} \tag{22.22a}$$

$$\boxed{Z_C = \frac{Z_1 Z_2}{Z_1 + Z_2 + Z_3}.} \tag{22.22b}$$

This is the transformation usually needed and is thus the one to remember.

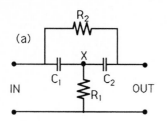

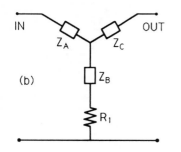

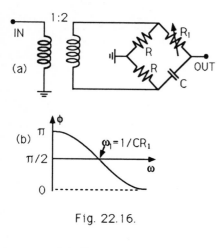

Fig. 22.15. The bridged–T filter
and an equivalent circuit.

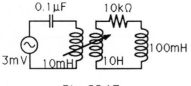

Fig. 22.16.

0.1 μF   10kΩ

100mH

3mV   10mH   10H

Fig. 22.17

What happens if $Z_3 = \infty$? What happens if it is zero?

Star to delta

The converse transformation is easily obtained by noting that

$$Z_A/Z_B = Z_2/Z_1 \qquad \text{and} \qquad Z_C/Z_A = Z_1/Z_3.$$

Substituting these into (22.22),

$$Z_A = \frac{Z_1^2(Z_A/Z_B)(Z_A/Z_C)}{Z_1(1 + Z_A/Z_B + Z_A/Z_C)} = \frac{Z_1 Z_A^2}{Z_B Z_C + Z_A Z_C + Z_A Z_B}$$

so

$$Z_1 = (Z_A Z_B + Z_B Z_C + Z_C Z_A)/Z_A \qquad (22.23a)$$

$$Z_2 = (Z_A Z_B + Z_B Z_C + Z_C Z_A)/Z_B \qquad (22.23b)$$

$$Z_3 = (Z_A Z_B + Z_B Z_C + Z_C Z_A)/Z_C. \qquad (22.23c)$$

You can construct mnemonics for these formulae. If you can re-member their general form, you can quickly check with special cases such as $Z \to 0$ or $Z \to \infty$, where results are obvious.

These transformations are useful not only in dealing with three-phase systems, but also in simplifying networks. As an example, consider the bridged-T filter of figure 22.15(a). Hitherto, this cir-cuit would have been analysed using Kirchhoff's laws and either mesh currents or the node voltage at X. Now we see that the ar-rangement $C_1 C_2 R_2$ at the top is a delta configuration drawn upside down, and this may be replaced by the star configuration of fig-ure 22.15(b), which is easier to analyse than the original network.

22.10   Exercises

1.   In figure 22.16(a), the transformer turns are in the ra-tio shown, and the input from a low-impedance source is $V_0 \sin \omega t$. Show that the output observed by a high-impedance device is $V_0 \sin(\omega t + \phi)$ and find how ϕ varies with R_1. (Ans: figure 22.16(b).)

2.   (QMC 1969) Assuming perfect coupling between the primary and secondary windings of the transformer of figure 22.17 and using suitable approximations, calculate the resonant frequency of the primary circuit, the current in the primary and the voltage appearing across the 100 mH inductor in the secondary circuit at this resonance. (Ans: $\omega = 10^6$ rad/s, 0.3 mA, 300 mV.)

3. Show that the two circuits of figures 22.18(a) and (b) are equivalent for suitable values of Z_i, Z_j and Z_k. (Ans: $Z_i = j\omega(L_1 + M)$, $Z_j = -j\omega M$, $Z_k = j\omega(L_2 + M) + Z_2$.)

4. Figure 22.19 shows the Heavyside bridge. Find the balance conditions and express M in terms of the other components when the bridge is balanced. (Ans: $PR = SQ$ and $R(L_2 + M) = Q(L_3 - M)$.)

5. Show that the maximum power P_m which may be delivered to a load by an AC voltage source V having output impedance R is $P_m = V_{RMS}^2/4R$. Show that this result is not affected if a transformer is inserted between source and load.

6. (QMC 1985) Find the magnitude and phase of the current labelled I_1 in figure 22.20. Does it lead or lag the voltage? Sketch the behaviour as a function of frequency. The voltage source V may be taken as $V_0 \cos \omega t$ or $V_0 e^{j\omega t}$ and the transformer may be assumed to be ideal. (Ans: current leads voltage by angle ϕ where $\tan \phi = t = \omega L_1/[R(\omega^2 C L_2 - 1)]$; $I_0 = V_0/[R(1 + t^2)^{1/2}]$.)

7**. For an imperfect transformer, $M^2 = k^2 L_1 L_2$. Show from equations (22.12) and (22.13) that the primary and secondary circuits may be represented for AC signals of frequency ω by the equivalent circuits shown in figures 22.21(a) and (b).

8. A star load with three equal arms of 25 Ω is connected across a three-phase supply with 110 V RMS phase voltage. What is the RMS line voltage? What is the peak current in each arm of the star and the total power dissipation? (Ans: 191 V, 6.22 A, 1.45 kW.)

9. A delta load with three equal arms of $(15 + 20j)$ Ω is connected across a three-phase supply with 415 V RMS line voltage. What is the peak phase voltage? What is the RMS line current in magnitude and how does its phase relate to (a) line voltage, (b) phase voltage? (Ans: 339 V, 28.75 A, (a) lags 23.1°, (b) lags 23.1°.)

10. (QMC 1987) A three-phase, 415 V system supplies a balanced delta load with impedances of $20\angle-40°$ Ω in each arm and a parallel star load with impedances of $(15-j25)$ Ω in each arm. Find the active and reactive power in each load and the magnitude of the total line current. Hint: take phase voltage with respect to neutral as the reference phase. Reactive power = mean power stored in reactive components; active power = power dissipated in resistance. (Ans: 19.8 kW, 16.6 kW, 5.07 kW, 43.8 A RMS.)

11. (QMC 1988) Show that the line current is equal to $\sqrt{3}$ times phase current for a three-phase delta-connected load and

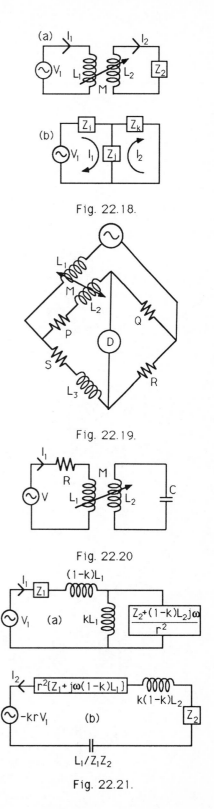

Fig. 22.18.

Fig. 22.19.

Fig. 22.20

Fig. 22.21.

draw a phasor diagram for the voltages and currents. A delta-connected three-phase system has a 10 Ω resistance in series with a 40 mH inductor in each phase of the load. If the load is supplied from a 415 V RMS 50 Hz supply, find the magnitude and phase of the line current and the power supplied to the load. A star-connected load, consisting of a 20 Ω resistance in each arm of the star is connected in parallel with the delta load. What is the magnitude and phase of the new line current? (Ans: RMS line current = 44.76 A, 51.5° behind the phase voltage; 20.03 kW; 53.05 A, 41.3° behind the phase voltage.)

12. A star load with resistances of 25, 30 and 35 Ω is connected between the three phases of a 110 V RMS supply and neutral. What is the RMS magnitude of the current in the neutral? If instead the neutral is left floating, what is its RMS voltage? (Ans: 1.09 A, 10.7 V.)

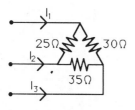

Fig. 22.22.

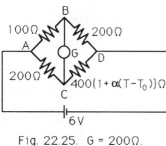

Fig. 22.25. G = 200Ω.

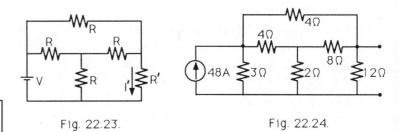

Fig. 22.23.   Fig. 22.24.

13. A 415 V three-phase supply is connected to the delta network of figure 22.22. What are the RMS currents I_1, I_2 and I_3? (Ans: I_1 = 26.4 A, I_2 = 24.8 A, I_3 = 22.3 A.)

14. (RHC 1974) Express the current I' of figure 22.23 in terms of R, R' and V. What is this current if $R = R' = 1\ \Omega$ and $V = 2$ V? (Ans: $I' = 12\ \mathrm{V}/(9R + 15R') = 1$ A.)

15. (QMC 1986) Reduce the circuit of figure 22.24 to its Thevenin equivalent. (Ans: $V_{\mathrm{EQ}} = 54$ V, $R_{\mathrm{EQ}} = 3\ \Omega$.)

16*. (RHBNC 1987) By applying a delta to star transformation to the network ABC in figure 22.25, find the resistance between the points A and D, assuming that $\alpha(T - T_0) \ll 1$. Hence obtain expressions for the potential difference between the points B and C and the current flowing through the galvanometer in terms of $(T - T_0)$. (Ans: $200[1 + (2/9)\alpha(T - T_0)]$, $V_{\mathrm{B}} - V_{\mathrm{C}} = -2\alpha(T - T_0)/3$, $I = \alpha(T - T_0)/300$ A.)

17**. (RHBNC 1988) Show that the network of figure 22.26(a) is equivalent to that of figure 22.26(b) if $Y_1 = Y_{\mathrm{B}}Y_{\mathrm{C}}/(Y_{\mathrm{A}} + Y_{\mathrm{B}} +$

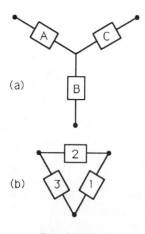

Fig. 22.26.

Y_C), where Y are admittances. Use the star to delta transformation for the components in the lower half of the figure to show that the Tuttle bridge of figure 22.27 is balanced if $2\omega^2 LC = 1 - \omega^2 C^2 R_L R = 4R_L/R$.

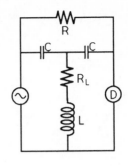

Fig. 22.27.

Appendix A

Thevenin's Theorem

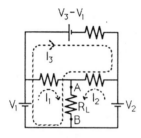

Fig. A1. Loop currents through terminals AB.

Any network can be described by loop currents passing from voltage sources to the terminals AB, hence through a load resistor R_L. An example is shown in figure A.1. In order to establish Thevenin's theorem, it is necessary to show that $V_{AB} = V_{EQ} - R_{EQ} I_{AB}$, where V_{EQ} and R_{EQ} do *not* depend on the load resistance R_L.

Around each current loop, there is a *linear* equation:

$$V_{AB} = V_1 - R_{11} I_1 - R_{12} I_2 - \cdots - R_{1n} I_n$$
$$V_{AB} = V_2 - R_{21} I_1 - R_{22} I_2 - \cdots - R_{2n} I_n$$
$$\vdots$$
$$V_{AB} = V_n - R_{n1} I_1 - R_{n2} I_2 - \cdots - R_{nn} I_n.$$

Because the voltage drops appearing on the right-hand side in the form $R_{ij} I_j$ stop short of the terminals AB, the load resistance R_L does not appear explicitly anywhere in the equations. They can be solved for currents (by simple elimination and substitution), with the results

$$I_1 = Y_{11} V_1 + Y_{12} V_2 + \cdots + Y_{1n} V_n - \gamma_1 V_{AB}$$
$$I_2 = Y_{21} V_1 + Y_{22} V_2 + \cdots + Y_{2n} V_n - \gamma_2 V_{AB}$$
$$\vdots$$
$$I_n = Y_{n1} V_1 + Y_{n2} V_2 + \cdots + Y_{nn} V_n - \gamma_n V_{AB}$$

where again none of the coefficients depend on R_L (since R_{ij} did not). Then $I_{AB} = \sum_i I_i$ can be written $B_1 V_1 + B_2 V_2 + \cdots + B_n V_n - C V_{AB}$. This is of the required form if $C = 1/R_{EQ}$ and $V_{EQ} = (B_1 V_1 + B_2 V_2 + \cdots + B_n V_n)/C$.

In Chapter 6, it is shown that there is a linear relation between V and I for capacitors and inductors when the notation of complex numbers is used. The linear relation between V and I is the foundation of Thevenin's theorem. Hence the proof carries over to include capacitors and inductors in AC circuits. Fourier's theorem expresses any waveform in terms of AC components, and Thevenin's theorem is therefore generally valid for any network where the relation between V and I is linear. This extends it to small signals in non-linear circuits, following the methods of section 1.11.

Appendix B

Exponentials

The exponential function $f(x) = e^x$ or $\exp(x)$ is defined by

$$df/dx = f \qquad \text{(B.1)}$$

$$f(x = 0) = 1. \qquad \text{(B.2)}$$

It is shown graphically in figure B.1. The slope of the curve is everywhere equal to its value, hence it diverges as $x \to \infty$:

$$e^x \to \infty \qquad \text{as} \qquad x \to \infty. \qquad \text{(B.3)}$$

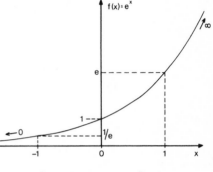

Fig. B.1. The Exponential Function

Indeed it diverges rapidly: for $x = 20$, $e^x = 4.85 \times 10^8$. Even politicians have heard of exponential growth! It will be shown below that $e^x e^{-x} = 1$; hence

$$e^x \to 0 \qquad \text{as} \qquad x \to \infty. \qquad \text{(B.4)}$$

The exponential is, however, finite for any finite value of x. It is everywhere positive: if it were to reach the x axis, df/dx would be zero and f would be 0 everywhere.

With the substitution $x \to \alpha x$ in (B.1)

$$\frac{1}{\alpha}\frac{d(e^{\alpha x})}{dx} = e^{\alpha x} \qquad \text{or} \qquad \frac{d}{dx}(e^{\alpha x}) = \alpha e^{\alpha x}. \qquad \text{(B.5)}$$

So, if $df'/dx = \alpha f'$,

$$f' = e^{\alpha x}. \qquad \text{(B.6)}$$

Equation (B.5) is important for differentiating exponentials. Conversely,

$$\int e^{\alpha x}\, dx = (1/\alpha)e^{\alpha x} + C \qquad \text{(B.7)}$$

where C is a constant.

The exponential function may be written as a power series (and is sometimes alternatively defined this way):

$$f(x) = \sum_{n=0}^{\infty} \frac{x^n}{n!} \qquad \text{(B.8)}$$

$$f(x) = 1 + x + \frac{x^2}{2!} + \frac{x^3}{3!} + \frac{x^4}{4!} + \ldots . \qquad (B.9)$$

Differentiating this series term by term gives (B.1). Equation (B.1) is also satisfied by

$$f(x) = \text{constant} \left(1 + x + \frac{x^2}{2!} + \frac{x^3}{3!} + \cdots \right)$$

and the condition (B.2) sets the constant to 1. From (B.9) with $x = 1$, e itself is given by 2.718 ….

An important property of the exponential is that

$$e^x e^y = e^{(x+y)}. \qquad (B.10)$$

This may be verified by multiplying the power series for e^x and e^y:

$$e^x e^y = \left(1 + x + \frac{x^2}{2!} + \frac{x^3}{3!} + \cdots \right) \left(1 + y + \frac{y^2}{2!} + \frac{y^3}{3!} + \cdots \right)$$

$$= 1 + x + y + \frac{x^2 + y^2}{2!} + xy + \frac{x^3 + y^3}{3!} + \frac{x^2 y + yx^2}{2!} + \cdots$$

$$= 1 + (x + y) + \frac{(x + y)^2}{2!} + \frac{(x + y)^3}{3!} + \cdots$$

$$= e^{(x+y)}.$$

An alternative demonstration is as follows. If $f = e^{\alpha x}$ and $g = e^{\beta x}$,

$$\frac{d(fg)}{dx} = f\frac{dg}{dx} + \frac{df}{dx}g = \beta fg + \alpha fg \qquad \text{by (B.1)}$$

$$= (\alpha + \beta)fg$$

so fg itself satisfies

$$\frac{d(fg)}{d[(\alpha + \beta)x]} = fg.$$

Hence, from (B.5),

$$fg = e^{(\alpha+\beta)x}.$$

A particularly important exponential in electrical and electronic circuits arises when a signal V (voltage *or* current) decays exponentially with time t:

$$V = V_0 e^{-\alpha t} \equiv V_0 \exp(-\alpha t).$$

The shape of the signal is sketched in figure B.2. At time $t = 0$, $e^{-\alpha t} = 1$ and $V = V_0$. When $\alpha t = 1$, V has fallen to V_0/e; this value of t is called the **decay time**. As $t \rightarrow \infty$, $V \rightarrow 0$. The signal obeys the differential equation

$$\frac{dV}{dt} = -\alpha V.$$

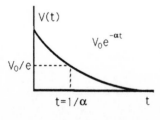

Fig. B.2. A signal decaying exponentially with time,

Logarithms

The inverse of the exponential is

$$x = \ln f. \tag{B.11}$$

From (B.2),

$$\ln 1 = 0. \tag{B.12}$$

From (B.1),

$$dx = d(\ln f) = df/f \tag{B.13}$$

$$\frac{d}{dx}(\ln f(x)) = \frac{1}{f}\frac{df}{dx}. \tag{B.13a}$$

Equation (B.13) appears frequently in integrations.
Figure B.3 shows a sketch of the logarithmic function.

$$\ln e = 1 \tag{B.14}$$

$$\ln(1/f) = -x \tag{B.15}$$

$$\ln f \to \infty \quad \text{as} \quad x \to \infty \tag{B.16}$$

$$\ln f \to -\infty \quad \text{as} \quad x \to 0. \tag{B.17}$$

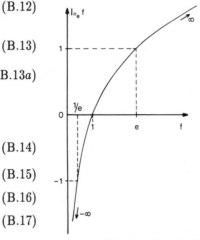

Fig. B.3. The Logarithmic Function

Complex exponentials

If $x = j\theta$ is substituted in (B.1),

$$\frac{1}{j}\frac{df''}{d\theta} = f'' \quad \text{or} \quad \frac{df''}{d\theta} = jf''. \tag{B.18}$$

This implies that f'' is complex, and that on the complex plane the slope of f'' is everywhere at 90° to its value. Hence, as θ varies, f'' describes a circle. From the definition (B.2), this is the unit circle. The solution (figure B.4)

$$f'' = \cos\theta + j\sin\theta \tag{B.19}$$

satisfies (B.5) since

$$df''/d\theta = -\sin\theta + j\cos\theta = jf'' = je^{j\theta}. \tag{B.20}$$

Complex exponentials satisfy the rule

$$e^{j\theta_1}e^{j\theta_2} = e^{j(\theta_1+\theta_2)} \tag{B.21}$$

as is evident from the product

$$e^{j\theta_1}e^{j\theta_2} = (\cos\theta_1 + j\sin\theta_1)(\cos\theta_2 + j\sin\theta_2)$$
$$= (\cos\theta_1\cos\theta_2 - \sin\theta_1\sin\theta_2)$$
$$\quad + j(\sin\theta_1\cos\theta_2 + \sin\theta_2\cos\theta_1)$$
$$= \cos(\theta_1+\theta_2) + j\sin(\theta_1+\theta_2)$$
$$= e^{j(\theta_1+\theta_2)}.$$

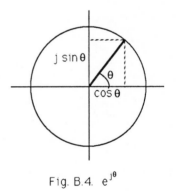

Fig. B.4. $e^{j\theta}$

Three further useful results are

$$e^{-j\theta} = \cos\theta - j\sin\theta \tag{B.22}$$

$$\cos\theta = \frac{1}{2}(e^{j\theta} + e^{-j\theta}) = 1 - \frac{\theta^2}{2!} + \frac{\theta^4}{4!} - \cdots \tag{B.23}$$

$$\sin\theta = \frac{1}{2j}(e^{j\theta} - e^{-j\theta}) = \theta - \frac{\theta^3}{3!} + \frac{\theta^5}{5!} - \cdots \tag{B.24}$$

Appendix C

PN Junctions and Bipolar Transistors

Expressions for $n(x)$ and $p(x)$

It is assumed that the reader has digested the qualitative features of diodes and transistors from Chapters 9 and 10. Here, mathematical details are provided. These are used in programs like SPICE for modelling transistor performance.

Currents are carried by electrons and holes which are freed by thermal excitation to the conduction band. Figure 9.14, reproduced from Chapter 9, shows the currents due to diffusion of electrons and holes and their drift under the action of the electrostatic field across the junction. To find these currents, it is necessary to find expressions for $n(x)$ and $p(x)$, the densities of free electrons and holes. These distributions are shown in figure 9.12. In the n region, $n(x)$ is high, but it falls precipitously across the depletion layer (note the logarithmic scale); $p(x)$ behaves in the opposite sense.

The probability that an electron is excited thermally to energy E is given by the Fermi–Dirac distribution

$$F_e(E) = \frac{1}{1 + \exp[(E - E_F)/kT]}.$$

It falls rapidly with increasing E. Since the number of donors is a factor of 10^6 or so less than the number of silicon atoms, the number of available levels in the conduction band far exceeds the number of excited electrons. It is then a reasonable approximation to assume they all enter levels at the bottom of the conduction band. Then, from figure 9.13,

$$E - E_F = \tfrac{1}{2}E_N + |e|(V_0 - V(x))$$

where $\tfrac{1}{2}E_N$ is the energy difference between donor levels and the bottom of the conduction band in the n region (figure 9.7), and V_0 is the potential difference across the full junction (figure 9.11(d)).

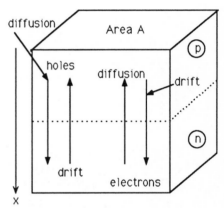

Fig. 9.14. Directions of diffusion and drift currents.

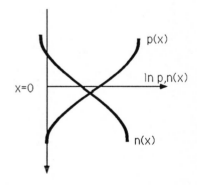

Fig. 9.12. p(x) and n(x).

359

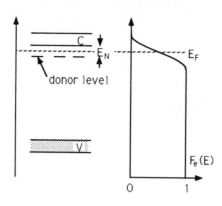

Fig. 9.7. Energy levels and $F_e(E)$
for an n-type material.

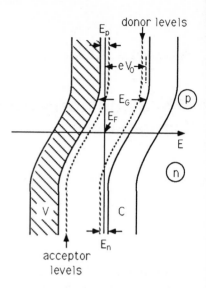

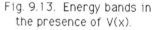

Fig. 9.13. Energy bands in
the presence of $V(x)$.

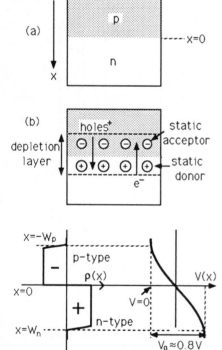

(a)

(b)

(c)   (d)

Fig. 9.11 (a) a pn junction, (b) the
diffusion of mobile electrons and
holes, (c) the charge distribution
$\rho(x)$, (d) the potential $V(x)$ across
the junction.

From this point onwards, the modulus sign around e will be dropped and it will be used to denote the *magnitude* of the electron's charge, 1.6×10^{-19} C. The signs of the charges on electrons and holes will be put into the equations explicitly. Then

$$n(x) = \frac{N_0}{1 + \exp\{[\frac{1}{2}E_N + e(V_0 - V)]/kT\}}. \tag{C.1}$$

The constant N_0 is expressed in terms of the density of donor atoms N in the n region by

$$N = \frac{N_0}{1 + \exp(\frac{1}{2}E_N/kT)}. \tag{C.2}$$

If the 1 in the denominators of equations (C.1) and (C.2) is neglected,

$$n(x) \simeq N \exp[e(V(x) - V_0)/kT]. \tag{C.3}$$

Likewise

$$p(x) \simeq P \exp(-eV(x)/kT). \tag{C.4}$$

The difference in sign inside the exponential comes from the opposite charges of the holes and electrons.

Expressions for E and V(x)
The density of mobile charges is known in terms of $V(x)$ and the problem is to find $V(x)$ itself within the depletion layer. For this

Gauss' theorem is applied:

$$\int \boldsymbol{E} \, \mathrm{d}S = \rho/\epsilon\epsilon_0 \tag{C.5}$$

where ρ is the charge density and $\boldsymbol{E}$ the electric field. If the cross-sectional area of the junction is A, Gauss' theorem applied to the material between x and $(x + \mathrm{d}x)$ gives

$$A\left(\boldsymbol{E}(x) + \frac{\mathrm{d}\boldsymbol{E}}{\mathrm{d}x}\,\mathrm{d}x - \boldsymbol{E}(x)\right) = A\frac{\mathrm{d}\boldsymbol{E}}{\mathrm{d}x}\,\mathrm{d}x$$
$$= \frac{eA}{\epsilon\epsilon_0}\,\mathrm{d}x(N + p(x) - P - n(x)). \tag{C.6}$$

On the right-hand side, note that donor atoms when ionised have positive charge and contribute with the same sign as $p(x)$; acceptor atoms have negative charge and contribute like $n(x)$. Above the junction, $N = 0$ and $n(x)$ is small. So the charge appearing in the bracket on the right-hand side of (C.6) is $p(x) - P$. Well above the junction this is zero. So $\boldsymbol{E} = 0$. However, as soon as V starts to rise, $p(x)$ drops precipitously. The essential results may be derived assuming that the charge distribution $p(x)$ has a well-defined edge at $x = -W_\mathrm{p}$, i.e. $p(x)$ drops to zero there, with the result that $\rho(x) = -P$ within the depletion layer on the p side. This is not strictly correct. The difference between P and $p(x)$ is responsible for the curved upper edge of $\rho(x)$ in figure 9.11(c), but $p(x)$ drops so rapidly that it can be neglected to a first approximation. Note that we are taking P to be constant with x, and this itself is an idealised model; in reality, the junction is not sharp.

From (C.6), $\mathrm{d}\boldsymbol{E}/\mathrm{d}x$ above the junction is then given simply by

$$\mathrm{d}\boldsymbol{E}/\mathrm{d}x \simeq -eP/\epsilon\epsilon_0$$

so

$$\boldsymbol{E} \simeq -ePx/\epsilon\epsilon_0 + \text{constant} = eP(-W_\mathrm{p} - x)/\epsilon\epsilon_0 \tag{C.7}$$

if $x = 0$ at the junction and $x = -W_\mathrm{p}$ at the top edge of the depletion layer ($\boldsymbol{E} = 0$). Then

$$V(x) = -\int \boldsymbol{E}\,\mathrm{d}x \simeq eP(\tfrac{1}{2}x^2 + W_\mathrm{p}x)/\epsilon\epsilon_0 + \text{constant}$$

$$V(x) = eP(x + W_\mathrm{p})^2/2\epsilon\epsilon_0 \tag{C.8}$$

if $V(x)$ is defined to be 0 at $x = -W_\mathrm{p}$. If we go back and allow for $p(x)$, $\boldsymbol{E}$ will tail off gradually for $x \leq -W_\mathrm{p}$, so $V(x)$ will depart from a parabola near there.

On the n side of the junction, $P = 0$ and similar algebra gives

$$\boldsymbol{E} = eNx/\epsilon\epsilon_0 + \text{constant}. \tag{C.9}$$

Neglecting $n(x)$, $\boldsymbol{E}$ drops to zero at a distance W_n from the junction and

$$V(x) = V_0 + eN(W_n - x)^2/2\epsilon\epsilon_0 \qquad (C.10)$$

within the depletion layer. The potential distribution of equations (C.8) and (C.10) is shown in figure 9.11(d).

Einstein's relations

The main text, Chapter 9, shows that drift currents ($\propto \boldsymbol{E}$) and diffusion currents ($\propto dp/dx$ and dn/dx) lead to hole current I_p and electron current I_n given by

$$I_p = eA(\mu_p p\boldsymbol{E} - D_p dp/dx) \qquad (9.13)$$

$$I_n = -eA(\mu_n n\boldsymbol{E} + D_n dn/dx). \qquad (9.14)$$

At equilibrium, both I_p and I_n are zero, otherwise the charge distributions would change with time. So from (9.13),

$$\frac{dp}{dx} = \frac{\mu_p}{D_p}p\boldsymbol{E} = -\frac{\mu_p}{D_p}p\frac{dV}{dx}. \qquad (C.11)$$

But from (C.4)

$$\frac{dp}{dx} = -\frac{e}{kT}p\frac{dV}{dx}. \qquad (C.12)$$

Equating these two expressions for dp/dx,

$$\mu_p/D_p = e/kT. \qquad (C.13)$$

This is Einstein's relation. He derived it originally for conduction in a metal. From (9.14) likewise

$$\mu_n/D_n = e/kT. \qquad (C.13a)$$

Width of the depletion layer

At the junction

$$V(x = 0) = ePW_p^2/2\epsilon\epsilon_0$$

and likewise for the other side of the junction

$$V_0 - V_a - V(x = 0) = eNW_n^2/2\epsilon\epsilon_0$$

if the applied voltage is V_a. Adding these two equations

$$V_0 - V_a = e(PW_p^2 + NW_n^2)/2\epsilon\epsilon_0. \qquad (C.14)$$

However, the total charges on the two sides of the junction must balance so

$$W_{\mathrm{p}}P = W_{\mathrm{n}}N. \qquad (C.15)$$

Substituting in (C.14),

$$V_0 - V_{\mathrm{a}} = \frac{e}{2\epsilon\epsilon_0}\left[\frac{W_{\mathrm{n}}^2 N^2}{P} + NW_{\mathrm{n}}^2\right].$$

$$W_{\mathrm{n}}^2 = 2\epsilon\epsilon_0(V_0 - V_{\mathrm{a}})P/[eN(N+P)] \qquad (C.16)$$

$$W_{\mathrm{p}}^2 = 2\epsilon\epsilon_0(V_0 - V_{\mathrm{a}})N/[eP(N+P)]. \qquad (C.16a)$$

The width of the depletion layer is proportional to $(V_0 - V_{\mathrm{a}})^{1/2}$ and shrinks under forward bias. Equations (C.16) also demonstrate that the width of the layer on the n side, W_{n}, increases when the p side is heavily doped and the n side is lightly doped; and vice versa.

Junction capacitance

The pn junction has opposite charges Q on the two sides of the junction and therefore forms a capacitor. It turns out that this capacitance is only important for reverse bias. For forward bias it is swamped by a second source of capacitance discussed below. As the applied voltage changes, equation (C.16) demonstrates that the width of the depletion layer is altered, hence the stored charge. The capacitance of the junction is

$$C = -\mathrm{d}Q/\mathrm{d}V_{\mathrm{a}}$$

where the minus sign allows for the negative sign of V_{a} for reverse bias. Now

$$Q = AW_{\mathrm{n}}N = A[2\epsilon\epsilon_0 NP(V_0 - V_{\mathrm{a}})/e(N+P)]^{1/2}$$

$$C = -\mathrm{d}Q/\mathrm{d}V_{\mathrm{a}} = A[\epsilon\epsilon_0 NP/2e(N+P)(V_0 - V_{\mathrm{a}})]^{1/2}. \qquad (C.17)$$

Because of the variation of C with V_{a}, a reverse biased diode may be used as a variable capacitor. It is called a **Varactor diode**. It is always used reverse biased, since for forward bias it has a low resistance in parallel.

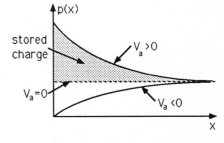

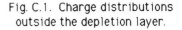

Fig. C.1. Charge distributions outside the depletion layer.

Storage capacitance

Consider the n region of the diode between the edge of the depletion layer and the terminal. The small hole density there $p(x)$ is given for $V_{\mathrm{a}} = 0$ by the thermal equilibrium due to

$$\mathrm{atom} \leftrightarrow \mathrm{electron} + \mathrm{ion}$$

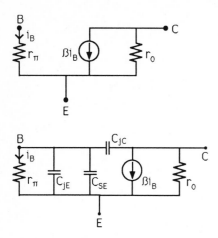

Fig. C.2 (a) hybrid π equivalent circuit, (b) including capacitance.

and hence by the flat broken line in figure C.1. For reverse bias, $p(x)$ is reduced at the edge of the depletion layer and for forward bias it is greatly increased. The hole distribution falls exponentially beyond the depletion layer as holes encounter electrons flowing in from the terminal and annihilate. For forward bias, the current is large so there is a substantial charge stored in this region, proportional to the current flowing.

The same is true for electrons on the p side. These charges Q give rise to a large **storage capacitance** (sometimes called diffusion capacitance) for forward bias. It increases linearly with current I. It is conventional to express Q in the form

$$I = Q/\tau$$

where τ is the time constant for the stored charge to change. For saturated operation of a diode or transistor, this time constant can be 10–300 ns and is the dominant limitation on switching speed unless special precautions are taken (Chapter 21).

Capacitance in the bipolar transistor

Chapter 17 derives an equivalent circuit for the transistor shown in figure C.2(a). To this it is necessary to add the capacitances shown in figure C.2(b). The capacitance C_{jE} is the junction capacitance of the base–emitter depletion layer, allowing for variations in its width with V_{BE}. For normal forward-biased operation it is dwarfed by the storage capacitance C_{SE} corresponding to charge stored in the controlling base region. It is also necessary to include the junction capacitance C_{jC} of the reverse-biased collector–base depletion layer; because of the Miller effect of Chapter 8, this can be very important, despite its low value, typically 1 pF.

For forward bias, a typical value of C_{SE} is 10 pF $\times I_E$ (mA). A fundamental characteristic of the transistor is the frequency f_T at which the current gain I_C/I_B falls to 1. This is

$$f_T = \beta/2\pi(C_{jE} + C_{SE})r_\pi.$$

Now $r_\pi \simeq 25\beta\ \Omega/I(\text{mA})$. So if C_{jE} is ignored, f_T is independent of current and β and roughly equal to 10^9 Hz. Its value may be used to find C_{SE} experimentally.

The Ebers–Moll model of the bipolar transistor

Chapter 9 considers normal operation of the npn transistor where the electron distribution in the base region is as shown in figure C.3(a). A model is now derived which deals with the other three possible modes shown in figures C.3(b)–(d).

It is simply necessary to trace the fate of electrons injected into the base from either emitter or collector. For the former,

$$I_{E1} = I_1[\exp(eV_{BE}/kT) - 1]$$

$$I_{C1} = \alpha I_{E1} \tag{C.18}$$

$$I_{B1} = (1 - \alpha)I_{E1}$$

where α is very close to 1. For forward bias, $\beta = \alpha/(1 - \alpha)$. Likewise, for electrons originating from the collector, figure C.3(c),

$$I_{C2} = -I_2[\exp(eV_{BC}/kT) - 1]$$

$$I_{E2} = -\alpha' I_{C2} \tag{C.19}$$

$$I_{B2} = -(1 - \alpha')I_{C2}.$$

A constraint on the parameters is that, if $V_{BC} = V_{BE}$, collector and emitter are at the same voltage, so there is no current direct from emitter to collector. Hence $I_{C1} + I_{E2} = 0$ or

$$\alpha I_1 = \alpha' I_2.$$

Adding these two sets of equations, (C.18) and (C.19),

$$I_E = I_1[\exp(eV_{BE}/kT) - 1] - \alpha' I_2[\exp(eV_{BC}/kT) - 1]$$

$$I_C = \alpha I_1[\exp(eV_{BE}/kT) - 1] - I_2[\exp(eV_{BC}/kT) - 1]. \tag{C.20}$$

$$I_B = I_E - I_C.$$

These are the Ebers–Moll equations. They are used (or elaborations of them) to model the performance of the transistor in SPICE. In the saturation region, they give the dependence of β on V_{BC}.

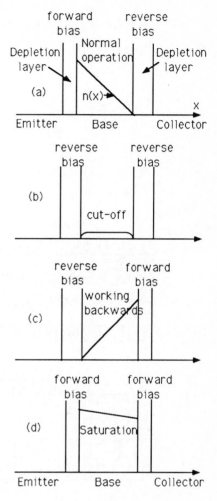

Fig. C.3. The electron density n(x) in the base region of an npn bipolar transistor for four modes of operation.

Appendix D

IC Layouts and Transistor Specifications

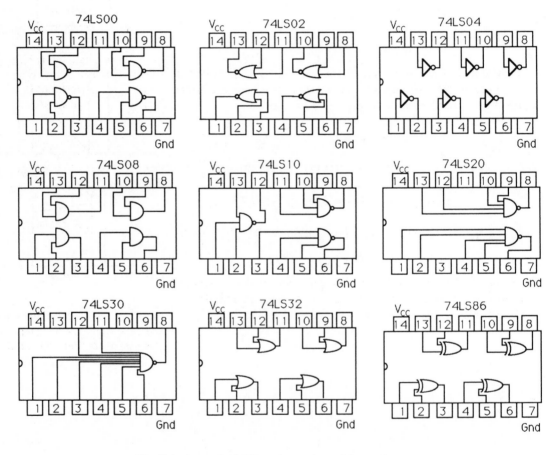

Fig. D.1. Layout of TTL gates, viewed from above.

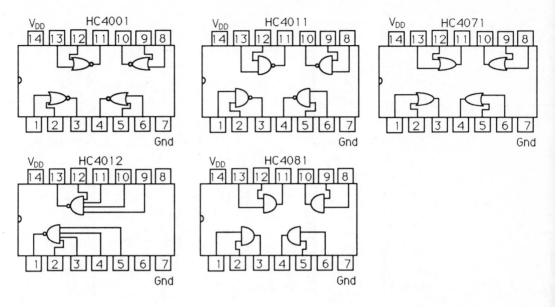

Fig. D2. Layout of CMOS NOR, NAND, OR and AND gates, viewed from above.

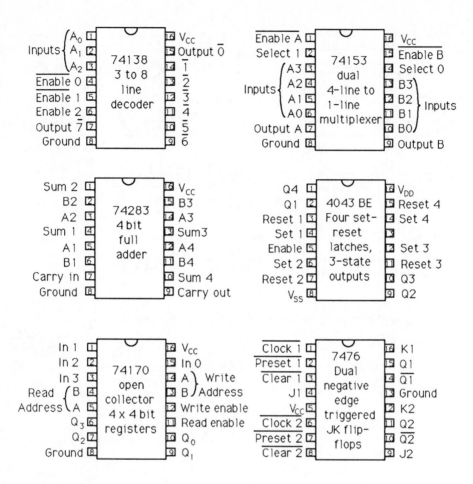

Fig D.3. Pin layouts of some useful ICs for projects. For the 74153, the select lines control both multiplexers. The 74170 will read and write simultaneously using different registers.

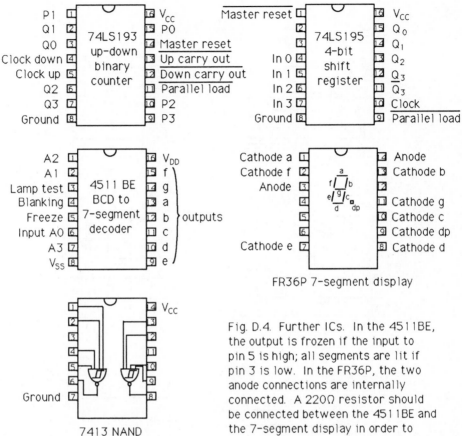

Fig. D.4. Further ICs. In the 4511BE, the output is frozen if the input to pin 5 is high; all segments are lit if pin 3 is low. In the FR36P, the two anode connections are internally connected. A 220Ω resistor should be connected between the 4511BE and the 7-segment display in order to limit the current.

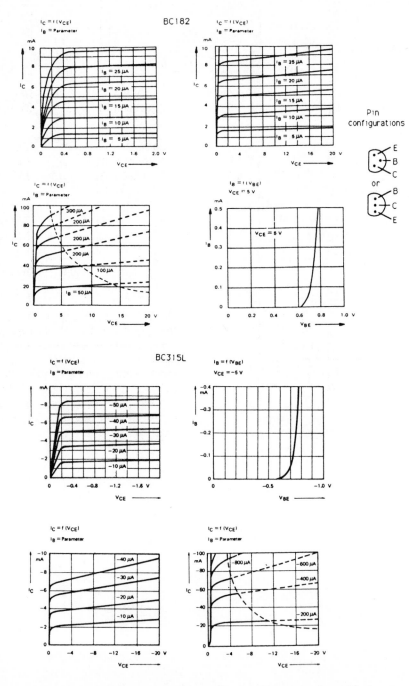

Fig. D.5. Characteristics of some bipolar transistors (Texas Instruments).

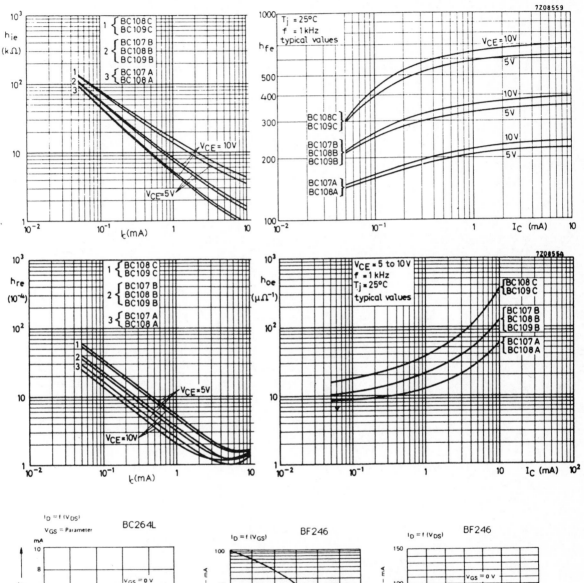

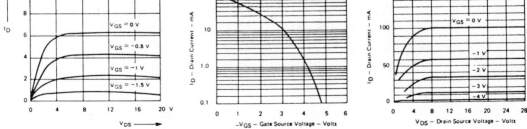

Fig. D.6. Characteristics of some bipolar and FET transistors.

Index